The Handbook
of Environmental Chemistry

Editor-in-Chief: O. Hutzinger

Volume 3 Part F: Anthropogenic Compounds

Detergents

Volume Editor: N. T. de Oude

With contributions by
F. Balk, J. G. Batelaan, J. G. Blumberg, R. S. Boethling,
L. Butterwick, P. Christophliemk, J. S. Falcone,
T. Feijtel, P. Gerike, H. L. Gewanter, P. A. Gilbert,
C. G. van Ginkel, W. E. Gledhill, D. Gleisberg,
J. L. Hamelink, M. S. Holt, H. L. Hoyt, H. B. Kramer,
D. G. Lynch, G. C. Mitchell, H.-J. Opgenorth,
H. A. Painter, M. Potokar, K. Raymond, R. J. Watkinson

With 48 Figures and 116 Tables

Springer-Verlag

Berlin Heidelberg New York
London Paris Tokyo
Hong Kong Barcelona
Budapest

Professor Dr. Otto Hutzinger
University of Bayreuth
Chair of Ecological Chemistry and Geochemistry
P.O. Box 101251, W-8580 Bayreuth, FRG

ISBN 3-540-53797-X Springer-Verlag Berlin Heidelberg New York
ISBN 0-387-53797-X Springer-Verlag New York Berlin Heidelberg

The Handbook of environmental chemistry.
Includes bibliographies and indexes. (Revised for volume 3-F)
Contents: v. 1. The natural environment and the biogeochemical cycles / with contributions by P. Craig ... [et al.] — — v. 2. Reactions and processes / with contributions by W. A. Bruggeman ... [et al.] — — v. 3. Anthropogenic compounds — / with contributions by R. Anliker. [et al.]   1. Environmental chemistry.   I. Hutzinger, O.
QD31.H335   574.5'222   80-16607
ISBN 0-387-09688-4 (u.s. : v. 1)

This work is subject to copyright. All rights are reserved, whether the whole or part of the material is concerned, specifically the rights of translation, reprinting, reuse of illustrations, recitation, broadcasting, reproduction on microfilm or in any other way, and storage in data banks. Duplication of this publication or parts thereof is permitted only under the provisions of the German Copyright Law of September 9, 1965, in its current version, and permission for use must always be obtained from Springer-Verlag. Violations are liable for prosecution under the German Copyright Law.

© Springer-Verlag Berlin Heidelberg 1992
Printed in Germany

The use of general descriptive names, registered names, trademarks, etc. in this publication does not imply, even in the absence of a specific statement, that such names are exempt from the relevant protective laws and regulations and therefore free for general use.

Typesetting: Macmillan India Ltd., Bangalore-25
Printing: Color-Druck Dorfi GmbH, Berlin; Binding: Lüdertz & Bauer, Berlin
52/3020-543210 – Printed on acid-free paper

Preface

Environmental Chemistry is a relatively young science. Interest in this subject, however, is growing very rapidly and, although no agreement has been reached as yet about the exact content and limits of this interdisciplinary subject, there appears to be increasing interest in seeing environmental topics which are based on chemistry embodied in this subject. One of the first objectives of Environmental Chemistry must be the study of the environment and of natural chemical processes which occur in the environment. A major purpose of this series on Environmental Chemistry, therefore, is to present a reasonably uniform view of various aspects of the chemistry of the environment and chemical reactions occuring in the environment.

The industrial activities of man have given a new dimension to Environmental Chemistry. We have now synthesized and described over five million chemical compounds and chemical industry produces about one hundred and fifty million tons of synthetic chemicals annually. We ship billions of tons of oil per year and through mining operations and other geophysical modifications, large quantities of inorganic and organic materials are released from their natural deposits. Cities and metropolitan areas of up to 15 million inhabitants produce large quantities of waste in relatively small and confined areas. Much of the chemical products and waste products of modern society are released into the environment either during production, storage, transport, use or ultimate disposal. These released materials participate in natural cycles and reactions and frequently lead to interference and disturbance of natural systems.

Environmental Chemistry is concerned with *reactions in the environment*. It is about distribution and equilibria between environmental compartments. It is about reactions, pathways, thermodynamics and kinetics. An important purpose of this Handbook is to aid understanding of the basic distribution and chemical reaction processes which occur in the environment.

Laws regulating toxic substances in various countries are designed to assess and control risk of chemicals to man and his environment. Science can contribute in two areas to this assessment: firstly in the area of toxicology and secondly in the area of chemical exposure. The available concentration ("environmental exposure concentration") depends on the fate of chemical compounds in the environment and thus their distribution and reaction behaviour in the environment. One very important contribution of Environmental Chemistry to the above mentioned toxic substances laws is to develop laboratory test methods, or mathematical correlations and models that predict the environmental fate of new chemical compounds. The third purpose of this Handbook is to help in the basic understanding and development of such test methods and models.

The last explicit purpose of the handbook is to present, in a concise form, the most important properties relating to environmental chemistry and hazard assessment for the most important series of chemical compounds.

At the moment three volumes of the Handbook are planned. Volume 1 deals with the natural environment and the biogeochemical cycles therein, including some background information such as energetics and ecology. Volume 2 is concerned with reactions and processes in the environment and deals with physical factors such as transport and adsorption, and chemical, photochemical and biochemical reactions in the environment, as well as some aspects of pharmacokinetics and metabolism within organisms. Volume 3 deals with anthropogenic compounds, their chemical backgrounds, production methods and information about their use, their environmental behaviour, analytical methodology and some important aspects of their toxic effects. The material for volumes 1, 2, and 3 was more than could easily be fitted into a single volume, and for this reason, as well as for the purpose of rapid publication of available manuscripts, all three volumes are published as a volume series (e.g. Vol. 1; A, B, C). Publisher and editor hope to keep the material of the volumes 1 to 3 up to date and to extend coverage in the subject areas by publishing further parts in the future. Readers are encouraged to offer suggestions and advice as to future editions of "The Handbook of Experimental Chemistry".

Most chapters in the Handbook are written to a fairly advanced level and should be of interest to the graduate student and practising scientist. I also hope that the subject matter treated will be of interest to people outside chemisry and to scientists in industry as well as government and regulatory bodies. It would be very satisfying for me to see the books used as a basis for developing graduate courses on Environmental Chemistry.

Due to the breadth of the subject matter, it was not easy to edit this Handbook. Specialists had to be found in quite different areas of science who were willing to contribute a chapter within the prescribed schedule. It is with great satisfaction that I thank all authors for their understanding and for devoting their time to this effort. Special thanks are due to the Springer publishing house and finally I would like to thank my family, students and colleagues for being so patient with me during several critical phases of preparation for the Handbook, and also to some colleagues and the secretaries for their technical help.

I consider it a privilege to see my chosen subject grow. My interest in Environment Chemistry dates back to my early college days in Vienna. I received significant impulses during my postdoctoral period at the University of California and my interest slowly developed during my time with the National Research Council of Canada, before I was able to devote my full time to

Environmental Chemistry in Amsterdam. I hope this Handbook will help deepen the interest of other scientists in this subject.

Otto Hutzinger

This preface was written in 1980. Since then publisher and editor have agreed to expand the Handbook by two new open-ended volume series: Air Pollution and Water Pollution. These broad topics could not be fitted easily into the headings of the first three volumes.

All five volume series will be integrated through the choice of topics covered and by a system of cross referencing.

The outline of the Handbook is thus as follows:
1. The Natural Environment and the Biogeochemical Cycles
2. Reactions and Processes
3. Anthropogenic Compounds
4. Air Pollution
5. Water Pollution

Bayreuth, February, 1992   Otto Hutzinger

Contents

Introduction XII

H. A. Painter Anionic Surfactants *1*

M. S. Holt, G. C. Mitchell, R. J. Watkinson The Environmental Chemistry, Fate and Effects of Nonionic Surfactants *89*

R. S. Boethling, D. G. Lynch Quaternary Ammonium Surfactants *145*

D. Gleisberg Phosphate *179*

P. Christophliemk, P. Gerike, M. Potokar Zeolites *205*

H. L. Hoyt, H. L. Gewanter Citrate *229*

K. Wolf, P. A. Gilbert EDTA—Ethylenediaminetetraacetic Acid *243*

W. E. Gledhill, T. Feijtel Environmental Properties and Safety Assessment of Organic Phosphonates Used for Detergent and Water Treatment Applications *261*

K. Raymond, L. Butterwick Perborate *287*

P. A. Gilbert TAED—Tetraacetylethylenediamine *319*

J. G. Bafelaan, C. G. van Ginkel, F. Balk Carboxymethylcellulose (CMC) *329*

H.-J. Opgenorth Polymeric Materials, Polycarboxylates *337*

H. B. Kramer Fluorescent Whitening Agents *351*

J. S. Falcone, G. Blumberg Anthropogenic Silicates *367*

J. L. Hamelink Silicones *383*

Subject Index *399*

List of Contributors

F. Balk
Akzo International Research
Department of Organic
and Polymeric Chemistry
P.O. Box 9300
NL-6800 SB Arnhem

Dr. J. G. Batelaan
Akzo International Research
Department of Organic
and Polymeric Chemistry
P.O. Box 9300
NL-6800 SB Arnhem

John G. Blumberg
PQ Corporation Research
and Development Center
280 Cedar Grove Road
Conchohocken, PA 19428
USA

Dr. R. S. Boethling
US Environmental Protection Agency
Exposure Evaluation Division
Washington, DC 20460
USA

Lucy Butterwick
Environmental Resources Ltd.
106 Gloucester Place
London W1H 3DB
England

Dr. Peter Christophliemk
Henkel KGaA
Postfach 1100
4000 Düsseldorf
FRG

James S. Falcone
PQ Corporation Research
and Development Center
280 Cedar Grove Road
Conchohocken, PA 19428
USA

Dr. T. Feijtel
Monsanto Company
800 N. Lindbergh Boulevard
St. Louis, Missouri 63167
USA

Dr. Peter Gerike
Henkel KGaA
Postfach 1100
4000 Düsseldorf

Dr. Herman L. Gewanter
Pfizer Inc., Chemical Division
Eastern Point Road
Groton, CT 06340
USA

Dr. Philip A. Gilbert
Unilever Research Laboratory
Quarry Road East
Bebington, Wirral
Merseyside L63 3JW
England

William E. Gledhill
Monsanto Company
800 N. Lindbergh Boulevard
St. Louis, Missouri 63167
USA

Dr. Dietrich Gleisberg
Hoechst AG, Werk Knapsack
5030 Hürth
FRG

Dr. Jerry L. Hamelink
Dow Corning Corporation
Midland, Michigan 48686-0994
USA

M. S. Holt
Shell Research Limited
Environmental Microbiology
Sittingbourne Research Center
Sittingbourne, Kent ME9 8AG
England

Hazen L. Hoyt
Pfizer Inc., Chemical Division
Eastern Point Road
Groton, CT 06340
USA

Hans B. Kramer
Chemiewinkel
University of Amsterdam
P.O. Box 20242
NL-1000 HE Amsterdam

D. G. Lynch
US Environmental Protection Agency
Exposure Evaluation Division
Washington, DC 20460
USA

G. C. Mitchell
Shell Research Limited
Environmental Microbiology
Sittingbourne Research Center
Sittingbourne, Kent ME9 8AG
England

Dr. Hans-Joachim Opgenorth
BASF AG
Unternehmensbereich Spezialchemikalien
6700 Ludwigshafen
FRG

Dr. H. A. Painter
Freshfield Analysis Ltd.
Caxton Villa
Park Lane
Knebworth, Herts SG3 6PF
England

Dr. Mattias Potokar
Henkel KGaA
Postfach 1100
4000 Düsseldorf
FRG

Karen Raymond
Environmental Resources Ltd.
106 Gloucester Place
London W1H 3DB
England

C. G. van Ginkel
Akzo International Research
Department of Organic
and Polymeric Chemistry
P.O. Box 9300
NL-6800 SB Arnhem

Dr. R. J. Watkinson
Shell Research Limited
Environmental Microbiology
Sittingbourne Research Center
Sittingbourne, Kent ME9 8AG
England

Introduction

The chapters in this volume cover the major raw materials used in consumer and industrial detergents. Most of the raw materials are used in both groups of products. The emphasis in the selection has been on laundry detergents for the consumer market for the simple reason that this category of products represents the largest volume of detergent chemical use.

The chemicals have been described by their different chemical classes because that is the most rational way to present the information. In reality, these chemicals are rarely, if ever, used in isolation. They are used in 'formulated' detergents, carefully balanced mixtures of ingredients, designed to meet the consumer needs. These needs vary between consumers and are a function of fibre type, white/colored fabrics, water hardness, temperature, etc.

Soil or dirt, has been defined as "matter in the wrong place", a definition so old that the origin appears to have been lost. It is convenient to classify soils in different categories, such as fatty/oily soil, bleachable stains, proteinaceous matter and so on. The assumption is that there is a specific ingredient to take care of a specific category of soil, e.g. a bleach will remove bleachable stains. In reality, soils always are mixtures of materials and it is rare that a single detergent ingredient can remove a single category of soil. Mixtures of chemicals will remove the mixtures of materials that constitute soil and do that more efficiently and effectively than single ingredients could. Detergents then are mixtures of chemicals, formulated to meet the varied needs of the consumers in the most economical manner.

Detergent products differ between continents due to differences in washing conditions. Characteristic washing conditions for two continents are shown in Table 1.

These differences in washing conditions have also resulted in formulation differences between the two continents. Table 2 provides examples of so-called heavy-duty detergents, that is detergents for the main wash. The consumer washes his laundry in trying to satisfy three needs: hygiene, esthetics and cleansing. Hygiene was an objective of mankind, long before the scientific basis existed to define how hygiene could be achieved. Many early attempts, such as the building of walls, the use of smoke and fire, and rituals were unsuccessful attempts to create hygienic conditions. Detergents for cleaning the body, fabrics in contact with the body, eating utensils, etc. now provide that hygiene in a convenient manner. The consumer is seen to respond to unsatisfactory cleaning

Table 1. Washing conditions comparison

	Europe	U.S.A.
Machine type	horizontal drum	vertical drum
Product concentration	0.7%	0.1%
Wash time (min)	up to 60	10
Wash solution (l)	18	50
Heating	built-in heater	none (hot fill, 50 °C max)
Loads washed per month	14	33
Drying: line/dryer (%)	86/14	22/78

Table 2. Typical compositions of detergents for consumer laundry products

		Europe Powders with P	Powders without	U.S.A. Liquids phosphate	Powders with P
Anionics	LAS	5–10	5–10	10–15	5–10
Nonionics	alkylethoxylates	2–10	0–10	10–15	2–10
Builders	see note	10–35	20–25	0–5	25–35
Bleaches	perborate	10–25	20–25	—	—
Activators	TAED	0–5	0–2	—	—
Polymers	CMC, acrylates	1–2	1–2	0–3	1–2
Brighteners	—	<0.3	<0.3	<0.3	<0.3
Silicate	—	2–6	2–6	<0.3	2–6
Silicones	—	<1	<1	<1	—

Note: Sodium triphosphate is used in detergents with phosphate. Zeolite is used in powders without phosphate. Also soap and citrate are used, both in powders and in liquids

by adding effort, such as additional washing steps, the use of auxiliary products or a higher washing temperature.

Esthetics is another motivation. Textiles are part of our surroundings which we try to make esthetically pleasing by selecting the form, color and texture of the material things that surround us. The consumer also has the esthetic aspects of textiles in mind when purchasing them or when deciding to discard them because color, feel or shape no longer meet the demands of the user. Part of formulating detergents then is the attempt to maintain the esthetic values of textile for as long as possible.

Finally: cleansing. It is an age-old need for man to cleanse himself. Ritual cleansing occurs all over the world from east to west, from north to south. This topic goes well beyond the scope of the current book, but it is relevant because man has recently discovered that cleansing himself results in polluting the environment. It is now generally recognized that any human action has some environmental effect. Hence, the demand for environmental information which this volume is trying to satisfy. The authors are experts who provide a summary of published as well as new data.

Introduction

Two chemicals are not included in this volume: nonylphenol ethoxylates and nitrilo triacetate because they are of limited interest only. Instead, a brief rationale is given below.

Nonylphenol with 9 ethoxy units (NPE) is a biodegradable material, but biodegradation slows down when only 2 or 3 ethoxy units are left. This slower biodegrading intermediate is also more toxic than the parent material. These findings made it clear that the use of NPEs must be restricted to levels that avoid accumulation of the semi-degraded material. The detergent industry did so by removing NPE from its laundry consumer detergents.

Sodium nitrilo triacetate (NTA) has a very checkered history. It started with an alarm about possible teratogenicity, which was rapidly shown to have been false. The next concern was carcinogenicity. NTA was tested in a bioassay at the Maximum Tolerable Dose (MTD). The MTD is that concentration at which the test animals will not show toxic effects during the test duration. NTA has a very low toxicity, i.e. the MTD was rather high. At these high concentrations, secondary effects occur due to an upset in the cation balance of the test animals; these was no genuine genotoxicity. The findings have been generally dismissed as irrelevant for the exposure resulting from detergent use; only the State of New York has banned NTA. Finally, a concern was expressed because of the ability of NTA to bind metals. There is no simple environmental criterion for the size of the effect that should be considered as adverse. The effects that have been noticed are a reflection of the sensitivity of the experimental technique used, rather than of demonstrated adverse environmental effects. Several countries (Germany, Netherlands, Switzerland) have accepted that levels of 3–5% NTA in all laundry detergents will not cause adverse effects; these countries have not taken a position about the safety of higher levels. Levels of 3–5% NTA are insufficient to replace phosphates in detergents, hence, they are not used for that purpose. In general, detergent manufacturers have shied away from using NTA because of the past controversy. Limited use occurs in Switzerland and careful monitoring programs have shown good removal in sewage treatment plants and from groundwater. Extensive use during almost two decades in Canada has not shown adverse environmental effects either.

Apart from these two materials, all other important detergent chemicals are reviewed in this volume. That includes phosphate, which for many years has been debated hotly as the possible solution to the eutrophication issue. Recently, it has become clear that good sewage treatment is essential and sufficient to eliminate the phosphate from municipal sewage. In some countries, the question has been asked if phosphate should be reintroduced in detergents because of its excellent human safety. However, good sewage treatment exists only in a few countries and the issue is still so politically sensitive that it seems unlikely that a major turn around will occur any time soon. Phosphate is still used in a large number of countries, which justified including a chapter on this material.

Anionic Surfactants

H. A. Painter
Formerly with Water Research Centre Medmenham UK, Freshfield Analysis Ltd., Caxton Villa, Park Lane, Knebworth, U.K. SG3 6PF

Symbols and Abbreviations 5
Introduction 6
   Types of Anionic Surfactants 7
Alkyl Aryl Sulfonates 8
   Structure 8
   Nomenclature 9
   Production Methods 10
   Amounts Produced 10
   Uses and Applications 11
   Legislation Concerning Anionic Surfactants 12
   Physical Properties 12
      Adsorption 14
   Chemical Properties 14
   Analytical Methods 15
      Colorimetric Methods 15
      Chromatography 16
      Spectroscopy 19
   LAS in the Environment 20
      Sewage 20
      Effluents from Biological Treatment Plants 22
      Sewage Sludges 23
      Soils 24
      River Water 24
      River Sediments 25
      Estuarine and Coastal Waters and Sediments 26
      Tapwater 26
      Mean Chain Length of LAS in Environmental Samples . . 27
   Environmental Fate 27
      Metabolic Pathways 27
      Intermediates 28
      Effect of Structure of LAS on Biodegradability 29
      Degradation in Laboratory Tests 30
      Removal of LAS in Sewage Treatment 32
      Removal of LAS in Rivers and Sediments 34
      Removal in Soil and Sub-Surface Sediments 34

The Handbook of Environmental Chemistry,
Volume 3 Part F, Ed. O. Hutzinger
© Springer-Verlag Berlin Heidelberg 1992

Effects on the Environment 35
   Sewage Treatment 35
   Natural Waters 36
   Soils 36
   Aquatic Organisms 36
   Terrestrial Plants 40
Effects on Small Mammals and Man 41
Alkyl (or Alcohol) Sulfates 41
   Structure 41
   Nomenclature 42
Production Methods 42
   Amounts Produced 43
   Uses and Applications 43
   Legislation 44
   Physical and Chemical Properties 44
   Analytical Methods 44
   Environmental Concentrations 45
   Environmental Fate 46
      Primary Biodegradability 47
      Ultimate Biodegradability 47
      Effect of Chain Length 48
      Sea Water 49
      Soil and Sludges 49
      Sewage Treatment 50
      Percolating (Biological) Filters 50
      Anaerobic Treatment 50
      Metabolic Pathways and Intermediates 51
   Environmental Effects 51
      Effects on Bacteria 51
      Effects on Algae 52
      Effects on Invertebrates 52
      Effects on Other Aquatic Organisms 53
      Effects on Fish 54
      Bioaccumulation in Fish 56
      Effects on Higher Plants 57
   Effects on Small Mammals and Man 57
Alkyl Ether Sulfates 58
   Structure 58
   Nomenclature 58
   Production Methods 59
   Production Amounts 59
   Uses and Applications 59
   Physical and Chemical Properties 60
   Analytical Methods 60
   Environmental Concentrations 61

- Environmental Fate 61
  - Primary Biodegradability 61
  - Ultimate Biodegradability 62
  - Effect of Structure on Biodegradation 62
  - Simulation and Field Tests 63
  - Anaerobic Biodegradation 63
  - Other Alkoxylate Sulfates 64
  - Metabolic Pathways and Intermediates 64
- Effects on the Environment 65
  - Bacteria 65
  - Algae 65
  - Invertebrates 66
  - Fish 66
  - Higher Plants 68
- Effects on Small Mammals and Man 68

Alkane Sulfonates (SAS) 69
- Structure 69
- Nomenclature 69
- Production Methods 70
- Amounts Produced 70
- Uses and Applications 70
- Physical and Chemical Properties 71
- Analytical Methods 71
- Environmental Concentrations 71
- Environmental Fate 72
  - Biodegradability 72
  - Metabolic Pathways 73
  - Microbial Growth 73
- Effects on the Environment 73
  - Bacteria 74
  - Algae 74
  - Daphnia 74
  - Fish 74
- Effects on Small Mammals and Man 75

α-Olefine Sulfonates 75
- Structure and Nomenclature 75
- Production Methods 76
- Amounts Produced 77
- Uses and Applications 77
- Physical Properties 77
- Analysis 78
- Environmental Concentrations 78
- Environmental Fate 78
  - Anaerobic Biodegradability 80
  - Metabolic Pathways 80

Effects on the Environment 80
    Algae. 80
    Invertebrates 80
    Fish 81
Higher Plants 82
Small Mammals and Man. 82
References 82

Summary

The various types of anionic surfactants are listed and the five most used types are dealt with in detail. More attention is given to linear alkyl benzene sulfonates (LAS), since they are used more than any other surfactant and more data have been reported on them. The structure, methods of production and application of each type are presented and relevant analytical methods are summarized. The legislation regulating their use is explained and their routes to and fate in the various environmental sectors are discussed. Their impact on the aquatic and soil environment has been assessed by comparing evidence of their concentrations in the various environmental sectors with those concentrations which produce observed effects on appropriate aquatic organisms. It is concluded that no environmental problems arise from the use of present-day anionic surfactants.

Symbols and Abbreviations

ABS	alkyl benzene sulfonate
AE	alcohol ethoxylate
AES	alcohol ethoxylate sulfate
AOS	α-olefine sulfonate
APES	alkyl phenol ethoxylate sulfate
AS	alkyl sulfate
BOD_n	biochemical oxygen demand after n days
COD	chemical oxygen demand
DOC	dissolved organic carbon
EC_n	concentration of chemical giving effect of n%
EC (EEC)	European (Economic) Community
EO	ethylene oxide
FAB	fast atom bombardment
g	gram
GC	gas chromatography
GC/MS	gas chromatography-mass spectroscopy
h	hour
IR	infra-red
ISO	International Organisation for Standardisation
k	first order reaction constant
kg	kilogram
L	litre
LAB	linear alkyl benzene
LAS	linear (secondary) alkyl benzene sulfonate
LC_n	concentration giving n% death of the population
LPAS	linear primary alkyl sulfate
M	molar
mM	milli-molar
MB	methylene blue
MBAS	methylene blue active substances
mg	milligram
MITI	Ministry of International Trade and Industry (Japan)
mL	millilitre
MW	molecular weight
n	normal, linear
ng	nanogram
nm	nanometre
NMR	nuclear magnetic resonance
NOEC	no observed effect concentration
OECD	Organisation for Economic Co-operation and Development
PAS	primary alkyl sulfate
R	alkyl or other group

SAS	secondary alkane sulfonate
SDA	Soap and Detergent Association
sec	secondary
TLC	thin layer chromatography
ThCO$_2$	theoretical yield of carbon dioxide
ThOD	theoretical oxygen demand
TPBS	tetrapropylene benzene sulfonate
UV	ultra-violet
μ	specific growth rate
μg	microgram
ϕ	benzene ring

Introduction

The large-scale replacement in laundry and cleaning formulations of soap powder by synthetic surfactants (surface active substances) around the end of the Second World War was made possible both technically and economically by the development of ABS, alkyl benzene sulfonates. The change-over was fairly rapid, starting in the U.S.A. and then in other developed countries. The first synthetic surfactants seem to have been developed in Germany during the First World War; these were short-chain alkyl naphthalene sulfonates which were good wetting agents but only moderately good detergents. In the late 1920s long-chain alkyl sulfates came on to the market and later long-chain alkyl aryl sulfonates with benzene as the aromatic nucleus and the alkyl portion from a kerosene fraction.

After the Second World War the kerosene product was soon replaced by an alkyl benzene produced from propylene tetramer coupled to benzene, and this product was present in more than half of the detergents used throughout the world in the period 1950–1965. However, as early as 1952 in the U.K., the tetrapropylene benzene sulfonate, TPBS, was found not to biodegrade efficiently. It was degraded by only about 50% in sewage treatment units and as a result excessive foaming often occurred in activated sludge aeration tanks, as well as in receiving rivers. The foaming was far worse than that caused by proteinaceous material in sewage prior to the introduction of synthetic surfactants and in extreme cases sewage-works operators were killed by asphyxiation after falling into foaming tanks from walkways made slippery by the foam. The efficiency of sewage treatment was impaired and the risk of greater aerial distribution of bacterial pathogens was enhanced when winds blew the foam from the tanks.

Because of its incomplete biodegradation in sewage treatment, the concentration of TPBS in river waters reached as high as 2 mg L^{-1}, and water tended to foam when coming out of the tap. While the concentrations in the public water supply were low (<0.5 mg L^{-1}) and even though anionic surfactants are toxic to man only at much higher concentrations, it was decided in the U.S. and by other authorities to set a limit of 0.5 mg L^{-1} in water supplies, partly for aesthetic reasons and partly because foaming does not occur below this level. For this and other reasons, the manufacturers sought other alkyl benzene surfactants which were more biodegradable for use as the major surfactant used in laundry detergents. Evidence existed that bacteria are able to metabolise straight, or linear, chain compounds more readily than branched chain isomers and this proved to be the case with ABS. The resulting linear alkyl benzene sulfonate, LAS, introduced in the early 1960s, has been shown to be more than adequately biodegradable and remains the "work-horse" surfactant to the present time.

Along with LAS, there are many other types of anionic surfactants available, which have their particular niche and application for domestic and industrial use. Because of its importance, more about LAS is given in this chapter than about the dozen or so other types; indeed, more is known about LAS.

Recent texts which include descriptions of anionic surfactants are Davidsohn and Milwidsky [1], Swisher [2], Gerike [3] and Schoberl, Bock and Huber [4]. Two earlier comprehensive reviews were produced by Little [5, 6].

Types of Anionic Surfactants

A surfactant consists essentially of a hydrophobic group and a hydrophilic group. This chapter deals with those synthetic surfactants having as their hydrophilic group an anion; most common anions are sulfonate and sulfate, and more rarely phosphate. (Soap, having the carboxy group as the hydrophile, is not discussed here). As many as sixty groups and sub-groups have been listed, of which only seven were not sulfonated or sulfated products [7].

Some of the commonly used types of anionic surfactants are listed below; only the first five listed are described in this chapter.

Alkyl aryl sulfonates (ABS, LAS)

$RC_6H_4SO_3^-H^+$ R is in the *para* position

Long chain fatty alcohol sulfates; alkyl sulfates (AS)

$RCH_2OSO_3^-Na^+$ (LPAS)
primary

$R_1CHOSO_3^-Na^+$ (AS)
|
R_2
secondary

Alkyl ether sulfate; alkyl ethoxylate sulfate; alcohol ethoxy sulfate (AES)

$R(OCH_2CH_2)_nOSO_3^-Na^+$ R is primary or secondary

Alkane sulfonate; aliphatic sulfonate; paraffin sulfonate

$RSO_3^-H^+$ R is primary or secondary (SAS)

α-olefine sulfonates (AOS) —a mixture of alkene and hydroxyalkane sulfonates

$RCH:CHCH_2SO_3^-Na^+$ $R.CHOH(CH_2)_nSO_3^-Na^+$

α-sulfo fatty acids and methyl esters

$$\underset{SO_3^-H^+}{RCH.COOH} \qquad \underset{SO_3^-H^+}{RCHCOOCH_3}$$

Sulfo-succinate esters

$COONa.CH(SO_3^-Na^+)CH_2.COOR$ mono-ester

$COOR.CH(SO_3^-Na^+)CH_2.COOR$ di-ester

Sulfo-succinamates

$(R_1R_2N)COCH_2CH(SO_3^-Na^+)CONR_1R_2$

Alkyl isethionates

$RCOOCH_2CH_2SO_3^-Na^+$

Alkyl taurides

$RCONR^1CH_2CH_2SO_3^-Na^+$

Phosphate esters

$ROPO(OH)_2$ $(RO)_2POOH$
mono-alkyl   di-alkyl

It will be seen from what is presented here that the present day anionic surfactants used in detergents and other products cause no problems in natural waters or in soil.

Alkyl Aryl Sulfonates

Structure

The surfactants of the alkyl aryl type are the most widely used since they have excellent detersive power, are of low cost since they are made of easily available materials and their formulations have attractive physical properties. They consist of alkyl chains, usually with a mixture of 10–15 C atoms, but principally C_{11} and C_{12}, attached to the benzene ring in the *para* position to the sulfonate group, SO_3H or SO_3Na, namely

$RC_6H_4SO_3^-H^+(Na)$

Sometimes toluene, xylene and naphthalene are used in place of benzene. The average chain length is normally around 12, though in some cases of polypropylene benzene sulfonates the chain length has averaged 13 or 14. The TPBS surfactants contain a high degree of branching in the chain and the number of isomers of the five or so homologues run into scores, but these polypropylene derivatives are not now in normal use because of the low degree of biodegradability of some of the components.

The convention has grown that ABS is used to describe the branched products, although it is sometimes used to describe any alkyl benzene sulfonate; the linear product is called LAS. LAS products are derived from linear alkyl benzenes in which the ring is attached to a C atom which is itself attached to two other C atoms. The benzene ring is attached to any of the C atoms from C2 to C6, but not to C1. That is, LAS is a mixture of secondary alkyl benzene sulfonates, e.g.

$CH_3CH(C_6H_4SO_3H)(CH_2)_9CH_3$ $CH_3(CH_2)_4CH(C_6H_4SO_3H)(CH_2)_5CH_3$

2-(4′-sulfophenyl)dodecane   6-(4′-sulfophenyl)dodecane

The primary isomer is not formed in the method of production and the aromatic ring is randomly distributed over the internal C atoms. The use of aluminum chloride as the alkylation catalyst in the manufacture of the parent hydrocarbons results in a higher proportion of 2-phenyl isomers than does the use of hydrofluoric acid.

The isomers exist in stereo-isomeric forms but apparently there is no significant difference in the environmental behavior between pairs of stereo-isomers [8].

Nomenclature

Individual LAS homologues are designated by the number of C atoms in the chain, namely C_{10}-LAS for decyl-benzene sulfonate, while for mixtures of homologues the range of C atoms in the chain is given thus, C_{10}–C_{12}-LAS indicates that decyl, undecyl- and dodecyl-benzene sulfonates are present. $C_{10}C_{12}$-LAS would indicate that only decyl and dodecyl homologues were present. If the proportions of each homologue are known, they are included as, for example, LAS [25:50:25, $C_{10}:C_{12}:C_{14}$]. Commercial samples of known average chain length are represented as $C_{11.6}$ and the average molecular weight is sometimes given.

Individual isomers have the number of the C atom in the chain joined to the ring given as 5-ϕ-C_{12}, indicating 5-(4-sulfophenyl)dodecane. The primary alkyl benzene sulfonate, not present in commercial products, is described by 1-ϕ-C_{12}. A homologue particularly rich in a given isomer is indicated as, for example, as "high 2-ϕ-C_{12} LAS".

Production Methods

The original ABS products were produced by a one-step Friedel-Crafts alkylation of benzene with tetrapropylene, involving addition of the benzene at the double bond of the olefine, using a catalyst such as aluminum chloride or hydrogen fluoride, followed by sulfonation with sulfuric acid or sulfur trioxide:

$$C_{12}H_{24} + C_6H_6 \longrightarrow C_{12}H_{25} \cdot C_6H_5 \xrightarrow{H_2SO_4} C_{12}H_{25} \cdot C_6H_4SO_3^- H^+$$

The much more environmentally acceptable LAS is synthesised from straight chain paraffins with an average chain length of 12 C atoms. The linear alkylate is produced, for example, from a petroleum fraction obtained by use of a molecular sieve. The alkane is mono-chlorinated and then used directly in a Friedel-Crafts reaction to alkylate benzene with an anhydrous aluminum chloride catalyst. Alternatively, a corresponding olefine is formed by dehydrochlorinating the chlorinated hydrocarbon and then using hydrogen fluoride as catalyst in a Friedel-Crafts reaction. Another method of forming the olefin is by thermal dehydrogenation.

Sulfonation is carried out with sulfuric acid, oleum of various concentrations or chlorosulfonic acid, but these are now being replaced by the use of gaseous sulfur trioxide obtained by the vaporization of liquid sulfur trioxide (sulfuric acid anhydride) or directly as converter gas from a contact sulfuric acid plant.

It is becoming standard practice to refine the raw alkyl benzene to a product with an average molecular weight of 233–245 and another with an average of 257–265. The first is roughly equivalent to a C_{12} and the second a C_{13}, though both are mixtures. The latter, as sulfonate, has in general better detergency and foams better in soft water, while the C_{12}-LAS has a lower cloud point and a lower viscosity in liquid formulations.

LAS as produced is acidic and can be neutralized after production or when mixed in formulations. The sodium salt is usually formed but for special purpose aluminum, calcium, or magnesium salts are produced as well as salts of amines and alkanolamines.

Impurities present in LAS in small proportions include non-linear alkyl benzene sulfonates, dialkyltetralin sulfonates and dialkyl-indane sulfonates.

Amounts Produced

Estimates of production and consumption of surfactants are difficult to make, since exact figures are not published because of intercompany competition and because of import-export movements. Berth and Jeschke [9] gave estimated regional consumption values of LAS for 1987:

	metric tonnes/annum
Western Europe	485 000
U.S.A. and Canada	380 000
Japan	160 000
Central and South America, Africa Middle East and S. E. Asia	500 000
Eastern Bloc States	300 000
Total (approx)	1 800 000

The same authors estimated a world-wide consumption of all forms of surfactant to be 15 million t/annum for 1987, of which 6.7 million t/annum were synthetic, and the remainder was soap. For W. Germany, the authors gave 85 000 t/annum, of LAS representing about 22% of total synthetic surfactant of 380 000 t/annum.

The application of LAS in W. Europe was about 54% in laundry detergents, 27% in cleaning agents, and the remaining 19% in other products. In W. Germany even more, 70%, went to laundry detergents and 15% to cleaning agents.

Uses and Applications

The uses and application of LAS, together with the mechanism of its action, have been described by Berth and Jeschke [9]. The main use is as cleaning agents in heavy-duty grade detergent washing powders and liquids. Its cleansing properties are excellent for this purpose and it is also of relatively low cost. LAS is used as its sodium salt as the sole surfactant in a formulation or in conjunction with other anionic, non-ionic or cationic surfactants. Other uses for the sodium salt are in dry surfactant preparations such as scouring powders and suspended blocks used to clean WC lavatory basins. Some industrial applications of the sodium salt are inclusion in agrochemicals and in the textile and cement industries.

When sodium is replaced by other cations, the properties of the resulting salt are suitable for use as emulsifier in pesticide formulations (Ca); in dry-cleaning detergents, degreasing agents for metal cleaning (amine); hand-cleaning gels (isopropylamine); cosmetic detergents, car shampoo, bubble baths, de-sizing (triethanolamine). Short side-chain LAS, e.g. toluene-, xylene-, cumene-sulfonates, find application as hydrotropes in heavy-duty liquid detergents by depressing the cloud-point.

Naphthalene sulfonates are used as agents for wetting, dispersing and emulsifying in applications such as vat dyes, plastics, paint.

Legislation Concerning Anionic Surfactants

There are general laws in many countries ensuring that the aquatic environment is managed in such a way that any avoidable impairment is prevented, for example, the Water Resources Law [10] in the Federal Republic of Germany (FRG) and the Control of Pollution Act [11] in the U.K. Thus, surfactants, though not mentioned specifically, would be subject to such scrutiny under these laws. However, by far the greater control over surfactants is executed by laws relating to the production and marketing of detergents. The specific relevant legislation in many countries stems from EC Directives and normally refers to surfactants in general present in washing and cleaning agents. The regulations, which relate only to primary biodegradability, apply to any surfactants which react in the methylene blue test for anionic surfactants or in the Dragendorff reagent test for non-ionic surfactants. LAS is not specifically mentioned. Thus the FRG [12] fixed by law a lower limit of 80% biodegradability on anionic surfactants and in the U.S.A. the limit set by SDA [13] was 90% using their semi-continuous activated sludge (SCAS) test method. These values are realistic in that they are high enough to achieve the environmental objectives but not so high as to be unattainable by industry.

The OECD co-ordinated the studies with various countries and recommended generally applicable methods for testing the primary biodegradability of surfactants [14, 15]; these proposals were adopted by the EEC in Directives [16]. These Directives, including amending Directives [17, 18], described the Die-Away and Confirmatory simulation test methods to be used and prohibited the marketing of detergents if the average primary biodegradability of the surfactants of any category present was less than 90% in the Die-Away test. However, to allow for normal tolerances in the measurement of biodegradability, taking account of the unreliability of the test methods, a rejection decision may only be taken where the results obtained with the Die-Away test determined by a single analysis is less than 80%. The more lengthy and more costly confirmatory test is applied only if the removal is less than 80%.

If a "new" surfactant (new as defined in the EEC Directive on Dangerous Chemicals [19]) is to be put on the market, it would be subject to the tests relevant to the yearly tonnage produced. As a consequence it would be subject to tests for ultimate biodegradability, rather than primary bio-degradability, and the whole range of toxicological and ecotoxicological tests.

Physical Properties

The key to the useful physical properties of any surfactant is the balanced strongly hydrophobic-strongly hydrophilic structure. In LAS the hydrophobic portion is the alkyl-aryl chain and the hydrophilic part is the sulfonate cation. The surfactant concentrates in the water surface of aqueous solutions to a higher value than in the body of the solution and lowers the surface tension. The accumulation in the surface is made use of in a method of separation of low concentrations of surfactant (not just LAS) from other organic matter in environmental samples. The method, called sublation, involves the bubbling of air or nitrogen through the sample on which ethyl acetate is placed. Surfactant accumulates on the surface of the bubbles which carry it into the ethyl acetate layer for further concentration and determination. Also, separation is sometimes achieved more crudely by bubbling air through an aqueous solution of surfactant, or an environmental sample, and collecting the foam in which the surfactant accumulates, leaving an aqueous phase with a lower concentration.

Another general property of surfactants is to lower the rate of re-aeration of solutions of the surfactant deficient in dissolved oxygen, when diffused air is used for re-aeration.

For LAS, the 2- and 3-phenyl isomers have been reported to be less effective as detergents than the internal isomers, 5- and 6-phenyl [1].

At low concentrations (about 1 mg L^{-1}) surfactants dissolve in water normally and each individual molecule or ion is present as a separate entity. As more is added a concentration is reached (100–1000 mg L^{-1}) depending on the salt content of the solution, the temperature, and the chemical nature and structure of the surfactant, at which micelles are formed. This concentration is called the critical micelle concentration (CMC). Micelles contain tens or hundreds of molecules which are oriented with their hydrophobic groups clustered together, the hydrophilic ends extending outwards. Beyond the CMC much more surfactant can be dissolved and the micelles increase in number, the number of single molecules staying roughly the same. Further additions cause the solubility limit to be exceeded and a new phase is formed, usually a hydrated liquid or solid. Micelles play a part in the cleaning action of surfactants but, as far as it is known, they have no bearing on environmental effects of surfactants.

The acid forms of LAS containing less than about 10% water are viscous liquids at room temperature and are pale yellow to dark brown depending on purity. The sodium salts are supplied as waxy solids, pastes or slurries depending on the water content. Iso-propylamine salts are produced as pale amber viscous liquids.

LAS and other alkyl benzene sulfonates have complex but characteristic infra-red absorption spectra by which they have been identified and their degradation followed. The three characteristic and intense absorption bands in the ultra-violet range exhibited by benzene are also shown by LAS, though the wavelengths differ somewhat from those exhibited by benzene itself. This property has been used to detect the biodegradability of LAS, especially in later stages involving ring fission, though it is necessary to bear in mind possible interferences from, for example, nitrate.

Adsorption

In common with other surfactants, LAS adsorbs onto surfaces and suspended solids and it was originally thought that it was removed in sewage treatment processes by adsorption on to sludge rather than by biodegradation. It was, however, soon established that only relatively small proportions (1.5 to 3%) were removed via adsorption on sludge in the treatment process [20, 21], although other authors have reported higher proportions removed, for example, 10% [22], 15% [23].

The adsorption obeys the Langmuir and Freundlich equations [24], over the range of concentration tested, 0.25 to 15 mg LAS L^{-1}. The Freundlich isotherm is $x/m = K_d \cdot c^{1/n}$.

where
   x/m = concentration of LAS on the solids,
   c = equilibrium concentration of LAS in the liquid phase,
   K_d = partition coefficient
   n = a constant, usually close to unity.

The value of K_d appears to vary with the organic content of the suspended matter. In seven sediments from four rivers, K_d varied between 6 and 91 $L\,kg^{-1}$; and the lowest value was for the sediment with the lowest C content and the highest value was that with the highest C content. Based on the organic C content, the average K_d was 1900 $L\,kg^{-1}$ [25]. Other river sediments had values, based on total weight, from 37 to 300 $L\,kg^{-1}$. Soils gave lower values, at 2 to 20 $L\,kg^{-1}$.

Primary sewage sludge had much higher values for K_d, in keeping with their higher organic carbon content, namely 600–1400 $L\,kg^{-1}$ [26], while activated sludge exhibited even higher values at 1000 to 5000 $L\,kg^{-1}$ [27, 28].

In general, adsorption on soils, sediments and sludge was higher for the longer chain homologues and higher for the 2- and 3-phenyl isomers than for the internal isomers. Games [29] reported that K_d for each homologue was about 3-fold higher than that for the homologue with one less C atom in the chain. Hand and Williams [30] found the increase to be about 2.8-fold and that the increase was about 2-fold from the 5-phenyl to the 2-phenyl isomers. The latter authors differed from Urano et al. [25]; Hands and Williams [30] found that adsorption correlated well with silt content but not with organic C content.

Chemical Properties

In common with other anionic surfactants, LAS reacts with large cationic molecules to form a salt or paired ion, which is insoluble in solvents such as chloroform. If a cationic dye is used, the coloured paired ion and the dye can be separated by extraction with chloroform. This reaction forms the basis of the most common analytical method for determining LAS and other anionic surfactants. LAS can be desulfonated by heating with concentrated phosphoric acid at 200–210 °C to yield the parent alkylbenzenes. Under more severe conditions (pyrolysis with or without the presence of P_2O_5) decomposition products are formed which yield distinctive patterns on gas chromatographic (GC) analysis. Alkyl phenols are produced from LAS by fusion with molten alkalies and these can be analysed by direct GC or after bromination or acetylation. LAS may also be converted to alkyl thiophenols by reduction, to sulfonyl chlorides by treatment with thionyl chloride and to sulfonate methyl esters by reaction with diazomethane.

LAS is stable under aqueous acid and alkaline conditions; this has useful practical application and also for distinguishing between alkyl sulfates and alkyl ether sulfates, which are hydrolysed by acid.

Analytical Methods

Methods for the determination of surfactants were reviewed by Llenado and Neubecker [31] and Llenado and Jamieson [32]. The properties of LAS, given earlier, indicate methods which have been used in its determination. Semiquantitative methods for all types of surfactants involved the measurement of heights of foam produced under standard conditions. However, foaming is not a linear function of concentration of surfactants and the presence of a second surfactant or polyethylene glycol gives non-additive heights of foam, often manyfold in excess of expected values. Similarly, the direct measurement of surface tension has been found to be of only limited application.

Colorimetric Methods

Colorimetric methods for the determination of anionic surfactants, including LAS, are all based on the formation of an ion-association complex (paired ion) with a cationic dye. The complex formed is soluble in an organic solvent such as chloroform whereas the dyestuff, in the salt form (e.g. chloride), is not, so that the coloured complex can be separated from the unreacted dye prior to measurement of the intensity of the colour.

Methylene blue has generally been adopted as the reagent dye and the standard mehtod adopted by the International Organisation for Standardisation (ISO) [33] is based on Abbott's [34] modification of the Longwell and Maniece method [35]. Briefly, the surfactant solution is mixed with an alkaline solution of methylene blue and the mixture is extracted with chloroform. The chloroform phase is backextracted with the acidified dye. The first extraction avoids negative interference due to proteinaceous matter present in environmental samples and the second extraction removes interferences of those substances, such as nitrate, chloride, which form methylene blue complexes of low chloroform-extractability. The absorbance of the final chloroform phase is determined at 650 nm. Calibration is made with solutions of a standard surfactant obtained in the purest state possible, and results of tests are expressed in terms of the standard used. For comparative purposes, results are usually expressed in terms of sodium dodecylbenzene sulfonate, the factors for converting other standards to the C_{12}-LAS being 0.781 for dodecane-1-sulfonic acid, sodium salt, 0.828 for dodecane-1-sulfuric acid, sodium salt, and 1.276 for dioctylsulfosuccinic acid, sodium salt [33].

Even though the effects of some positive and some negative interferants are eliminated by this method, not all are removed to allow a direct methylene blue determination of the synthetic anionic surfactants in environmental samples. For this reason the entities detected by this method are more correctly referred to as "methylene blue active substances" or MBAS. For domestic waste water, the MBAS response can generally be taken as an acceptable over-estimate of the synthetic anionic surfactants present, but the synthetic material may contribute only a small proportion of the total MBAS in surface waters, (see Sects pp 20–22, 24–26).

Methods for cleaning up environmental samples, prior to analysis for further elimination of interferants include sublation into ethyl acetate and adsorbing on to adsorbents such as Amberlite XAD-2, followed by elution. Extraction with methanol of sludges, sediments and soils has been successfully used to determine MBAS in solid samples [36] and the efficiency of extraction is reported to be improved by the use of ultrasound [37]. The limit of detection of the standard method carried out manually is given as 0.05 mg L^{-1} for a 100 ml sample. Other cationic dyes have been used in place of methylene blue but have not become established. Some proposed substitutes were claimed to be more sensitive and others faster and simpler to use than methylene blue.

The cationic dye method is not specific for LAS, but by prior acid hydrolysis of the sample any alkyl sulfates and alkyl ethoxy sulfates will be removed.

Chromatography

However, the chromatographic methods for LAS are specific and, in addition, can give measures of not only the various homologues present but also of the isomers.

Thin Layer Chromatography (TLC)

TLC is suitable for separating the various types of surfactants in commercial detergent formulations and also for resolution of homologous hydrophobes but not the various hydrophilic classes. TLC has been used more for non-ionic surfactants and for qualitative or semi-quantitative estimates.

Gas Chromatography (GC)

GC cannot be carried out directly on LAS since they are not sufficiently volatile, and it is necessary to produce a volatile derivative while preserving the original homologue and isomer pattern. In theory GC should permit the separation and measurement of each component homologue and its isomers. In practice the degree of resolution achieved depends on factors such as stationary phase, column efficiency and operating conditions—temperature, carrier gas, flow-rate—and, of course, the objective of the analysis. Prior to forming a volatile derivative it is necessary to concentrate and isolate LAS from the environmental sample and this is done by adsorption, solvent extraction or sublation, or combinations of these. Other clean-up procedures involve cation exchange chromatography and hydrolysis.

The next stage was originally desulfonation by heating with phosphoric acid. It has been found that the conditions such as type of apparatus, temperature, duration, are critical for both quantitative formation and retention of the derived linear alkyl benzenes (LAB), so that checking of any procedure with known standards is essential [38, 39]. Many authors have shown that the LAB's produced by desulfonation of LAS retained the homologue pattern of the original LAS.

Other derivatisation techniques used include the production of sulfonyl chlorides by addition of thionyl chloride and the products can be further converted to the methyl esters [40, 41, 42]. The corresponding fluorides have been formed using PCl_5 and KF [43] and the methyl esters of LAS have been converted to the corresponding thiols [44].

In the GC analysis, capillary columns are more effective than packed columns and a wide range of operating conditions can be tolerated for the separation of the LAB's derived from the LAS's. Capillary column GC is nowadays almost invariably synonymous with high resolution GC and has been the technique of choice of present day investigators.

Complete resolution of all isomers does not appear to have been achieved even with the high efficiency capillary columns. Even the complex procedure of Eganhouse et al. [45], involving an initial TLC separation, permitted the complete resolution of only 3 pairs of LAB isomers. Better resolution in a shorter time (less than 40 min per analysis) than indicated by previous workers was achieved by Waters and Garrigan [38] who employed temperature-programming; even so, the concentration of some pairs of isomers had to be expressed jointly. The overall recovery for their complete procedure using a C_{12}-LAS internal standard was 91%. The use of a mass spectrometer as a detector (GC-MS) provided a powerful means of identification; this technique has been used by authors such as Hon-Nami and Hanya [40, 41] and McEvoy and Giger [42].

Although the limits of detection are such that much less than 1 μg of LAS can be detected, for reasons of handling 100–1000 μg are usually preferred. By fine tuning, Waters and Garrigan [38] were able to desulfonate as little as 50 μg total

LAS and starting with 8L of river water they were able to detect much less than 10 $\mu g\, l^{-1}$ for total LAS and down to about 1% of individual components.

High Performance Liquid Chromatography (HPLC)

As with GC, HPLC—or high pressure liquid chromatography—is in theory capable of resolving all the homologues and isomers of LAS. Present techniques are so versatile that the possibilities range from obtaining a single chromatographic peak representing all the LAS present through to single peaks representing all homologues and their isomers. Although the degree of resolution of isomers that can be obtained with HPLC is not as great as with high efficiency GC columns, HPLC has the advantage over GC that no derivatization is necessary. It is superior to TLC in being less time-consuming and of much greater efficiency. Interference in HPLC is due to the same causes as in GC so that clean-up may be necessary. The same techniques have been used as with GC, including isolation of the MB complex and separation by resin adsorption or ion-exchange. For sludges, sediments and soils, LAS can be extracted with methanol or, to improve selectivity, with methanol and methylene blue.

In earlier work Tanaka et al. [46] obtained all the C_{12}-LAS isomers in one peak using a column of Zorbax SIL with a mobile phase of acetic acid-hexane-isopropanol, having pre-treated the environmental samples on Amberlite XAD-2 resin column. Reverse-phase ion-pair HPLC, used by Gloor and Johnson [47], gave similar high resolutions of LAS isomers and homologues whether ammonium chloride or trimethylammonium chloride was used; tributylammonium chloride gave lower degrees of resolution because its larger size dominated the retention time of the ion-pair. The column was Micropak MC, a monomolecular layer of C_{18} silane bonded to the surface of 10 μm silica. They found that the relative homologue composition of a standard LAS was almost identical by the HPLC and GC methods.

Ion-pairing was used to overcome poor efficiency and low chromatographic selectivity on ion-exchange columns, as well as distorted and split irreproducible shapes. These faults have been overcome by adding a strong electrolye, e.g. Na_2SO_4 instead of ion-pairing [48]. The ion-pair mobile phase method was also applied by Linder and Allen [49] using tetrapropylamine phosphate and tetrahydrofuran, the detector being UV fluorescence. The resolution was chosen to produce single peaks for each LAS homologue and was suitable for detecting aromatic intermediates having at least 2 C atoms in the side chain.

A Japanese group [50] first used a porous styrene/divinylbenzene copolymer as packing, and with perchlorate in methanol-water as the mobile phase, to separate the homologues and the 2-phenyl isomers from the other phenyl isomers using UV fluorescence detector. Later [51] using a porous spherical octadecyl silanized silica gel—LiChrosorb RP-18—of average diameter 5 μm and an aqueous acetonitrile-perchlorate mobile phase, the same group achieved excellent separation of the phenyl isomers. An important refinement was the addition of a high concentration of sodium dodecyl sulfate to reduce loss of LAS by absorption. The method was used to identify the intermediate 3-(4'-sulfophenyl)butyrate in biodegradation tests.

De Henau and his colleagues [52, 53] have improved Nakae's method [51] by using linear gradient elution with 0.15 M perchlorate in 49 to 63% aqueous acetonitrile in 25 min; they preferred UV rather than UV fluorescence for detection. A better separation of homologues was obtained with gradient elution than with isocratic systems. By using a C_1-SAS column instead of the normal C_{18} column, with a mobile phase of sodium perchlorate in aqueous tetrahydrofuran, Castles and Ward [54] were able to elute each homologue in a compact single peak.

For marine samples, Kikuchi et al. [55] rejected hitherto existing methods as being unsuitable, too time-consuming or needing too many steps. They used the Nakae et al. [50, 51] chromatographic system but shortened the extraction system. LAS was extracted from filtered aqueous samples, from methanol extracts of homogenised sediments or fish tissue by adsorption on to mini-columns of Bond Elut C_{18} RP silica. Methanolic eluates were concentrated by nitrogen blow-down and then chromatographed. The limit of detection for each LAS component was about 0.1 μg L^{-1} for aqueous samples and 0.03 μg g^{-1} (dry weight) for sediments; recovery was 80% and 87% respectively.

De Henau et al. [52] used a similar pre-treatment procedure, but containing more steps, to remove interfering non-ionic matter and to ensure higher degrees of recovery of LAS from columns. First, they evaporated down to dryness and took up the residue in aqueous acetonitrile. The amounts of samples taken were normally 5 ml sewage, 10 ml effluent, 100 ml river water, 1 g dry sludge and 10 g dry sediment. LAS was extracted from solids by refluxing, while Kikuchi et al. [55] manually extracted three times. The time per sample was considerably shorter than for GC, but the resolution was not so good. The detection limit was about 0.05 μg for each homologue with a relative standard deviation of about 6%. Recovery of LAS from spiked samples was 94% for water, 87% for sediment and 84% for sludge.

Spectroscopy

Spectroscopic methods generally have been used to identify and characterize unknown compounds rather than to determine them quantitatively.

Hellmann [56] followed the degradation of LAS by using infra-red absorption (IR) measurements but the method normally plays only a confirmatory role; there is also some doubt about the number of absorption peaks necessary for full identification [57, 58]. IR can also be used to estimate the relative amounts of TPBS and LAS from the intensities of absorption in the 7.1–7.3 μm region in which they exhibit different absorptions.

Ultra-violet (UV) absorption, used to detect LAS isomers and homologues in many HPLC separations, can also be used to quantify LAS as a class. Uchiyama [59] extracted MBAS from aqueous samples with methylene blue and

1,2-dichlorethane. The extract was acidified and the liberated LAS was taken back into water; the absorbance in the UV at 222 nm was determined. UV methods can generally be used in biodegradability studies, indicating the removal of the aromatic ring, providing that the presence of interferants, such as nitrate, is taken into account [60].

In a short review of the use of nuclear magnetic resonance (NMR), Swisher [2] concluded that direct application of the technique to biodegradation tests and environmental samples was still some way off, though it was a useful research tool.

LAS in the Environment

The routes to the environment of LAS, as with most chemicals, vary from country to country. In many countries the major route is via sewage treatment works where LAS is subjected to physical and biological treatment, the effluent being discharged to a nearby river. Other routes are direct discharge of sewage to rivers, lakes and the sea. In some countries more LAS passes directly to surface waters than to sewage treatment works.

Sewage

Until fairly recently the concentration of synthetic anionic surfactants in environmental samples has been determined by use of the methylene blue method. Although the methods used are designed to reduce the effects of interferants, the resulting MBAS values still include substances other than synthetic anionic surfactants (see p 16). A further problem is that anionic surfactants other than LAS are also present and these, too, will be included in the MBAS values. However, LAS will usually contribute 70–90% of the MBAS [9]. In the 1980s LAS-specific methods have been widely applied but whichever method is used, the concentrations in untreated sewage vary from place to place depending on local circumstances. The determining factors are the per capita daily consumption of water, whether the sewage includes industrial waste waters and the length of the sewer. Consequently better comparisons are obtained when the results are expressed as g LAS or MBAS per capita per day.

Table 1 provides values from 6 countries for LAS and MBAS in sewages; the lowest values were 2–4 mg L^{-1} in four countries up to about ten-fold higher in Spain and the U.K. An intermediate value was reported from Stuttgart, FRG, of 11.9 mg L^{-1} representing a study lasting one year. These values may be taken to be representative, since they refer either to composite samples from one sewer, or to many samples from a sewer over a long period (up to 6 years), or from a number of sewers.

Table 1. Concentration of LAS and MBAS in sewage

Country	Concentration (mg L^{-1})	Ref.
U.S.A.	3.7 (1.1)	Rapaport et al. [62]
	3.8–6.5	Osburn [39]
	5.25*, 3.5*†	Sedlak and Booman [61]
	3.73, 2.93†	
Canada	2.0 (0.6)	Rapaport et al. [62]
Switzerland	2.4 (0.9)	Giger et al. [23]
West Germany	4.0 (0.5–12.4)	Matthijs et al. [53]
	11.9*	Wagner [63]
Spain	14.7 (9.4–21)	Berna et al. [22]
U.K.	7.4* (6.7–8.6)	Painter and King [64]
	20.8* (15.7–24.5)	Painter et al. [65]
	12.1* (10.2–13.3)	Stiff [66]

* MBAS as dodecyl benzene sulfonate, remainder are LAS
† Known to be settled sewage; remainder unsettled
Values in brackets are ranges or standard deviations

The detailed study by Sedlak and Booman [61] covered three 24-h periods and showed that 71% of total MBAS (5.25 mg L^{-1}) in raw sewage was LAS (3.73 mg L^{-1}) and in settled sewage 84% was LAS (namely 2.93 mg LAS L^{-1} out of 3.5 mg MBAS L^{-1}). Values for sewage in other countries were; Japan 22 mg MBAS L^{-1}, Israel 9–11 mg MBAS L^{-1} and China 5–15 mg MBAS L^{-1}.

Table 2 shows less variation in the LAS or MBAS daily usage per capita, namely 1.8 to 2.7 g cap^{-1} d^{-1} for values obtained by analysis, than in the concentrations. Only values which appeared to be reliable have been included. In the three cases in which comparison is possible, the short-falls between the "sales" value and the analytical value are 20% in the U.S.A., 23% in the U.K. but a much larger discrepancy of 54% for the Spanish values. For W. Europe as a

Table 2. Daily amounts of LAS or MBAS used per person

Country	Water usage (L cap^{-1} d^{-1})	LAS usage (g cap^{-1} d^{-1})	Ref.
U.S.A.	560	2.6ˣ, 2.1	Rapaport et al. [62]
West Germany	185	2.2	Wagner [63]
Spain	—	5.6ˣ, 2.6 (1.7–4.0)	Berna et al. [22]
U.K.	208†	2.5*	Stiff [66]
		3.5*ˣ, 2.7*	18th Rep. STCSD [67]
Japan	120†	2.3*	Nara et al. [68]
		1.8*	Kondoh and Aoi [69]

* MBAS; ˣ calculated from sales, remainder by analysis
† Vol. of sewage produced
Values in brackets are range for 5 works, over a 1–2 y period

whole the LAS used in 1984 was 460 000 tonnes so that with a population of 320 millions the daily per capita usage is about 3.98 g. Table 2 gives the European load to be about 2.5 g cap^{-1} d^{-1}, which represents a "loss" of 35%. Biodegradation in the sewer is a likely contributing factor; the rate of "loss" of MBAS in raw (unsettled) sewage was about 0.2 mg L^{-1} h^{-1} [70] so that for a sewage containing 5 mg MBAS L^{-1} to "lose" one third of its LAS the retention time in the sewer would have to be about 8 h.

Effluents from Biological Treatment Plants

It is common experience that effluents from sewage treatment units contain much less LAS/MBAS than found in the in-going sewage. A representative selection of reported concentrations is given in Table 3. Effluents from activated sludge units contained less than 0.1 mg LAS L^{-1}, range 0.02 to 0.09, except for Spain where the value was 0.2 mg L^{-1}. Effluents from biological (trickling or percolating) filters contained more at 0.14 to 0.76 mg LAS L^{-1}, and 0.2 to 1.0 mg MBAS L^{-1}. These differences are at least partly due to the higher absolute volumetric and BOD loadings on filters than on activated sludge and also to the higher rate at which filters are loaded in U.S.A. compared with Europe. Further, there is a greater possibility of adsorption in the activated sludge process.

The contribution of LAS to total MBAS was lower than in sewage. Sedlak and Booman [61] reported 11–25% and Osburn [39] reported 35–60% for U.S.A. effluents, while in Germany Matthijs and de Henau [53] found 18–53%.

Table 3. Concentration of LAS and MBAS in treated sewage

Country	Concentration (mg L^{-1})	Ref.
U.S.A.	0.05 (0.04)	Rapaport et al. [62]
	0.6† (0.3)	„
	0.14†–0.60†	Osburn [39]
	0.02, 0.05††	Sedlak and Booman [61]
Canada	0.09 (0.05)	Rapaport et al. [62]
Switzerland	0.09 (0.12)	Giger et al. [23]
W. Germany	0.59*	Wagner [63]
	0.07 (0.05–0.11)	Malthijs and De Henau [53]
	0.76† (0.6–0.9)	„
Spain	0.20 (0.13–0.25)	Berna et al. [22]
U.K.	0.2*–1.0*ˣ	Painter et al. [64, 65]
	0.1–0.3@	Waters and Garrigan [38]

* MBAS
† Filter effluent, remainder from activated sludge
†† High-rate effluent
ˣ Both types of effluent
@ Unidentified effluent
Values in brackets are ranges or standard deviations

Sewage Sludges

Up to about 20% of the LAS/MBAS entering treatment works is associated with the sludge settling in the primary settlement tanks. Few data exist for the LAS content of primary (or raw) sludge and these are given in Table 4 with some earlier MBAS values. Also presented are some of the larger number of values for anaerobically digested sludge, as well as values for other types of sewage sludge. Most values relate to grab samples; those of Sedlak and Booman [61] represent composite samples taken over a period of several months.

Table 4 shows that both primary and anaerobically digested sludges, which are the types applied to agricultural land as fertilizer, contained 2–12 g LAS kg^{-1}, most in the range 4–10 g kg^{-1}. These values are close to that predicted by Gilbert and Pettigrew [71] and also that predicted from the adsorption partition coefficient of LAS (see p 14).

Aerobically digested sludges contained lower amounts at 2.1–4.3 g LAS kg^{-1}, indicating possible aerobic degradation. Activated sludge contained even less, at 0.09 to 0.86 g kg^{-1}, clearly demonstrating degradation. The work of Sedlak and Booman [61] was most detailed. Their raw sludge contained 5.3–6.3 g LAS kg^{-1}, whereas their (secondary) activated sludge, from a unit operated with a sludge retention time of 3.2 d, contained only 0.41 g LAS kg^{-1} and the sludge from their high rate unit, sludge retention time of 0.8 d, contained 0.86 g LAS kg^{-1}. The authors' anaerobically digested sludge contained 6.66 g LAS kg^{-1}, compared with 4.25 g LAS kg^{-1} after aerobic digestion. These digested sludges when dried on open air drying beds had significantly reduced contents of LAS of 0.15 g LAS kg^{-1}.

Table 4. Concentration of LAS in sewage sludges

Type	Source	Concentration ($g\,kg^{-1}$)	Ref.
Primary	1 works U.S.A.	5.3–6.3	Sedlak and Booman [61]
	10 U.K. works	7.3* (4.7–11)	Bruce et al. [72]
Anaerobically digested	49 grab samples at 5 U.S.A. works	4.66 (1.54)	Rapaport et al. [62]
	1 works in U.S.A.	6.7	Sedlak and Booman [61]
	7 Swiss works	5.9 (3.1–11.9)	Giger et al. [23]
	24 Austrian works	4.2 (1.2)	,, ,,
	10 German works	4.9 (1.3–9.9)	Matthijs and De Henau [24]
	4 Spanish works	9.7 (7.0–12.8)	Berna et al. [22]
	10 U.K. works	9.9* (8.5–16)	Bruce et al. [72]
Aerobically digested	1 U.S.A. works	4.3	Sedlak and Booman [61]
	5 Austrian works	2.1 (1.8)	Giger et al. [23]
	1 Swiss works	2.9	McEvoy and Giger [42, 73]
Activated sludge	2 samples, 1 U.S.A. works	0.41, 0.86	Sedlak and Booman [61]
	1 Japanese works	0.09	Yoshimura [74]
Air-dried sludge	2 samples, 1 U.S.A. works	0.15, 0.16	Sedlak and Booman [61]

* MBAS. Values in brackets are ranges or standard deviations

Soils

In a U.S.A. field treated from 1972 with anaerobically digested sludge at about twice the normal annual rate, analysis shortly after the last application in 1979 showed little, if any, penetration of LAS had occurred below the top 15 cm (< 3 mg LAS kg^{-1}). The top 7.5 cm contained 28(13–47) mg LAS kg^{-1} [62].

De Henau et al. [52] found 0.9 to 1.3 mg LAS kg^{-1} in German soil and 2.2 mg LAS kg^{-1} in U.K. soil; they predicted the concentration to be 7 mg LAS kg^{-1}, while Gilbert and Pettigrew [71] calculated a value of 16 mg LAS kg^{-1} in U.K. soil, indicating some removal.

The soil in 42 fields in the U.K. which had not received sludge for at least six months contained less than 0.5 mg LAS kg^{-1}; in another 7 fields the concentration was 1.1 to 2.5 mg kg^{-1} [75]. Immediately after dosing with sludge, the concentration in 7 soil samples taken from 5 fields ranged from 2.5 to 40.3 mg LAS kg^{-1} (median 25 mg LAS kg^{-1}); these values fell to 'control' values within 21 to 122 days.

River Water

Published data on the concentration of LAS in river water and other surface waters are sometimes difficult to interpret because of lack of information concerning the degree of pollution of the sampled river. Consequently, values as disparate as 0.005 to 6.9 mg MBAS L^{-1}, and up to 1.6 mg LAS L^{-1} have been reported as being present in river waters. Often the intention of a study has been to determine the fate of LAS downstream of a known discharge of effluent or sewage in order to establish the kinetics of degradation, so that values for concentration of LAS reported in the literature are not derived from random samples. Some of these values are presented in Table 5, together with examples in which the information on pollution is more precisely stated. It may be concluded from these and other data that for samples of water not grossly polluted by sewage and effluent discharges, the most likely range of concentration of LAS is 0–0.04 mg L^{-1}. For MBAS the corresponding range is 0.005 to 0.2 mg L^{-1}, but the MBAS is known to contain material other than LAS and other than synthetic anionic surfactants. Even where higher levels occur, the evidence is that the downstream concentration soon falls to levels approaching those representative of 'unpolluted' sites.

The position of river water in Japan is complicated since in some parts of the country only the waste waters containing excreta are treated at sewage works, the laundry washing water being discharged untreated to rivers. Because of this, values as high as 2.5 mg LAS L^{-1} [76] and 2.6 mg MBAS L^{-1} [82] have been reported. On the other hand, the mean value for a number of rivers in the Hiroshima prefecture was only 0.018 mg LAS L^{-1} [83]. In five instances in Japan the ratios of LAS to total MBAS were found to be 10, 40, 50, 60–70 and 37–83%.

Table 5. Concentration of LAS and MBAS in river waters

Country and Ref.	Site	Concentration (mg L^{-1}) "Unpolluted"	Polluted	Type
U.S.A.				
Rapaport et al. [62]	Just below discharge	—	0.099 (0.085)	LAS
	<5 km below	0.063 (0.03)	—	
	>5 km below	0.041 (0.03)	—	
Hennes and Rapaport [77]	8 rivers about 1 km below	0.01–0.04	—	LAS
W. Germany Fischer [78]	4 rivers	0.075 (0.002)	0.2–0.5	MBAS
	many sites	0.03–0.04	—	*IL-MBAS
Hennes and Rapaport [77]	11 rivers 1 km below discharge	0.01–0.09	—	LAS
Gerike et al. [79]	R. Rhine	0.017 (0.009–0.035)	—	LAS
U.K.				
Waters and Garrigan [38]	1 river:	0.039	—	LAS
	2 rivers:	—	0.2–0.5	MBAS
	1 river: above discharge	0.012 (0.008–0.019)	—	LAS
	just below	—	0.08 (0.01–0.17)	LAS
	5–16 miles below	0.04 (0.008–0.095)	—	LAS
Spain				
Moreno Danvilla [80]	1 river three sites	<0.1–0.19	—	MBAS
Belgium				
Inst. d'Hygiene [81]	189 sites	0.005–0.2	0.2–6.9	MBAS

*"Interference limited" MBAS

In the U.S.A. LAS contributed only 10% of total MBAS in a river 55 miles downstream from a discharge [39], while in W. Germany the proportions were 43–70% (R. Neckar), 54% (R. Main), 50% (R. Ruhr) and 15–35% (R. Rhine) [78]. The value for the Rhine has also been reported as 28% [79]. In the U.K. and the Netherlands the proportions were 26% and 18%, respectively [38].

A model has been developed using results of an extensive monitoring programme in the U.S.A., which has used percentage removals of LAS at various stages in the aquatic environment to predict the variability of concentrations of LAS in rivers below the mixing zones caused by effluent discharge [77]. Predicted concentrations agreed well with those observed.

River Sediments

Concentrations of LAS in river sediments from three countries depended on the sampling site. Sediments taken upstream of discharges of sewage or sewage effluent contained 1 to 10 mg LAS kg^{-1} (dry solids), while in the vicinity of the discharge values as high as 100–200 mg LAS kg^{-1} were observed.

Generally sediments from Japanese rivers contained higher concentrations, presumably because of the lack of treatment of laundry wastes. For example, the mean for 9 rivers in the Tokyo region was 107 (1–567) mg LAS kg^{-1} [84], although for 20 rivers in Nagano province the median was as low as 0.5 mg LAS kg^{-1} [85].

Sediments taken from the close vicinity of a discharge in a river in the U.S.A. contained 174–275 mg LAS kg^{-1}, while downstream samples contained less than 12 mg/kg^{-1} [29, 62]. At 10 sites in W. Germany the range of LAS content was 1.5 to 10 mg kg^{-1}; at four other sites the range was 25 to 174 mg kg^{-1} [86]. In sediments distant from discharges, the proportion of LAS of total MBAS was 17–25% in Japanese rivers [87], in the U.S.A. the value was 27–38% [39]. Sediments containing a mean of 107 mg LAS kg^{-1}, also contained concentrations of the parent hydrocarbons of 3.6 mg LAB kg^{-1}, range 0.01 to 15 [84].

Estuarine and Coastal Waters and Sediments

Few data seem to have been published on the LAS content of waters and sediments in estuaries and the sea. Data from Japan indicate that in coastal waters some distance from a discharge the LAS content was in the region of 0.0008 mg L^{-1} [55]. Nearer to outfalls concentrations as high as 0.03 mg LAS L^{-1} were recorded. In Tokyo Bay the concentrations ranged from <0.003 to 0.014 mg LAS L^{-1}, corresponding to 0.03 to 0.07 mg MBAS L^{-1} [41]. In an unpolluted bay in France concentrations of 0.01 to 0.09 mg MBAS L^{-1} were found compared with 0.09 to 0.54 mg MBAS L^{-1} near the mouth of the R. Rhone [88]. The waters of the Baltic Sea were found to contain 0 to 0.09 mg "anionic surfactants" L^{-1} [89]. In sediments taken from the open sea, no LAS could be detected; nearer to an outfall 3–17 mg LAS kg^{-1} was reported [90].

Tapwater

Only a small number of values for the concentration of LAS in potable water were found. Samples from The Netherlands contained 0.003 mg LAS L^{-1}, equivalent to 27% of the total MBAS present, and in U.K. samples the values were 0.007 mg LAS L^{-1} and 33%, respectively [91]. Some 42 samples from 26 regions of Japan contained less than the limit of detection of 0.001 mg LAS L^{-1} [92] while Osaka tapwater contained 0.01 to 0.07 mg LAS L^{-1} [93]. Italian well-water contained from undetectable amounts to 0.008 mg LAS L^{-1} [94].

Mean Chain Length of LAS in Environmental Samples

The C_{10} to C_{13} homologues were found in all environmental samples examined, with C_{11} and C_{12} dominating; sometimes C_9 and C_{14} were also present, and on occasion C_{15} was also detected. The average chain length is in the C_{11}–C_{12} region; rarely lower and higher means have been reported—10.9 in Japanese river waters [95] and >12 in U.S.A. river sediments and anaerobic sludges [62]. In their extensive study with large numbers of samples, Rapaport et al. [62] showed a significant difference, at the 0.01 level, between the mean chain length of LAS in sewage (12.0) and the activated sludge effluent (11.8). Similarly, the mean chain length for sludges (12.5) was significantly higher than the value for sewage (12.0). However, in some individual treatment plants no difference was observed between the chain length distributions in sewage and treated effluents. The chain length was also higher in river sediments (11.8 to 13) than in river water (10.9 to 11.2); the difference is statistically significant. These differences have been ascribed to the difference in the extent to which the homologues are adsorbed on to solid particles and to the differences in their rates of biodegradation.

Environmental Fate

Metabolic Pathways

Very many studies have been made of the biodegradation of LAS and these have shown that the surfactant is well biodegraded under a wide variety of aerobic conditions (see pp 30–35). Much of the evidence is given by Swisher [2] and has been recently reviewed by Painter and Zabel [96] and, with particular reference to metabolic pathways, by Cain [97] and Schöberl [98]. Knowledge of the pathways is not fully complete; presumably this is partly due to the large number of homologues and isomers of LAS and partly due to the versatility of bacteria. The consensus of opinion from the available evidence is that the main pathway (Fig. 1) involves, first of all, attack on the alkyl side chain by ω-oxidation (on the terminal CH_3-group) followed by β-oxidation to form a series of sulfophenyl alkanoic acids. The methyl group further from the aromatic ring is attacked first to form a carboxylic acid and β-oxidation results in carboxylic acids with 2 carbon atoms fewer, although there is evidence that α-oxidation also occurs. It is thought that this oxidation proceeds until 4 or 5 C atoms remain in the side chain to form, for example, sulfophenyl butyric acid. These intermediates are then converted, in steps, to the 2,3-dihydroxy derivatives which then undergo meta-fission of the ring to form alpihatic sulfo-keto unsaturated dicarboxylic acids. It is only at the next stage that the sulfo-group is split off to form sulfate and the keto-unsaturated dicarboxylic acids which are readily converted to catabolites,

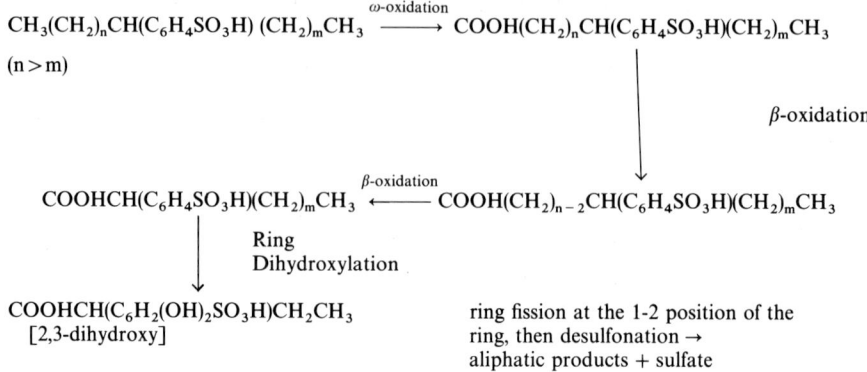

Fig. 1. Postulated metabolic pathway of LAS

such as glyoxylic acid, which are introduced into the normal metabolic pathways, Krebs and glyoxylate cycles.

Other alternatives which may occur if other, perhaps rarer bacterial species are present, are:

(a) α-oxidation of the side-chain,
(b) formation of dialkanoic acids of the type, sulfophenylsuccinic acid and sulfophenylmalonic acid; the former is converted to sulfophenylbenzoic acid and the latter to sulfophenylacetic acid,
(c) either hydrolytic or reductive desulfonation before the alkyl side chain is oxidized [99]; but the overwhelming evidence is that side-chain oxidation predominates as the first site of attack: although sulfo-aromatic acids have been found, there are no reports of the presence of the corresponding phenylalkanoic acids,
(d) unsaturated aromatic acids e.g. sulfocinnamic acid have been detected and do not fit into the proposed pathway.

Intermediates

The only aromatic intermediates identified from cultures grown on LAS are the short-chain sulfophenyl mono-alkanoic acids, including 2- and 3-(sulfophenyl)-butyric, 3- and 4-(sulfophenyl)valeric, 4- and 5-(sulfophenyl)hexanoic, and 5- and 6-(sulfophenyl)heptanoic acids [100]. Accumulation of acids with the phenyl group in the middle of the chain or having the original LAS alkyl chain length did not occur. The short-chain acids accounted for almost all of the organic matter present; other organic chemicals identified include aliphatic compounds containing the sulfonic group.

The *Pseudomonas* sp. investigated by Divo and Cardini [100] could not degrade these intermediates but on addition of a mixed culture the aromatic intermediates rapidly and completely degraded. Similarly, another pure culture degraded and grew on LAS and sulfophenyl-undecanoate but could not break the ring nor degrade the seven or so short chain aromatic acids; on adding a second species, which could not degrade LAS, the intermediates rapidly disappeared [74]. Many species can carry out the initial degradation steps of LAS but it appears that few, if any, are able to mineralize the surfactant. Most LAS-degraders oxidized the surfactant at only low concentrations, about 30 mg LAS L^{-1}, but some degraded LAS at concentrations as high as 1000 mg LAS L^{-1}. The rates of removal of LAS by four isolates, individually and collectively, were lower than the rate with activated sludge but when the four isolates were grown in "floc" form the degradation rate increased to that with the sludge [101] indicating that adsorption might play a part.

Effect of Structure of LAS on Biodegradability

There is ample evidence that the primary biodegradability, but not necessarily mineralization, of homologues of LAS (C_{10} to C_{14}) in general increases as the chain length increases. For the isomers of a given homologue the greater the distance between the sulfonic group and the more distant terminal methyl group on the alkyl chain the faster the degradation. Swisher [2] has called this the "distance principle". These phenomena have been observed both in batch experiments [e.g. 102] and in laboratory-scale continuous activated sludge units [e.g. 103].

Some exceptions to these general rules are brought about by factors such as:

— the concentration of the LAS homologue or isomer,
— possible inhibitory effects of LAS,
— concentration of suspended solids,
— the state of acclimatization of the inoculum,
— the presence of LAB's in the LAS used, if desulfonation-GC is used for analysis.

An example is the rate of primary biodegradation in die-away tests of the isomers of C_{12} LAS. When present as a mixture of all five isomers the order, in decreasing magnitude, was 2-, 3-, 4-, 5-, 6-, but when present separately, at 30 mg L^{-1}, these changed to 3-, 2-, 4-, 6-, 5 for primary biodegradation and for ring opening the order was 3-, 2-, 6-, 5-, 4 [104]. The longer chain homologues of LAS degraded to CO_2 more slowly than the shorter chain homologues in shake flask systems inoculated with sewage or soil. This was attributed [105] to the higher inhibition of bacteria of the higher chain homologues, added at 30 mg L^{-1}, since the effect was reduced or eliminated on further addition of LAS during the first week.

The evidence from the distribution of homologues and isomers in environmental samples and in commercial LAS is conflicting on this point. Simko et al. [106] found the same distribution in sewage and effluents from a full scale activated sludge plant and Waters and Garrigan [38] also observed no effect in homologue distribution but did find a predominance of the internal isomers of C_{10}–C_{13} homologues in effluents. Four groups of Japanese workers confirmed both rules; for example, Abe and Seno [107], confirmed both while Utsumoniya et al. [108] could not confirm the homologue rule.

These rules were based on experiments with LAS at or around 10 mg L^{-1}. Larson and De Henau [109] investigated the removal of LAS homologues and isomers in various environmental compartments at much lower concentrations (10 to 100 μg L^{-1}) using ^{14}C-labelled materials, measuring the rate of ^{14}CO$_2$ production. They found neither clear indications of significant differences between homologues and isomers, nor trends one way or the other. The mean 1st order decay constant, k, for the homologues C_{10} to C_{14} was 0.81 ± 0.17 d^{-1} in river water and 1.12 ± 0.15 d^{-1} in river sediment, which are equivalent to half-lives of 20 ± 2 h and 15 ± 2 h, respectively. Also, homologues degraded at rates independent of the presence of one another. Results of this investigation confirm the complete biodegradation of LAS; the yield of ^{14}CO$_2$ was about 80% of theoretical, nearly all of the remaining ^{14}C being present as biomass on the membrane after the culture medium had been filtered.

Similar experiments with activated sludge [110] gave half-lives of C_{10}- to C_{14}-LAS of 1.5–2.2 d—little difference between homologues—with 60–70% ^{14}CO$_2$ evolved in 21 d; the values were similar to those for ^{14}C-labelled benzoic and phenylacetic acids. No explanation for the differences between the results at 10 μg L^{-1} and 10 mg L^{-1} are yet forthcoming.

There is apparently no difference in the rates at which enantiomorphic forms of LAS degrade [8].

Degradation in Laboratory Tests

In reviewing the evidence for biodegradability of LAS a number of factors must be borne in mind, no matter what type of test is applied. Methods range from die-away tests in synthetic media with various inocula or in river water, tests simulating waste water treatment and lastly, tests for inherent biodegradability which allow a greater chance for biodegradation by using less severe conditions and more time.

Die-Away Tests

Die-away tests make use of a wide range of bacterial cell density as inocula. The original OECD die-away test, incorporated in the EEC Directives on Detergents, employed only a very low cell density of about 0.5 to 2.5×10^2 cells mL^{-1} and assessed only primary biodegradability. However, when applied to the assessment of ultimate biodegradability of organic chemicals generally, many chemicals known to be readily biodegradable did not degrade, though they did so in other well-known tests, namely AFNOR, Sturm, MITI [111] and ISO

[112] methods (see Painter and King, [113], for further discussion on methods). The latter four methods employ cell densities up to 10^5 to 10^6 mL^{-1} and are widely accepted as passing only those chemicals which "biodegrade rapidly and completely" in a wide variety of aerobic environments, and do not give "false positives" [e.g. 114, 115, 116].

These differences in results with different cell densities explain why some reports give low removals of LAS-DOC, though accompanied by high removals of MBAS, using low cell densities. For example, Leidner et al. [117] found only 55% DOC removal after 56 d and Painter [118] reported only 39–63% after 19–55 d; in both cases >95% MBAS was removed. At higher cell densities, the removal of DOC was higher-81% [119], 75% [120], 90% [121]—as were the proportions of ThOD and ThCO$_2$-98% ThOD [122] and 75% ThCO$_2$ [123]. Removal of the ring was shown by 90% removal of UV absorbance [124]. The higher cell density tests give more reproducible results by ensuring a greater diversity of metabolic activity not always guaranteed in tests with low cell density. River water die-away tests gave variable results because of the variation in cell density, but high values were often obtained, for example, 85% DOC removal [125, 126].

Another factor causing low results is the concentration of test chemical used. The original OECD method employed only 5 mg MBAS L^{-1}, but many workers have used considerably higher concentrations of LAS than this. For example, at 10 mg LAS L^{-1} 85% DOC was removed but only 5% was removed when 100 mg LAS L^{-1} was tested, indicating that the higher concentration was inhibiting.

Simulation and Inherent Biodegradability Tests

There is clear evidence that LAS can be mineralized since DOC removal in the inherent biodegradability tests was high; in the semi-continuous activated sludge (SCAS) test the removal was 93% [119], 98% [115], 82–93% [120] and in the Zahn-Wellens test 95% was removed [120]. Also, UV-absorbance was removed in SCAS tests from three forms of C_{12}-LAS by 85–94% [127], showing high removal of the ring. Similar high removals (93–98%) were reported by Neilsen et al. [128], who tested ^{14}C-labelled LAS by the SCAS method, followed by a modified Sturm test.

However, the fate of organic C and the benzene ring in a given organic chemical cannot be readily determined in a full-scale plant and resort has to be made to laboratory simulation tests. Several authors have used various forms of the OECD Confirmatory Test, employing a control Husmann or Porous Pot unit receiving only synthetic sewage and a second identical, unit receiving LAS in addition [129]. Table 6 indicates that DOC removal ranged from 73 to 96%. Two exceptions to these high results are those of Brown [130]—33 to 41% DOC—and Hrsak et al. [131]—8 to 10% of UV-absorbance. Brown [130] also found that glucose was not well removed, only about 50% DOC being removed. Both groups of workers used the original OECD synthetic sewage recipe, which

Table 6. Removal of LAS in laboratory simulation tests (activated sludge)

	Removal (%)			Ref.
MBAS	DOC		UV absorbance (or other)	
—	80–85		79–83	Swisher [57]
97	80		—	Painter et al. [132, 133]
91–97	78–89		83 (COD)	Brown et al. [134]
96	85		90	Moreno Danvila [80]
100	—		80–90	Petresa [135]
90–95	73 [95, metabolism method]		—	Gerike and Jasiak [136]
—	82±2		—	Gerike et al. [137]
95–97	93.5* (6d) 96.6* (10d)		—	Gilbert and Kleiser [120]
92	33–41		—	Brown [130]
64–94	—		8–10	Hrsak et al. [131]

*In domestic sewage with extra 15 mg LAS L^{-1} added
Values in brackets are sludge retention times

could have been deficient in phosphorous and this deficiency caused low results [138].

Besides this P-deficiency, two other factors can lead to lower removals. The use of domestic sewage can lead to higher removals of DOC, for example, Gilbert and Kleiser [120] found an increase from 53±8 to 88±9% DOC removal of an alcohol ethoxylate when domestic sewage replaced P-amended synthetic sewage. The same authors found 93.5% and 96.5% DOC removal of an additional 15 mg LAS L^{-1} added to domestic sewage at 6 and 10 d sludge retention times, respectively. Secondly, the simulation test is restricted to 9 weeks, yet it has been found that the percentage removal of LAS-DOC gradually increased beyond this period [132]. Also, during the first few months of operation sulfophenyl alkanoic acids were identified in the effluents at 10–100 μg L^{-1}, but after a few more months the intermediates could not be detected.

^{14}C-labelled LAS studies [139, 140, 141] have confirmed that 80–90% of both ring and chain-C are removed as $^{14}CO_2$ and biomass under conditions simulating waste-water treatment.

Removal of LAS in Sewage Treatment

It is the practice to express the efficiency of sewage treatment works as a percentage removal of COD, BOD, etc., and this has been extended to MBAS and LAS. On kinetic grounds, the practice is faulty, since the removal is not zero order throughout the range of concentrations found in sewage treatment but changes to 1st order at lower concentrations. (This problem is overcome for river water—see p 34). The practice is also misleading since it is the concentration in the effluent—and hence in the receiving stream—which is the important factor in determining effects in the environment, although percentage removal does give a rough index by which to judge the amenability of chemicals to treatment.

Aerobic Conditions

Data, such as that given in Tables 1 and 3, show that in properly operated activated sludge treatment the removal of LAS is greater than 95%, usually 98 to 99%, while for biological filters the removal is lower at 80–91%. The corresponding removal of MBAS is not so high at 93–98% by activated sludge and 85–98% in filters. In grossly overloaded or poorly operated treatment plants removals lower than 60% have been reported. Some of the references cited in Tables 1 and 3 indicate that in full-scale treatment temperature had little or no effect on the removal of LAS. Using model activated sludge units, it has been shown that there was a lack of effect between 9 and 20 °C, but at 5–8 °C 90% MBAS was removed compared with 97% at 20 °C [142].

Gerike [3] observed that, in Germany, the efficiency of sewage treatment plants in removing LAS correlated well with expected results based on laboratory tests. In Switzerland 99% was removed in 8 treatment works, about 15% being adsorbed on activated sludge. Berna et al. [22] found an average of 98% removal of LAS at 5 treatment works in Spain, with about 10% adsorbed on to sludge. These authors also calculated 1st order rates, from influent and effluent concentrations of LAS and the retention time of sewage in the aeration tanks, to be between 0.1 and 1.2, mean $0.7 \, h^{-1}$. In the U.S.A. 95–99% of LAS was removed in activated sludge units with about 3% adsorbed on to sludge [61], but rather less removed in biological filter, 73–87%. In Europe, the removal of LAS in biological filters was higher than in U.S.A. presumably because filters are normally loaded at about 4.5 m^3 sewage $m^{-3} d^{-1}$ compared with only 0.5 m^3 sewage $m^{-3} d^{-1}$ in Europe.

The above removal or elimination values could have been achieved by non-biological processes; they do not indicate ultimate biodegradation or mineralization. However, it is difficult to prove conclusively that LAS (or indeed any other chemical) is ultimately biodegraded or mineralized in the environment. Only one study has been published, by Leidner et al. [117], on the removal of LAS and the production of known aromatic intermediates—in four Swiss full-scale treatment plants. Re-examination of the Leidner data [120] showed that the removal of LAS plus all known residues was as high as 92, 99.7, 99.7 and >99.7% in the four Swiss works. These values can be taken as representing ultimate biodegradability since the residues determined included all known intermediates.

Anaerobic Conditions

It is clear that LAS is not degraded in anaerobic sewage treatment [26, 61, 143]. However, an indication was given by Ward [110] that a pre-aerobic treatment of

LAS causes changes in the molecule which permitted subsequent degradation under anaerobic conditions.

Removal of LAS in Rivers and Sediments

From an earlier section (see Table 5) it may be inferred that LAS was degraded in rivers and this is backed up by results of die-away tests using 10 mg LAS L^{-1} in river water in which the oxygen uptake was 70% of ThOD [144].

Tests using low concentrations (5–500 $\mu g\,L^{-1}$) of ^{14}C-ring labelled LAS in river water showed that 68–82% of the theoretical $^{14}CO_2$ was produced with 1st order constants (k) dependent on the source of the water. In water taken above an effluent outfall k was 0.05 d^{-1}, or a half-life (t_{50}) of 14 d, and with sediment it was 0.26 d^{-1} $t_{50} = 2.8$ d. In river water taken below the effluent outfall, k was 0.5 d^{-1} $t_{50} = 1.4$ d, and with sediment k rose to 0.95 d^{-1}, $t_{50} = 0.7$ d [145, 146]. Thus, even water above the outfall contained microorganisms capable of degrading LAS to CO_2; the rate increased significantly when sewage effluent or sediment was present.

Ventullo and Larson [147] found that micro-communities exposed in tubes in a river to 1 mg LAS L^{-1} greatly increased, by 80-fold, the rate at which ^{14}C-LAS was converted to $^{14}CO_2$; that is from 1.5 ng $cm^{-2}\,h^{-1}$ to 134 ng $cm^{-2}\,h^{-1}$. Similar increases in activity were found in ground-water and sub-surface sediments [148].

It may be concluded that LAS is ultimately biodegraded extensively in a variety of aquatic environmental compartments and the rates exhibit 1st order kinetics at trace ($\mu g\,L^{-1}$) concentrations in natural systems. The rates are more than adequate to allow LAS to be biodegraded within the residence times in the specific compartments.

Removal in Soil and Sub-Surface Sediments

LAS reaches the soil by two main routes—by discharge of septic tank effluents and by application of various forms of sewage sludge to agricultural land.

In a field study of a septic tank tile field [149] t_{50} values were determined of LAS added to core samples taken at various distances from the septic tank located in the grounds of a Canadian University. The t_{50} values in water samples tended to decrease with increasing distance from the tank, whereas half-lives in soil samples tended to increase. Near the tank t_{50} was 11–19 d rising to 30 d in soil (but still about 11 d in water samples) some distance away. The concentration of LAS fell from 10 000 $\mu g\,L^{-1}$ to 30 $\mu g\,L^{-1}$ over a horizontal distance of 10 m. In an earlier section (p 24), concentrations of LAS in soil in fields receiving sludge indicated that LAS was well degraded with time. This has been confirmed by experiments in which ^{14}C-LAS was added to soil samples and also from field studies in which LAS-amended sludge was added in situ.

In laboratory tests the half-life for $^{14}CO_2$ evolution was 8–10 d, with 60% of theoretical $^{14}CO_2$ evolved in 25 d rising to 80% after 100 d; in control tests added ^{14}C-acetate was removed faster but yielded similar proportions of $^{14}CO_2$, namely, 40% after 60 d and 60% after 100 d [120].

In similar tests, [150], the half-life was 18 to 26 d, mean 22 ± 3 d in soils from one source which had received sludge for at least 5 years, and 18–19 d for a soil from another source. In the first soil the lag before degradation was only 1 to 4 d, but in the second soil the lag was 15 to 20 d. Two years later, t_{50} for the second soil had fallen to 1.8 to 2.1 d. Later, Larson and Bishop [151] reported the half-lives in 7 soils to be 1.5–3.3 d, even though the soils came from sites with no known history of exposure to sludge and surfactants.

From the paper of Giger et al. [152] in which a decrease of LAS from 45 mg LAS kg^{-1} to 5 mg kg^{-1} in soil from a Swiss field was reported over 104 d, Ward and Larson [150] calculated the t_{50} to be 4.4 d. In an extensive study of 5 fields in the U.K., the t_{50} for LAS ranged from 7 to 22 d; these agreed well with laboratory studies and monitoring findings [75].

The half-life for LAS in sub-surface sediments beneath a laundromat waste treatment pond was about 4 d down to 2 m rising to about 18 d at 13–20 m [151]. The average $^{14}CO_2$ yield was 33% and the fate of the remaining ^{14}C was not investigated. ^{14}C Benzylamine tested at the same time yielded only 31% $^{14}CO_2$, but had a t_{50} of 3.3 d.

Effects on the Environment

Because of the routes by which LAS enters the environment, LAS could affect the various stages of sewage treatment, free swimming and benthic organisms in rivers, estuaries and the sea and also microorganisms and plants in soil. Laboratory tests have been carried out on various aquatic organisms to ascertain what effects LAS has and, in some instances, laboratory tests have been backed up by field observations.

Sewage Treatment

Foaming problems in aeration tanks at sewage treatment works and in rivers associated with the poorly degraded TPBS disappeared with the introduction of LAS. The re-aeration rate of water bodies is known to be lowered by the presence of surfactants when diffused air is used, but there is no significant effect on the performance of the activated sludge process unless the air supply is only just sufficient, when the degree of nitrification may be reduced. With mechanical aeration, the aeration rate is not lowered and may in some cases be increased.

No reports have been seen of any harmful effects of LAS on the large-scale aerobic treatment of sewage and laboratory-scale tests have shown no harmful effects up to 30 mg MBAS L^{-1} [71, 142]. Concentrations up to 70 mg L^{-1} could be tolerated if the increase was not made in one step [153].

The oxygen uptake rate by activated sludge was not inhibited at concentrations of LAS up to 100 mg L^{-1} and in die-away tests of glucose-LAS mixtures, 15 mg LAS L^{-1} did not inhibit the removal of glucose but the surfactant did not begin to degrade until the glucose had been removed [124]. The EC_{10} of LAS against *Pseudomonas putida* was 50 mg L^{-1} [4].

Bench-scale anaerobic treatment of sewage sludge was inhibited at concentrations of LAS above about 20 g MBAS kg^{-1}. Up to 15 g kg^{-1} ($\equiv$ 25 mg LAS L^{-1} in unsettled sewage) there was no effect on gas production, but between 15 and 20 g kg^{-1} reliable digestion is sometimes impaired and above 20 g kg^{-1} a more serious inhibition is observed, especially if other potentially inhibitory compounds are present [72, 154]. If time for gradual acclimatization is allowed, concentrations as high as 30 mg MBAS kg^{-1} can lead to successful gas production. Present concentrations of LAS in primary sludges are around 5–6 g kg^{-1} so do not present problems for anaerobic digestion, and none have been reported.

Natural Waters

The metabolic activity of epilithic bacterial communities grown on plates suspended in a river above and below a sewage effluent outfall was unaffected by concentrations of LAS up to 0.1 mg L^{-1}. At 1 mg LAS L^{-1} the activity was significantly lowered at both sites, though there was little effect on total biomass. However, chronic exposure to LAS for 21 d brought about a complete recovery of microbial activity and, in some cases, even an increase [155]. Similar results were reported for bacterial communities in a lake.

Soils

LAS inhibited nitrification in soils but at concentrations much higher than currently found in soils. About 50% inhibition occurred at 500 mg LAS kg^{-1} and about 10% at 100 mg kg^{-1} [156], compared with current concentrations of 1–45 mg kg^{-1}. Litz et al. [157] found no effect on soil respiration rate by the addition of 5 or even 50 g LAS m^{-2} equivalent to about 30 and 300 mg LAS kg^{-1}, respectively.

Aquatic Organisms

Kimerle [158] has recently reviewed the data on the ecotoxicity of LAS and considered that the available data are as extensive and comprehensive as exist for any chemical. He concluded that the demonstrated exposure of LAS in water, sediments and sludge-amended soils provide substantial margins of safety for aquatic and terrestrial organisms. In general, the higher homologues are more toxic than the lower ones; while this is of academic interest, it has little bearing on environmental effects since commercial products have average chain lengths of C_{11}–C_{12}. For example, in W. Germany the average chain length is 11.6 (MWt = 348) and therefore the relevant toxicity data to be used in evaluating environmental hazards are those relating to $C_{11.6}$ or C_{11}–C_{12}. Intermediate degradation products have been shown to be considerably less toxic than the parent LAS. Besides the normal experimental reasons for erroneous results, another source of error is the lack of monitoring the concentration of the toxicant in the test solution. Not only can the actual concentration be different from the nominal, degradation can also occur to give falsely high LC_{50} values (for LAS).

A summary of toxicities to various organisms is given in Table 7.

Algae

Toxic effects on algae are expressed as an EC_{50} (mg L^{-1}) on the growth rate or biomass produced usually over a 72 to 96 h period. The range of EC_{50} values for some dozen or so species is very wide, most falling within 0.9 to 100 mg LAS L^{-1} [96] which compares with the range of 10 to 300 mg LAS L^{-1} for growth rate tests given by Schöberl et al. [4]. For example, a pure C_{13} homologue gave EC_{50}

Table 7. Summary of toxic concentrations of LAS to various organisms

Organism	Toxicity EC/LC$_{50}$ (mg L^{-1})
Bacteria	
heterotrophic	50*
anaerobic	20^x
nitrification (soil)	0.5†
Algae	
fresh water	1 to 100
marine	0.1 to 10
Daphnids	1 to 15
Other invertebrates	
ceriodaphnia	1.5 to 5
flatworms	1.8
chironomids	3
amphipods	3
nematodes	10 to 100
isopods	270
Fish	1 to 15
Terrestrial plants	⩾ 100

* EC_{10}
^{x} Maximum acceptable concentration; g kg^{-1}
† g kg^{-1}

values of 1.4 mg L^{-1} for *Navicula pelliculosa*, 5 mg L^{-1} for *Microcystis aeruginosa* and 116 mg L^{-1} for *Selenastrum capricornutum*. For a given species the EC$_{50}$ decreased with increasing chain length; for example, the EC$_{50}$ values for *M. aeruginosa* were 32–56, 10–20 and 0.9 mg L^{-1} for C$_{11.1}$, C$_{11.6}$ and C$_{12}$ respectively.

Field studies [159], rarely reported, indicated that indigenous species of algae were not affected until concentrations of C$_{11.8}$-LAS reached 27–108 mg L^{-1} (no intermediate concentrations were tested), which compared favourably with "first observed effect" levels from laboratory tests of about 1 mg C$_{11.8}$-LAS L^{-1} for *S. capricornutum* and 0.1 mg L^{-1} for *M. aeruginosa*. Photosynthesis in lake water by natural algal populations was inhibited by 50% by 3.4 mg C$_{12}$-LAS L^{-1} and by 1.9 mg C$_{13}$-LAS L^{-1}.

The few data available on toxicity to macrophytes indicate effects in the region of 1–5 mg LAS L^{-1}. EC$_{50}$ values for effect on duckweed (*Lemna minor*) were; dry weight 3.6 mg L^{-1}; root length, 4.9 mg L^{-1}; growth rate, 4.8 mg L^{-1}; frond counts, 2.7 mg L^{-1} [160].

Some of the lowest values reported for EC$_{50}$ were for the marine algae *Gymnodidium breve*, the "red tide" organism; the 24 h-EC$_{50}$ values were 0.025–0.125 mg C$_{13}$-LAS L^{-1} [161]. Reported values for other marine species were 0.2, 2, 4 and 10 mg L^{-1}.

Invertebrates

The most common test organism is *Daphnia magna* and was shown by at least 10 different workers to have 48 h-LC$_{50}$ values in the range 1 to 10 mg L^{-1} [96]; for a commercial product, C$_{11.8}$-LAS, the LC$_{50}$ was 2.8 to 6.8 mg L^{-1}. These values are lower than those given by Schöberl et al. [4] of 8.9–14 mg L^{-1}, which referred to a product used in W. Germany having an average chain length of 11.6; for a product with chain length greater than 11.6 the LC$_{50}$ value was lower, at 2–5 mg L^{-1}. One very low value of 0.013 mg L^{-1} (unstated type of LAS) for the LC$_{50}$ for *D. magna* was probably the result of poor technique, including fish illness, since 40% of the control population died during the test [162, 163].

The hardness of the test water plays a minor part in determining the LC$_{50}$. At a hardness of about 25 mg CaCO$_3$ L^{-1} the 48 h-LC$_{50}$ was 5 to 7 mg LAS L^{-1} while at 150 mg CaCO$_3$ L^{-1} the value was 2.5 to 4 mg L^{-1}, but further increase in hardness did not lower the LC$_{50}$. The 21 d-LC$_{50}$ values ranged from 1 to 4.1 mg L^{-1} [164].

The longer chain homologues were more toxic to *D. magna* than the shorter; Moreno Danvila [165] reported the 24 h-LC$_{50}$ values to be 0.1, 2, 8, 30 and 55 mg L^{-1} for C$_{14}$, C$_{13}$, C$_{12}$, C$_{11}$, and C$_{10}$, respectively. The 21-d "no observed effect concentration" (NOEC) for a commercial C$_{11.8}$ LAS was 1.9 mg L^{-1} in laboratory water and 4.5 mg L^{-1} for river water, thus demonstrating some mitigation of toxicity by material in field waters [166].

Intermediates of LAS, sulfophenylalkanoates, were much less toxic than LAS itself with LC$_{50}$ values 5000–6000 mg L^{-1}, while sulfonates of dialkyltetralins which occur in LAS as impurities are between one half and one tenth as toxic as LAS [167].

There are fewer data for other freshwater invertebrates. LC_{50} values, expressed as mg L^{-1}, have been reported as 1.4–5.3 for cerio-daphnia; 1.8 for flatworms; 2–3 for chironomids; 3–3.3 for amphipods; 16, 93 for nematodes; 23 for midges and 270 for the isopod, *Asellus* [166, 168, 169].

Of the marine invertebrates tested, the embryo-larvae of oysters are the most sensitive with 48 h-LC_{50} of 0.025 mg L^{-1} [170]. Other values for oyster larvae, however, were higher at 0.05, 0.1, 0.5 and 7.4 mg L^{-1} [158]. For other invertebrates the values (mg L^{-1}) are:

mysids, *Mysidopsis bahia*:	0.45; 1.4 (NOEC 0.9)
clams, *Tapes philippinarum*:	10.5
grass shrimps:	13.8
pink shrimps, *Penaeus duororum*:	19–154
blue crabs:	29.9

At 28 d the highest NOEC for the mysid shrimp was between 0.16 and 0.38 mg L^{-1} [166].

No discernible effect of LAS in effluent or in the bed of a small stream on the invertebrate fauna was observed, even though the concentration of LAS in the sediments was sometimes as high as 41 mg kg^{-1} [171].

Fish

The range of LC_{50} values for fish was similar to that for *D. magna*; most values were in the range 3 to 10 mg LAS L^{-1}, although for some species values as low as 0.72 mg L^{-1} were reported [96]. The range agrees with that given by Schöberl et al. [4]. For some species the individual values reported were relatively close together, e.g. fathead minnow (3.4 to 4.2 mg L^{-1}) while for other species they were disparate, e.g. guppy (1 to 10 mg L^{-1}).

Some of these differences can be attributed to differences in water hardness, temperature and in the type of LAS tested. For example, the LC_{50} for goldfish decreased from 15 mg LAS L^{-1} in very soft water to 5.7 mg L^{-1} in harder water [172]. The 96 h-LC_{50} for goldfish decreased from 16 mg L^{-1} at 15 °C to about 11 mg L^{-1} at 20 °C; rainbow trout also showed a decrease, from 6 to 4 mg L^{-1}, whereas the toxicity to golden orfe was little affected by temperature [3].

There was an approximately 10-fold difference in toxicity between homologues separated by 2 carbon atoms, and the 2- and 3-phenyl isomers were more toxic than the inner isomers. For example, Macek and Sleight [173] found the LC_{50} for the fathead minnow to be 0.5 mg L^{-1} for C_{13}, 6.6 mg L^{-1} for C_{12}, 27.9 mg L^{-1} for C_{11} and 57.5 mg L^{-1} for C_{10}; a mixture of C_{11}–C_{14} had an LC_{50} of 4.5 mg L^{-1}.

As with daphnids, fish were less susceptible to the sulfophenylalkanoic acid intermediates of LAS than to LAS itself. The NOEC values for the survival of fathead minnow fry were more than 52 mg L^{-1} for the C_{11} acid and more than

1400 mg L^{-1} for the C_4 acid [174]. LAS itself had a 14 d-NOEC on fry survival between 10 and 50% of the 96 h-LC_{50}. For commercial preparations of LAS the proportions were 14 to 33% or an average of 0.63 mg L^{-1} [175]. The 30 d-NOEC for survival of fathead minnow fry was found to be 1–2 mg LAS L^{-1} [166]. These values support the conclusion of Gerike [3] that the highest concentration of LAS for the life cycle of this species is 0.25 mg L^{-1}. However, the rainbow trout is more sensitive and the toxic limit would be less than 0.12 mg L^{-1} [176].

Other studies have reported non-lethal effects such as gill damage, swimming behaviour, physiological and enzyme changes at concentrations from about 0.01 mg L^{-1}. However, such effects are not considered in hazard assessment, probably because the body of validated information on such effects caused by a sufficient number of chemicals is not yet sufficiently large.

Bioaccumulation in Fish

The octanol-water partition coefficients predict bioconcentration factors of 100–1000, with chain length being an important factor. However, experimental values of only 10–100 have been reported, the difference from predicted values resulting from the metabolism and elimination of LAS from the fish. Excess concentration of LAS occurs in the gall bladder and ^{14}C-labelled metabolites were observed in the bile when ^{14}C-LAS was used.

Sediment Organisms

LAS adsorbs on to sediments (see p 14) with partition coefficients of 50–1000. Organisms inhabiting sediments (*Chironomous riparius*) were found to have NOEC and first observed effect values as high as 319 and 993 mg LAS kg^{-1}, respectively. The corresponding NOEC for LAS in solution was as low as 2.4 mg L^{-1}, indicating that only a small fraction of LAS adsorbed in the sediment was biologically active [177]. Bressan et al. [178] reported similar results for marine benthic vertebrates.

Terrestrial Plants

The toxicity of LAS to terrestrial plants, such as barley, beans, cucumber, lettuce, tomatoes, has been extensively studied for direct foliage and hydroponic routes of exposure. Acute EC_{50} and NOEC values ranged from 10 to 1000 mg L^{-1}, but most values were at or above 100 mg L^{-1}. Fewer studies have been made with LAS-amended soils. Windeatt [179] found no reduction in growth of sorgum, sunflower and mung beans at 100 mg LAS L^{-1}; the LAS was C_{11}, C_{12}, C_{13}, MWt of 343. The 21 d-EC_{50} values were 167, 289, and 316 mg kg^{-1}, respectively. Mieure [180] spiked soil with anaerobically digested sludge and grew 10 species of plants. The 30 d NOEC values were 397 and 407 mg kg^{-1} for two different

LAS, respectively; for the earthworm the 14 d-EC_{50} was greater than 500 mg kg^{-1}. These values for LAS concentrations are to be compared with soil levels of 1–4 mg LAS kg^{-1}.

Effects on Small Mammals and Man

The acute oral toxicity of LAS in rodents is (LD_{50}) 500– 2000 mg kg^{-1} body weight; this is classified as moderately toxic, along with other household materials such as sodium chloride and sodium bicarbonate. In chronic studies lasting 2 years, LAS doses up to 5000 mg L^{-1} in the diet caused no pathological responses. Also at concentrations commonly employed in washing waters (1000–10 000 mg LAS L^{-1}) no hazards-irritation, sensitization-resulted from direct percutaneous exposure. Single doses of LAS to animals and man were rapidly and completely metabolized and excreted [181].

Estimates of surfactant intake from all sources by humans have been given as 1–10 mg anionic surfactants per day [182] and 0.3–3 mg per day [183]. These estimates seem to be excessive, for example, 0.3–1 mg is given as the amount of total anionic surfactants taken in from water. However, the estimates are orders of magnitude lower than doses to animals which resulted in no observable changes. Similarly, no other effects (carcinogenicity, mutagenicity, etc.) were observed in rodents when LAS was fed, or applied at relatively high amounts over periods of up to 2 years.

One exception is the report by the Mikami group in Japan [184] that embryotoxic and teratogenic effects were observed in rats allegedly caused by dermal application of LAS. As a result, four laboratories (including Mikami's) in Japan collaborated to investigate the alleged effect more thoroughly; they all concluded that there was no such effect and that the original report could have been due to an artefact.

The number of accidental exposures of humans to products containing LAS has been relatively small. No fatalities have been recorded and the number of incidents resulting in "moderately severe" symptoms was small, for example, 2% of 600 cases [185].

Alkyl (or Alcohol) Sulfates

Structure

Most of the alkyl sulfates (AS) employed in laundry and other products are linear primary alkyl sulfates (LPAS) but some linear and branched secondary sulfates (LSAS) are also used.

Primary alkyl sulfates have the structure

RCH_2OSO_3H

and secondary sulfates have the structure

$R_1R_2CHOSO_3H$.

The hydrophobic group usually contains 12–18 carbon atoms, although C_8 and C_{10} esters are used as wetting and foaming agents. The normal product is the sodium salt, but products with various other cations are produced, including ammonium, magnesium, mono-, di-, tri-ethanolamine and cyclohexylamine.

The commercial LPAS contain a range of homologues which reflect those in the natural fatty acid precursors and/or the breadth of the cuts collected in the distillation of the alcohols. However, supposedly pure products sold by supply companies, for example "sodium dodecyl sulfate" frequently contain large amounts of homologues other than the one suggested by the label [2]. Because of the method of production of secondary sulfates, the sulfate group is found at all positions along the chain on the alkane, except at the ends.

Nomenclature

Alkyl sulfates are generally described by the designation AS and linear primary isomers by LPAS. The number of C atoms in the alkyl chain is described by a subscript and mixed alkyl chain lengths are designated by a range of numbers, with the ratio of the chain lengths being given in brackets. Thus, Na *n-pri*-C_{10-12} (20:80)-AS is the sodium salt of a linear primary alkyl sulfate consisting of 20% C_{10} and 80% C_{12}. The abbreviation "ave" is used to describe a product derived from a broad-cut alcohol, for example, C_{12ave}-AS, while sodium dodecyl sulfate, often abbreviated to SDS, is C_{12}-AS.

Production Methods

The first alcohol sulfates came to the market in Germany in the early 1930s when fatty alcohols became available from the catalytic dehydrogenation of fatty acids. The alkyl sulfates are produced from the alcohols by conventional sulfation with either sulfur trioxide or chlorosulfonic acid, followed by neutralisation of the acid ester with the required base, for example, NaOH or NH_4OH. Neutralisation must be carried out as soon as possible to prevent acid hydrolysis of the alkyl sulfate.

The hydrophobic moiety is usually linear, such as lauryl or coco alcohols, tallow alcohol, Ziegler alcohols or linear 'oxo' alcohols. Most are completely linear but some are slightly branched, containing some secondary methyl branching. Highly branched AS are derived from tetrapropylene stock.

Ziegler alcohols are formed by polymerising ethylene using aluminium trialkylate as catalyst, followed by an oxidation process. These alcohols are even-numbered and are substantially the same as those derived from the natural saturated fatty acids from coconut oil and tallow. Coconut oil contains mainly C_{12} acids, with significant amounts of higher and lower homologues and tallow alcohol is about two-thirds C_{18}, the remainder being mainly C_{16} and C_{14}. There are small differences in minor components, the main one being that Ziegler alcohols contain small amounts of singly branched alcohols.

Linear 'oxo' alcohols are formed by catalytic addition of carbon monoxide and hydrogen to the double bond of an olefine. With a linear α-olefine addition may occur at either end of the double bond to give a mixture of linear and singly branched primary alcohols:

$$-C-C-C-C-OH \quad \text{and} \quad \begin{array}{c} -C-C-C \\ | \\ C-OH \end{array}$$

The conventional linear oxo alcohol is about equal parts linear and methyl branched, though with changed reaction conditions as much as 80% or more linear can be produced; this product is called Dobanol.

The development of linear paraffins for the production of LAS made linear secondary alcohols feasible as surfactant hydrophobes. The –OH group was introduced, either by reaction with oxygen or by chlorination followed by hydrolysis, into any of the internal carbon atoms in the chain. Secondary AS are more readily prepared by reacting the parent olefine with sulfuric acid; the sulfate ester group enters all positions except the chain ends.

Amounts Produced

Matson [186] reported the world-wide consumption of AS in 1976 to be 82 000 tonnes; of this 36 000 t were used in U.S.A., 400 000 t in W. Europe and 6000 t in Japan.

Schöberl et al. [4] reported that in W. Germany some 10 000 tonnes of AS were consumed in 1987 as compared with 87 000 tonnes of LAS.

From data given by Gilbert and Pettigrew [71], it may be calculated that the per capita use of primary AS in the U.K. was 0.3 g d^{-1} compared with 3 g LAS d^{-1}.

Uses and Applications

AS have mainly been used as wool-washing agents and as active ingredients in both light- and heavy-duty laundry formulations. They are also widely used in a variety of specialty products, including toothpastes, hair shampoos, antacids, cosmetics and in certain foods, without any indication of hazard to human health. Other uses are froth flotation in pharmaceuticals, pigment dispersion, bubble baths, car shampoos, hand-cleaning pastes and fire-fighting foams. In some forms they are used as emulsifiers for insecticides and in printing ink manufacture.

Matson [186] reported that in the U.S.A. the use of AS in laundry applications was decreasing in favour of an AS-alkyl ethoxy sulfate (AES) blend.

Legislation

No specific legislation exists for AS but they are governed by EC directives concerning anionic surfactants generally (see p 12). Briefly, the primary biodegradability of anionic surfactants must be shown to be greater than 80% before being placed on the market.

Physical and Chemical Properties

The sulfate group is more hydrophilic than the sulfonate group, but AS are of more limited stability, particularly under acid conditions. Straight chain AS are similar in performance properties, feel and emolliency characteristics to soaps of corresponding chain length. The higher the chain length, the higher the temperature required to attain maximum detergency and wetting effects. Commercial AS are supplied usually as the sodium or ammonium salt as nearly colourless solutions, sometimes viscous, in concentrations of 25–40% or as white or lightly coloured pastes, needles, powders or granules. Salts of mono-, di- and triethanolamine, singly or mixed, are also available. Purified sodium laurylsulfate (90–99%) is available specifically for use in toothpastes and shampoos. AS are stable in hard water and the Mg salt of C_{12}-AS gives voluminous foam of low water content—hence its use in carpet cleaners. However, AS hydrolyse to sulfate and the corresponding alcohol in hot alkaline and acidic media, unlike the sulfonates which are stable under these conditions. This difference in stability is used to distinguish sulfate surfactants from sulfonate surfactants by applying the MB test before and after hydrolysis.

Analytical Methods

Qualitative, or semi-quantitative, separations of the various anionic surfactants can be effected by, for example, the use of one- and two-dimensional paper and thin-layer chromatography [e.g. 187].

As with other sulfate and sulfonate anionic surfactants, AS react with large molecular cationic dyestuffs, notably methylene blue, to form coloured complexes, or paired ions, soluble in a solvent such as chloroform but not very soluble in water. However, unlike LAS and aliphatic sulfonates, AS are readily hydrolysed by acids and this affords a means of determining the concentration of AS. The concentration of MBAS before and after hydrolysis is determined by a standard procedure and the difference gives the concentration of AS plus AES, which are also hydrolysed by acid [188]. The nature of the sulfate surfactants can be deduced by examination of the hydrophobe liberated by hydrolysis, usually by GC or IR methods.

For other methods, the surfactants are normally extracted by sublation with ethyl acetate and concentrated to small volume finally being taken up in methanol. Subsequent separation/detection techniques are applied to the concentrate, although some further clean-up may be necessary, such as adsorption-desorption. For example, Arisue et al. [189] hydrolysed the extracted anionics to the corresponding alcohols, silylated them and applied GC, using flame ionization detection.

Smedes et al. [190] used HPLC to separate C_{10}–C_{18} homologues of AS on columns of Hypersil SAS or ODS with water-acetone mixtures as the mobile phase. An acetone gradient had to be used if more than five successive homologues were present. Detection was by ion-pairing with acridinium chloride and monitoring the complex (in chloroform) fluorimetrically. The detection limit was 1 to 5 ng and 1 ppb could be detected in aqueous samples under the conditions used.

Again using HPLC, Nakamura [191, 192] separated *prim.* C_{12}, C_{14}, C_{16} and C_{18}-AS on a reversed phase column using 85:15 methanol:water containing 0.4 M NaCl or M $NaClO_4$; without the salts the homologues eluted together at the beginning. The HPLC used by Irgolic and Hobill [193] involved a 10-μm styrene-divinyl benzene copolymer column with a mobile phase of 4:1 methanol:0.05 M ammonium acetate. Detection was effected by a sulfur-specific inductively coupled argon-plasma emission spectrometer. LAS, AES and alkane sulfonates, as well as AS, could be separated into homologues and isomers, with detection limits of about 15 μg S, corresponding to a solution with a concentration of 0.3 ng S L^{-1}.

Lyon et al. [194], Righton and Watts [195] and Rivera et al. [196] have applied fast atom bombardment (FAB) mass spectrum ionization to sublated or normal extracts of surfactants, including AS.

Environmental Concentrations

As with LAS and other anionic surfactants there is no limit placed on the concentration of AS as such in the environment—treated effluents, river water, potable water. There is a limit of not more than 0.5 mg MBAS L^{-1} in drinking water in the U.S.A.

Few data on the concentration of AS in the aquatic environment could be found; no extensive surveys appear to have been made as with LAS. However, Gilbert and Pettigrew [71] used production data for detergents and personal care products and for per capita daily use of water in the U.K. to calculate that the concentration of prim. AS in unsettled sewage would be about 1 mg L^{-1}, about 0.7 mg L^{-1} being derived from detergents. They also calculated that biologically treated sewage would contain about 0.01 mg AS L^{-1} and, for the worst possible case of 1:1 dilution, in river water the concentration would be about 0.005 mg L^{-1}. (A value for the biodegradability was assumed). The corresponding estimated concentration in fresh sewage sludge was 0.8 g AS kg^{-1} and 0.02 mg kg^{-1} in soil immediately after application of 50 m^3 anaerobically digested sludge per hectare ploughed to a depth of 10 cm. In the case of soil, allowance was made for the known anaerobic biodegradability of AS, but not for biodegradation in soil.

In two Japanese sewages [188] the yearly mean concentrations of MBAS were 8.2 (range 5.1 to 10.6) and 8.4 (6.5–14.0) mg L^{-1}, of which 17 and 15%, respectively, were AS plus AES; this represents 1.4 and 1.3 mg (AS plus AES) L^{-1} expressed as MBAS. No AS or AES could be detected in the biologically treated sewages.

Using FAB-MS all the homologues of AS from C_{10} to C_{19} were identified in sewage effluents and in some river waters but none were detected in drinking water [195]. The determinations were only semiquantitative; in sewage effluents the concentration of AS was no more than a few hundred $\mu g\ L^{-1}$ and only low $\mu g\ L^{-1}$ or less in river waters. In sewage samples the predominant AS were C_{12} and C_{13}.

Environmental Fate

As a class, AS has long been recognised as being readily biodegradable; the linear primary AS (LPAS) undergo complete primary biodegradation within a few days, sometimes within one day, in shake culture or river water. Secondary and moderately branched AS are also quite readily biodegraded, but more highly branched sulfate esters are degraded only at considerably lower rates. In earlier work results could have been somewhat blurred by the use of products of ill-defined composition; commercial linear products often contained undefined quantities of non-linear impurities, which could make the "linear" product appear to be less degradable than it really was.

LPAS exceeds all other anionic surfactants in the rate of primary and ultimate biodegradation indicating the widespread distribution of microbial sulfatase enzymes.

Primary Biodegradability

Many authors have reported rapid and near complete (95–100%) removal of MBAS in LPAS in river water and in media inoculated with low or high bacterial cell density. For example, Maurer et al. [197] found complete removal of 5 mg C_{16}-AS L^{-1} in river water, while Borstlap and Kooijman [198] reported 100% MBAS removal of C_{12}-AS in 1 d in a medium inoculated with 500 mg SS L^{-1}. A minority of authors reported longer times for high removals of LPAS. Miura et al. [199] found that it took 4–5 d to achieve 100% removal of 100 mg C_{12}-AS L^{-1} in an inoculated medium, while Itoh et al. [200] and de Fulvio [201] reported the necessary period to be 10 d. These differences presumably reflect differences in the cell densities, the microbial activity and the concentrations of AS employed.

Ziegler- and coco-lauryl AS were found to degrade by 100% MBAS in 10 d in inoculated medium [202] while *prim.* C_{12}–C_{14}-AS and *prim.* C_{12}–C_{18}-AS were removed by 98–99% in inoculated media [203]. The latter author also reported that the nature of the cation had little effect on the rate of biodegradation of the AS, but Sales et al. [204] found that triethanolamine C_{12}-AS degraded more easily in sea water than did the sodium salt.

More recently, Gerike [3] and Schöberl et al. [4] reported that C_{16-18} fatty alcohol sulfate and C_{12-15} oxo-alcohol sulfate degraded by 99% MBAS in the OECD screening die-away test.

In Europe, the product called Teepol, originally derived from cracked wax olefines (though not at present) and consisting of mixed linear secondary AS, was rapidly degraded, but not as rapidly as LPAS. For example, Degens et al. [205] obtained 96% MBAS removal from Teepol in river water after 5 d and Hammerton [206] reported 89% removal in river water in 4 d. A linear C_{18} 9-sulfate (i.e. secondary) was completely degraded (MBAS) in 7 d in an inoculated medium [207]. However, branched primary and secondary AS are much more difficult to degrade, as reported by Hammerton [206, 208] that for 3-methyl, 5-dimethylhexyl-prim-AS, only 4% MBAS was removed in 21 d in river water and only 37% of 3,9-diethyltridecyl 6-sulfate (i.e. secondary) was degraded in 79 d. In another study [209] a number of C_{12}–C_{15}-AS derived from various sources and with varying proportions of linear components were subjected to the OECD die-away test. The time required for the removal of 95% MBAS ranged from only 1 d for a coconut oil-AS, containing 99% linear material, through 3 d for an oxo wax-cracking-AS containing only 50% linear components, to as long as 12 d for an AS derived from tetra propylene containing less than 5% linear material.

Ultimate Biodegradability

There is much evidence from die-away tests for the ultimate biodegradability of LPAS and linear secondary AS. For 100 mg C_{12}-AS L^{-1} Miura et al. [199]

reported the removal of DOC and ThOD to approach 100% within 10 to 15 d, while Itoh et al. [200] found 70% DOC removal and 70% $ThCO_2$ evolved after 10 d. Using the MITI test with 30 mg SS L^{-1} of activated sludge as inoculum, Urano and Saito [122] obtained an oxygen uptake equivalent to 60 to 67% ThOD from 3 to 100 mg C_{12}-AS L^{-1} after only 3 d and 100–105% ThOD after 14 d. (It is probable that nitrification may have been responsible for the higher than theoretical oxygen uptake values.) They reported the 1st order reaction constant at 100 mg AS L^{-1} of $8.8 \times 10^{-2} h^{-1}$, higher than those for AES ($5.1 \times 10^{-2} h^{-1}$) and AOS ($2.0 \times 10^{-2} h^{-1}$). In a saprobic river water, AS degraded fastest of five surfactants examined with an MBAS half-life of about 0.5 d and 20 mg AS L^{-1} was completely converted to CO_2 in 12 d [210]. In Tama (Japan) river water, 20 mg C_{12}-AS L^{-1} was completely removed in 3 d (MBAS), while DOC was completely removed in 20 d; 20 mg glucose L^{-1} took 13 d under the same conditions [211]. Gerike [3] used the modified OECD test (28 d) and showed that 88% DOC was removed from C_{16-18}-AS; in the Closed Bottle test 91% of ThOD was obtained. In the latter test, the biodegradation of C_{12-15} oxo-AS was 86% ThOD. Schöberl et al. [4] reported that C_{12-18} AS and C_{12-15} oxo-AS were degraded by 88–96% (DOC), 63–95% (ThOD) and 64–96% ($ThCO_2$). Steber et al. [212] confirmed that poorly soluble tallow AS was ultimately degraded by using ^{14}C-labelling.

Fewer data are available for secondary AS; the COD of a linear secondary C_{16}-AS was removed by 95% in 5 d [213] and the oxygen uptake by a linear secondary C_{10-13}-AS in 22 d was 77% of ThOD. Branched chain AS compounds which were not well degraded as judged by MBAS removal were obviously not ultimately biodegraded.

Effect of Chain Length

The scanty evidence available suggests that there is little effect of chain length on ease of biodegradation. In Warburg respirometers, Winter [214] found that C_{10} and C_{18} homologues of a series of LPAS were degraded by about 50% (ThOD) in 24 h, while C_{12}, C_{14} and C_{16} were removed at a higher rate of around 70%. In shake flask studies using three linear secondary AS derived from kerosene, Gebril and Abou-Zeid [215] found that the half-life (t_{50}) decreased with increasing chain length. For the C_{11} homologue the t_{50} was 1 d, for C_{13} it was 0.9 d and for C_{15} it fell to 0.7 d. On the other hand, Flores Luque et al. [216] indicated that in sea water the rate of degradation decreased (that is, t_{50} increased) with increasing chain length.

Another study [5, 166] involved nine nearly completely linear AS examined by the BOD_5 and Sturm (CO_2) methods. Table 8 shows that all nine were readily biodegraded; the CO_2 evolved was between 74 and 95% $ThCO_2$ and for glucose the value was 80–85% under the same conditions. There is no consistent effect of chain length but the methyl branching in the C_{11} and C_{15} homologues appears slightly to lower the % $ThCO_2$, though the % ThOD shows little difference.

Table 8. Biodegradability of alkyl sulfates of varying chain lengths

Carbon chain length	% of ThOD (based on 5 d)	% ThCO$_2$ (based on 28 d)
11	73	93
11*	73	89
12	63	85
13*	66	86
12–13	61	74
12–14	57	95
15	67	95
15	63	81
16–18	60	83

*Slight methyl branching

Sea Water

AS are readily biodegraded (MBAS) in sea and estuarine water. In an average of 1 to 3 d, 75% removal of C_{12ave}-AS was observed in water taken from Chesapeake Bay, U.S.A. [217]. In sea water containing 25 g added sediment per 100 mL, primary biodegradation of 10 mg C_{12}-AS L^{-1} was 1st order (k = 1.25 × 10^{-3} h^{-1}) and 75% MBAS was removed in 1 d at 25 °C [218]. Vives-Rego et al. [219] used no sediment and reported t_{50} for 20 mg Na-C_{12}-AS L^{-1} to be 0.26 to 0.34 d at 22 °C in sea water; the numbers of bacteria increased from about 8×10^6 mL^{-1} initially to about 2.5×10^7 mL^{-1} at the peak of growth.

Soil and Sludges

No data on degradation in aerobic sludge were found but soil perfusion experiments showed that AS was well degraded in soil. C_{12}-AS was degraded in the early stages of perfusion to inorganic compounds and AS degraded faster than LAS and much faster than ABS [220].

Sewage Treatment

In simulations of the activated sludge process of sewage treatment more than 99% MBAS and more than 94% DOC were removed from n-C_{12}-AS up to loadings of 100 mg AS L^{-1} [221]. Above this concentration—very unlikely to be reached in practice—nitrification was inhibited by about 25%, a reduction in the degradation of organic N occurred, the effluent became very turbid and protozoa disappeared from the system. Similarly, Fischer and Gerike [222] found 99% removal of MBAB after 1 d and subsequently 107 ± 6% of COD in the OECD

Confirmatory test while Gerike [3] reported $97\pm7\%$ removal of DOC in the coupled units test (sludge exchange) and Schöberl et al. [4] reported 94–99% DOC removal without sludge exchange. Brown [130] reported 100% removal of MBAS from n-C_{12}-AS but only 41% of DOC; this low value was probably due to the use of a P-deficient synthetic sewage. Mixed primary homologues were also well degraded; for example, Gerike and Jasiak [223] found $97\pm7\%$ DOC removal for a C_{12-14} mixture and Moreno Danvila [224] reported 83% COD removal of Alfonic 1412-S.

Branched AS were not well degraded in activated sludge simulations; only 12% MBAS of a tetrapropylene C_{13}-AS was removed [225] and 35–55% MBAS was removed from a C_{13}-AS derived from tributene [203].

Percolating (Biological) Filters

In field trials at a biological filter installation, primary AS derived from either coconut alcohols or DOBANOL-25 were removed to the extent of 96–98% MBAS and these removals were maintained even during winter months [226]. At two large-scale sewage treatment works (types of processes not specified) in Japan, Oba et al. [188] showed that the sulfate surfactants (AS and AES) were completely removed (approx 100%) using a far infra-red analytical technique. No other large-scale investigations were found.

Anaerobic Treatment

Several workers have reported that MBAS disappears rapidly when AS is subjected to anaerobic conditions of sewage sludge treatment. For example, Pitter [227] reported 88–97% MBAS removal in 32 d from C_{12}-AS and Maurer [228] obtained 100% MBAS removal in 30 d from both hydrogenated tallow-AS and a C_{18}-9, 10 C_{12}-AS. These examples show that hydrolysis of the sulfate ester linkage is readily broken but indicates nothing about the degradation of the alcohol moiety of AS. However, Oba et al. [229] similarly reported complete removal of MBAS from a coco-AS in 7 d and also found 39% removal of COD in the same period, indicating some breakdown of the alcoholic portion. Itoh et al. [230] using aquatic anaerobic sediments, showed that carbon dioxide and methane were produced from AS (type not specified) and the rate of production was enhanced when glucose was added. The same authors [231] reported that AS was degraded faster, anaerobically, than all the other surfactants they tested. Wagener and Schink [232] confirmed the degradation of AS to CO_2 and CH_4 and showed that sulfate was converted to sulfide. Steber and Wierich [233] reported 95% ultimate degradation of stearyl sulfate (using ^{14}C-labelled AS) under anaerobic conditions and Birch et al. [234] found 88% degradation of the same compound in an anaerobic screening test (based on CO_2 and CH_4 production).

Dodgson et al. [235] reported the anaerobic degradation of AS with 6 or more C atoms in the presence of nitrate by facultative anaerobes, which used the same sulfatase enzymes found in aerobically grown cells.

Metabolic Pathways and Intermediates

Far more species of bacteria have been found to be capable of mineralizing AS than in the case of LAS; this may be due to the relative abundance of organic sulfates in nature. AS containing six or more C atoms in the chain are degraded by initial enzymic hydrolysis of the ester to produce sulfate and the corresponding alcohol. Those with five or fewer are not degraded in this manner. In primary AS the alcohol is readily oxidised via an aldehyde to the carboxy acid, which is then further oxidised by β-oxidation and the citric acid cycle; for secondary AS ketones and hydroxy-ketones are the intermediates. None of these intermediates were detected in activated sludge simulations.

There is little or no evidence that microbial oxidation of the chain occurs before hydrolysis of the ester bond (though this does occur in mammals) except in the short-chain AS. No sulfatases could be found which hydrolysed methyl and ethyl sulfates [236], which are thought to degrade initially by oxidation of the methyl group, as in the case of secondary propyl sulfate [237]. This sulfate is degraded via D-lactate-2-sulfate and hence to D-lactic acid, after release of sulfate.

Much attention has been paid to the various sulfatases which show little specifity for chain length above C_6 but show marked specificity as between primary and secondary sulfatases. The secondary sulfatases exhibit strong specificity for D- or L-isomers. Some sulfatases break this O–S bond while others break the C–O linkage of the C–O–S ester group.

Environmental Effects

Effects on Bacteria

The reporting of the effect of AS on bacteria, both pure culture and mixed cultures (activated sludge), has been limited to *prim.* C_{12}-AS. Microbial oxygen uptake, growth and enzyme activity have been used as end-points. The EC_{50} values for oxygen uptake of activated sludge were all greater than 100 mg L^{-1}: 170 mg L^{-1} [238], 135 mg L^{-1} [239] and >500 mg L^{-1} [240]. A freeze-dried bacterial preparation (Polytox^R) showed an EC_{50} for respiration of 470 mg L^{-1} [238].

The growth of pure cultures was inhibited by 50% at concentrations of *prim.* C_{12}-AS of 10 mg L^{-1} for *Escherichia coli*, 43 mg L^{-1} for *Spirillum volutans* and 65 mg L^{-1} for *Pseudomonas fluorescens* [241], but the growth of sewage-derived mixed cultures was less inhibited with EC_{50} of 400 mg L^{-1} [240]. The marine bacterium, *Photobacterium phosphoreum* MICROTOX$^{(R)}$, was the most sensitive microorganism tested with typical EC_{50} values of 1.5–3.5 mg L^{-1} after 5 minutes and 0.7–1.2 mg L^{-1} after 15 minutes contact [238]. Indigenous bacterial populations in sea-water were also fairly sensitive; using thymidine incorporation the EC_{50} was 4.4 mg L^{-1} and for the glucose metabolism test it was 2.1 mg L^{-1} [242].

In sewage treatment the normally expected concentrations of AS have no effect on microbial action, but at concentrations higher than 100 mg L^{-1} adverse effects on nitrification and removal of organic nitrogen compounds occurred [221]. In anaerobic digestion of sludge, gas production was only slightly retarded at 4.2% AS (w/w on dry sludge) a much higher concentration than occurs in practice and recovery was rapid.

Effects on Algae

Generally, algae are more sensitive to AS than are bacteria; there is wide variation in the results for algae which may be due to the lack of information on the nature of the AS used in some cases. Gerike [3] and Schöberl et al. [4] give a value of 60 mg L^{-1} for algae in general; Gerike states that it is NOEC (a no observed effect concentration) whereas it is not described by Schöberl et al. [4]. Other values do not agree with this low toxicity; for example, for *Selenastrum capricornutum* Beaubien et al. [243] reported an EC_{50} of 4 mg L^{-1}, and Nyberg [244] found 30 mg L^{-1}. However, other evidence tends to support the NOEC value of 60 mg L^{-1}; growth of *Chlorella vulgaris* was unaffected by 50 mg L^{-1} and the algae eventually tolerated 200 mg L^{-1} [245], and *Poterioochromas malhamensis* showed no adverse effect at 30 mg L^{-1} [246] but was completely inhibited at 89 mg L^{-1}. All the above workers, except Schöberl et al. [4] who list C_{12-18}-AS and C_{12-15} oxo-alcohol-AS, used Na-C_{12}-AS or did not describe what was tested.

Ukeles [247] tested MgC_{12ave}-AS which at 10 mg L^{-1} did not affect 10 marine algal species but 100 mg L^{-1} did, and for a further two species 1 mg L^{-1} had no effect but 10 mg L^{-1} did. Roberts et al. [248] reported EC_{50} values for C_{12}-AS for three estuarine algae, *Pseudoisochrysis paradoxa*, *Skeletomena costatum* and *Prorocentrum minimum*, to be 1.3, 2.3 and 1.3 mg L^{-1} respectively, using a test based on assimilation of ^{14}C-$NaHCO_3$.

Although there were varying values even the lower ones are much higher than the concentration of AS will be in surface waters.

Effects on Invertebrates

Table 9 shows the LC_{50} values for *Daphnia magna* to be fairly wide ranging but the range given by Gerike [3] and Schöberl et al. [4] of 5 to 70 mg L^{-1} is

Table 9. Acute toxicity of AS to *Daphnia magna*

AS	LC_{50} (mg L^{-1}) 24 h	48 h	Ref.
C_{12}	—	6.2, 8.1, 9.0	LeBlanc [250]
	—	27.8	Moore et al. [251]
	10	—	Beaubien et al. [243]
	80	—	Lundahl and Cabridenc [249]
C_{12} Ziegler	13.5	6.3	Lundahl et al. [202]
C_{12-14} natural	6.3	2.5	„ „
C_{12-18}	5–70	—	Schöberl et al. [4]
C_{12-18} OXO	5–70	—	„ „

probably representative. Lundahl and Cabridenc [249] showed that the higher homologues were more toxic to the organism; the LC_{50} values for n-$C_{4, 8, 9, 10, 12}$ and C_{13} were 8200, 4350, 2300, 800, 80 and 42 mg L^{-1}, respectively. LeBlanc [250] could find no increased resistance to n-C_{12}-AS after exposure of the daphnids for 20 h at 3 mg L^{-1} neither did significant resistance develop in daphnids cultured in the presence of AS for 15 generations. Tests over 10 d through four generations showed 100% survival at 2 mg L^{-1}, 40% at 4 mg L^{-1} and no survival at 6 or 8 mg L^{-1}. There were no adverse effects of AS on reproduction at any treatment containing surviving daphnids for the entire four-generation exposure. Lundahl and Cabridenc [252] showed that during degradation of C_{12}-AS the toxicity first increased, as measured by mobility, reaching a maximum at 30 h and then rapidly fell to almost a negligible value. The explanation given was that it was caused by the formation of dodecanoic acid by enzymic hydrolysis from C_{12}-AS and its further rapid degradation; dodecanoic acid is more toxic than C_{12}-AS, but is very short-lived.

Effects on Other Aquatic Organisms

Toxic effects of AS on other organisms are summarized in Table 10; most tests were done with NaC_{12}-AS. The effects ranged from insignificant inhibition by a *sec.* C_{10}-AS towards mosquito pupae up to 48 h-LC_{50} on larvae of clams at 0.35 mg C_{12}-AS L^{-1}. However, most of the LC_{50} values for adult organisms were over 5 mg L^{-1}. Although the flat worm, *Dugesia gonocephalla*, was unaffected at 0.6 mg C_{12}-AS L^{-1}, the regenerative capacity of the worms was reduced at half of this concentration. The low sensitivity of the grass shrimp, *Paleomonetes pagio*, to C_{12}-AS was explained as being a result of their relative inactivity and inhabiting bottom deposits.

Whereas most correlations between toxicity and chain length show an increasing toxicity with increasing chain length, Bode et al. [253] reported the budding in *Hydra attenuata* to be more affected by C_{10}-AS than by C_{12-16}-AS. C_{16}-AS had little effect at 688 mg L^{-1} (2 mM) and C_{14}-AS had no effect at

Table 10. Effect of AS on other aquatic organisms

Organism	Type of AS	LC$_{50}$ (mg L^{-1}) or other function	Ref.
Hydra attenuata	C$_{12}$	58 (death in 10 d)	Bode et al. [253]
	C$_{10}$	55 (death in 1 d)	
Copepods			
Eurytemora affinis	C$_{12}$	2.6 (96)	Roberts et al. [248]
Arcartia tonsa	C$_{12}$	0.6 (96)	
Worms:			
Dugesia gonocephalla	C$_{12}$	0.6 (LC$_0$ 30 d)	Patzner and Adam [254]
Arenicola marina	C$_{12}$	15.2 (48 h)	Conti [255]
Shrimps:			
Artemia salina	C$_{12ave}$	3.6 (24 h)	Price et al. [256]
Palaemonetes pugio	C$_{12}$	52–162 (96 h)	Tatem et al. [257]
Neomysis americana	C$_{12}$	5.7, 8.8 (96 h)	Roberts et al. [248]
Mysidopsis bahia	C$_{12}$	6.1, 7.1 (96 h)	,, ,,
Mosquito pupae	Na C$_{12}$	78	Piper and Maxwell [258]
(C.p *quinque fasciatus*)	NH$_4$ C$_{12}$	55	,,
	TEA C$_{12}$*	102	,,
	Na C$_{10}$	44	,,
	Na EH†	>200	,,
Molluscs:			
Corbicula fluminea	C$_{12}$	16.2 (5 d)	Graney and Giesey [259]
Tresus capax (larvae)	C$_{12}$	0.35 (48 h)	Cardwell et al. [260]
Mercenaria mercenaria (eggs)	AS	1 (62% reduction of development)	Hidu [261]
Crassostrea virginica (eggs)	AS	1 (61% reduction of development)	,,
Crassostrea gigas (larvae)	AS	1 (48 h)	Cardwell et al. [260]

* TEA = triethanolamine; †EH = 2-ethyl-hexyl

63 mg L^{-1} (0.2 mM), whereas C$_{12}$-AS caused death in 10 d at 58 mg L^{-1} (0.2 mM) and with C$_{10}$-AS death resulted in only 24 h at 55 mg L^{-1} (0.2 mM). The authors suggested that the decrease in toxicity with alkyl chain length was due to reduced solubility in water and resultant loss of surfactant activity at the test temperature (20 °C) but this needs to be confirmed.

Effects on Fish

Schöberl et al. [4], quoting mainly previously unpublished data for C$_{12-18}$ fatty alcohol AS and C$_{12-15}$ oxo-alcohol AS, gave a range of LC$_{50}$ values, presumably over 96 h, for fish generally of 3 to 20 mg L^{-1}, a range given earlier by Gerike [3]. Table 11 gives a selection of values and shows that in general this range is justified. The least toxic was C$_1$-AS (mono-methyl sulfate, not a surfactant) at 1000 mg L^{-1} against bluegill and the most toxic were NH$_4$ branched C$_{15}$-AS at 2.1 mg L^{-1} [5, 166] and C$_{12ave}$-AS against minnow at 1.4 mg L^{-1} [262]. Two other values require mention: for guppy the 24 h LC$_{100}$ was 2 mg L^{-1} for C$_{12-15}$-AS and the effect-concentration slope was so steep that the LC$_{50}$ appeared to be

Table 11. Toxicity of AS to fish

Species	AS	Hardness (mg CaCO$_3$ L^{-1})	Time (h)	LC$_{50}$ (mg L^{-1})	Ref.
Goldfish (*Carassius curatius*)	nC_{12}	100	6	60.0	Gafa [172]
	nC_{13}			18.3	
	nC_{14}			5.6	
	nC_{16}			>300	
	C_{14} branched			7.8; 49.1	
	C_{12}–C_{16}			12.0	
	C_{12}–C_{15}			7.8	
	Dobanol $C_{12,15}$			7.8	
Bluegill (*Lepomis macrochirus*)	NaC_1	35	96	1000	Procter and Gamble, Little [5, 166]
	NH_4C_{11}			26 (16)	
	NaC_{12}			4.8	
	NH_4C_{12}			20.3	
	NH_4C_{13} branched			18.4	
	NH_4C_{15}			5.2 (2.1)	
	NH_4C_{16}			22	
	NH_4C_{12}–C_{14}			3.2	
	C_{16}–C_{18}			76	
Bluegill	NaC_{12}	125	96	4.5	Bishop and Perry [160]
Killifish (*Oryzias latipes*)	NaC_{12ave}	dist H$_2$O	24	70	Kikuchi et al. [264]
	NaC_{14}			5.9	
	NaC_{16}			0.78	
Rainbow trout	NaC_{12}	350	24	4.6	Fogels and Sprague [266]
Zebra fish	"			7.8	
Flag fish	"			8.1	
Phoxinus phoxinus	C_{12}	spring H$_2$O	24	30.5	Lundahl and Cabridenc [249]
Minnow	C_{12ave}	65	96	1.4	Verna [262]
(*Macrones vittatus*)	C_{12ave}			1.5	
	TEA salt				
Lebistes reticulatus	C_{12}–C_{15}	—	24	20*	Van Embden et al. [263]
Cyprinodon variegatus	NaC_{12}	sea H$_2$O	96	4.1	Roberts et al. [248]
Menidia menidia	"	100	96	2.8	"

* LC$_{100}$; Values in brackets are for a branched isomer

not much less than $2\,mg\,L^{-1}$ [263] and for killifish the 24 h-LC_{50} was $0.78\,mg\,L^{-1}$ of C_{16}-AS [264].

Differences between certain values in Table 11 can be due to different species, but probably more likely due to different times of exposure and to the hardness of the water. Tovell et al. [265] showed that Na-C_{12ave}-AS was more toxic to goldfish and rainbow trout in hard (300 mg $CaCO_3\,L^{-1}$) than in softer (60 mg L^{-1}) or distilled water; in 70 mg AS L^{-1} 100% of the goldfish population died within 100 minutes at 300 mg $CaCO_3\,L^{-1}$ while in distilled water survival was longer than 24 h. In hard water the fish accumulated more AS than in soft; in distilled water, goldfish contained 9.6 μg AS g^{-1} while at 300 mg $CaCO_3\,L^{-1}$ the content of AS in fish was 85 μg g^{-1} and the authors considered this to be the reason for the higher toxicity in hard water. It is difficult to explain the $0.78\,mg\,L^{-1}$ of C_{16}-AS in distilled water reported by Kikuchi et al. [264]. In contrast to the situation with *Daphnia magna*, the effect of chain length is not clear or pronounced as judged by the studies of Gafa [172] and the Procter and Gamble Co. [166] (Table 11). However, Kikuchi et al. [264] reported that the 24 h-LC_{50} values for killifish in distilled water decreased by a factor of about 10 for an increase in C chain of 2; that is, C_{16} was 10 times as toxic as C_{14}, which was about 10 times as toxic as C_{12} (Table 11). Further work would be necessary to reconcile these differences.

In longer term tests the threshold LC_{50} values for zebra fish and flagfish were 7.9 and $6.9\,mg\,L^{-1}$ for Na-C_{12}-AS, respectively. For rainbow trout no threshold was evident but the 10 d LC_{50} was $2.8\,mg\,L^{-1}$ [266].

Reported sub-acute effects of AS include depressed olfactory sensitivity in white fish (*Coregonus clupeaformi*) at $0.1\,mg\,L^{-1}$ of C_{12ave}-AS [267] and a lowering of the coagulation time of the blood of *Saccobrancus fossilis* exposed for 60 d to $1.1\,mg\,L^{-1}$ of Na-C_{12}-AS, [268].

Bioaccumulation in Fish

Exposure of carp to solutions of ^{35}S-labelled Na-C_{12ave}-AS led to equilibrium after about 24 h and certainly after 72 h, giving bioconcentration factors (BCF) of about 4 for whole body, independent of initial aqueous concentration. BCF values for the hepatopancreas and gall bladder were 50 and 700, respectively. On transfer to fresh water the depuration rate was such that half of the ^{35}S was removed in 3 d [269]. Accumulation of C_{16}-AS was higher than that of C_{14}-AS which in turn was higher than C_{12}-AS, which is the same order as for toxicity [270].

Effects on Higher Plants

The length and dry weight of corn plants grown from seeds watered with 10, 100 and $1000\,mg\,L^{-1}$ of AS were greater than those of controls [271]. Barley seeds steeped in aqueous solutions of Na-C_{12ave}-AS for 24 h and then allowed to germinate showed inhibition at 288 mg L^{-1} (11%) and above [272]. Watering potted rice plants with a solution of 50 mg AS L^{-1} produced yellowing of leaves, inhibition of water uptake by roots and of photosynthesis, but did not affect plant height, number of shoots or dry matter production [273]. The aquatic plant, *Lemna minor*, showed EC_{50} values in flow-through tests of 18 mg C_{12}-AS L^{-1} for root length and up to 44 mg L^{-1} for growth rate [160].

The water hyacinth, *Eichhornia crassipes* was exposed for 16 d to solutions of Na-C_{12}-AS from 0 to 100 mg L^{-1}. At the highest concentration 42% of the leaves died and at 25 mg L^{-1} 25% of leaves died [274].

Effects on Small Mammals and Man

It is well recognized that AS are safe to humans when used at the recommended levels by the various Food Agencies. Broad cut Na-C_{12}-AS is permitted for use as an emulsifier (up to 1000 ppm in egg white solids), as a whipping agent up to 5000 ppm in gelatin, and in fruit drinks up to 25 mg L^{-1}. AS is also sanctioned for use in food packaging materials. Taken with their ease of biodegradation and general low order of toxicity, it is concluded that AS do not pose a hazard to human health.

The acute oral toxicity of AS to rodents is 1000–4000 mg kg^{-1} body weight (LD_{50}) and for the commercial product in practice the LD_{50} is 5000 to 15 000 mg kg^{-1}. In chronic tests lasting 1 year, 1% of AS in the diet of rats led to no adverse effects. Single doses to animals and man were rapidly and completely metabolized and excreted. Using ^{35}S-labelled AS, urinary metabolites from C_{12}-AS in rats were 4-OH-butyric acid and its sulfate ester, and the S-butyrolactone. Using C_{11}-AS, the major metabolite was propionic-3-sulfate and in lower concentration, pentanoic acid-5-sulfate. These products indicate that initial ω-oxidation of the AS occurs rather than hydrolysis to form sulfate.

In skin tests, applications of 0.1% solution of C_{12}-AS and C_{16}-AS to rabbits was without effect; at 1% irritation resulted after occluded exposure for 7 d, but in uncovered exposure 1% did not cause irritation. Several studies showed that AS was not a skin sensitizer. Instillation of 1% AS to rabbits eyes caused little or no irritation.

Rabbits, rats and mice showed no teratogenic effects when AS was dosed at 300 mg kg^{-1} daily for 10 d during pregnancy, but reduced litter size and foetal loss were observed in mice at 600 mg kg^{-1}, a dose level which caused severe maternal toxicity in all three species. No evidence of carcinogenicity caused by AS has been recorded. The same considerations apply to AS as to LAS (see p 47) in respect of estimated and safe daily intakes of anionic surfactants.

Alkyl Ether Sulfates

Structure

Alkyl ether sulfates, or alkyl ethoxy sulfates, are primary sulfate esters derived from alkyl ethoxylates and have the general structure:

$R(OCH_2CH_2)_nOSO_3^- Na^+$

or, more generally, alkyl alkoxylate sulfates $R(OC_mH_{2m})_nOSO_3^- Na^+$, where m is 2, 3, 4.

Normally R is a mixture of straight-chain alkyl groups containing in the region of 12 to 14 C atoms and sometimes up to 16 C atoms. R can also be alkyl phenol, as in alkyl phenol ethoxy sulfate, e.g.

$C_9H_{17}C_6H_4(OCH_2CH_2)_nOSO_3^- Na^+$

The alcohols on which the surfactant is based are normally primary, but can also be secondary.

The number of alkoxylate groups is normally smaller, $n = 1$ to 4, than in the corresponding nonionic surfactants, alkyl ethoxylates and alkyl phenol ethoxylates. These lightly ethoxylated alcohols are more complex that at first may be thought, since the ethylene oxide (EO) content is only a mean and appreciable portions of the alcohol remain unethoxylated. About 25% remains unreacted when 2 moles EO are added and 18% for 3 moles. Thus, the sulfated product will contain some alkylated sulfate (AS).

Nomenclature

Alkyl ether sulfates have other names, such as alkyl ethoxy sulfate, alkyl poly ether sulfate, alcohol ethoxy sulfate, alkyl polyglycol ether sulfate, and even alketh sulfate.

The alkyl group is designated by

C_{12} = straight-chain dodecyl
$C_{12-14}(40:60)$ = 40% C_{12} plus 60% C_{14}
C_x = unknown number of C atoms.

Branched chain primary alcohols such as 2-butyl, 2-hexyl ethyl (branched dodecyl) is given as $(C_4)(C_6)CC-$. Secondary alcohols are prefaced by *sec-*, for example, *sec-*C_{12}-AES.

The number of alkoxylate groups is designated by the subscript n as in E_n, P_n, B_n, for ethoxylate, propoxylate and butoxylate, respectively.

Production Methods

A fatty alcohol—natural, Ziegler-type or oxo-type—is first condensed with ethylene oxide so that not more than 4 molecules of ethylene oxide are added. An alkaline catalyst such as KOH, K_2CO_3 or sodium methylate, is used or for a narrow range of ethoxylates a catalyst of the alkaline earth type is probably used. The product at this stage with n not greater than 4, is not a surfactant. The ether is then sulfated using chlorosulfonic acid or sulfur trioxide and neutralized with caustic soda. Alkyl phenols are also subjected to the same process, except that sulfonation with sulfur trioxide leads to simultaneous sulfonation of the benzene ring and to disulfation, that is, sulfation of the –OH group and sulfonation of the ring of the same molecule. Complete exclusion of ring sulfation is achieved by the use of sulfamic acid, but this results in the formation of the ammonium salt.

Alkyl phenol ES are less complex since ethoxylation of alkyl phenols follows a different path from that with alcohols, all the phenol being ethoxylated before multiple ethoxylation starts.

Production Amounts

Gerike [3] gives the usage in 1985 for W. Germany of all anionic surfactants to be 138 000 tonnes, of which some 20 000 tonnes was alkyl ether sulfates. AES is one of the fastest growing surfactant in terms of annual production and usage; in 1966 1600 tonnes per year were produced in the U.S.A. while in 1980 the amount had risen to 108 000 tonnes per year.

Uses and Applications

AES are being used increasingly in combination with other anionic surfactants and non-ionic surfactants in household dishwashing. They are used to increase the detergency of liquid detergents based on LAS or *n*-paraffin sulfonates; about 20% of total active matter is replaced by AES.

AES, especially the monoethanolamine salts, are used in hair shampoos, bubble baths and shower-bath preparations. Other uses are in wallboard manufacture, textile industry and as a drilling auxilliary. They are used with AS and fluorosurfactants in the production of high expansion fire-fighting foams. The ammonium salts are used as air-entraining agents in the concrete industry for the production of light-weight cements. Blends of special 'cuts' of fatty acid AES are used to produce shampoos for babies and children and also in the cosmetic industry in make-up, skin cleansers and body lotions.

Alkyl phenyl ES (APES) are used as components of light-duty household detergent liquids, emulsion polymerisation and as bases for toilet preparations.

Physical and Chemical Properties

In concentrated form, about 70%, AES is supplied as a colourless or slightly yellow to straw coloured viscous liquid or free-flowing, pumpable paste or gel. At these concentrations it is thixotropic. At concentrations of 25 to 30% it is a colourless or yellow-straw coloured liquid.

Disadvantages of non-ionic surfactants of the EO condensate type (see later chapter) which are excellent detergents are that they produce weak or unstable foams and exhibit "invert solubility"—they are more soluble in cold than in hot water. This causes cloudiness under some circumstances. Alkyl sulfates (AS) are high foamers but their sodium salts produce clear solutions only in relatively low concentrations. The sulfated ethers (AES), however, have very low cloud points even in relatively high concentrations. They also have better wetting properties than other anionics and their foaming properties are considerably higher.

In common with other anionics, AES react with large molecular cationic dyestuffs, notably methylene blue, to form coloured complexes preferentially soluble in chloroform. Since AES are readily hydrolysed by acids (and alkalies) prior hydrolysis of an environmental sample removes AES, and AS, but not sulfonates, such as LAS. This forms the basis of a method for determining AES plus AS.

Analytical Methods

AES plus AS can be estimated from the difference between the MBAS content of a sample before and after acid hydrolysis of the sample, as indicated earlier for AS [188].

TLC methods can be used; for example, two-dimensional separations were obtained by using developing solvents, first a mixture of ethylacetate, benzene, ethanol and water and secondly, a mixture of 4-methyl-2-pentanone, propanol, acetic acid and methyl cyanide [275]. Spots were extracted and examined by UV, IR and NMR.

AES can also be separated by HPLC; Smedes et al. [190] used a C_{18} reversed-phase column and a water-acetone gradient to separate C_{12-14}-AE_3S into a series of C-chain homologues, each of which was partially resolved into individual ethoxamers of that homologue. Also, Irgolic and Hobhill [193] separated AES using a column of 10 μm styrene-divinyl benzene copolymer, and a mobile phase of 4:1 methanol: 0.05 M aq. ammonium acetate. Detection was by a sulfur-specific indirect-coupled-argon-plasma emission spectrometer.

GC analysis was used by Arisue et al. [189]. The sample was sublated with ethyl acetate and the various surfactants were separated from the concentrate by ion-exchange column chromatography. AES was eluted, hydrolysed and the product was silylated and chromatographed using flame-ionization detection.

Neubecker [276] first hydrolysed AES with methanolic HCl, which was said not to hydrolyse AS thus distinguishing AES from AS, and examined the liberated ethoxylate by HBr cleavage, followed by GC.

Fast atom bombardment (FAB) mass spectrum ionization has been applied to the detection of AES.

Environmental Concentrations

No surveys of the concentration of AES in the environment have been seen; Swisher [2] reports no values for AES. Since AES responds in the methylene blue test, any values reported for MBAS would also include any AES present. Gilbert and Pettigrew [71] estimated concentrations of AES, and other anionic surfactants, from tonnage produced, water usage and known biodegradability and from other assumptions. The predicted concentrations of AES were (mg L^{-1}): settled sewage 3.5; sewage effluent 0.1; river water 0.05; fresh sludge 0.4% (W/W); soil 0.1 mg kg^{-1}. Oba et al. [188] determined the yearly mean concentration of AES plus AS in two sewages and effluents in Japan. Of the 8.2 and 8.4 mg L^{-1} MBAS present in the sewages, the concentration of AES plus AS was 1.4 and 1.3 mg L^{-1} respectively: AES plus AS could not be detected in the effluents.

Rivera et al. [196], using FAB-MS, reported the presence in raw and potable waters of nonyl phenol ethoxy sulfates with less than 10 EO groups and *iso*-octyl phenol ethoxy sulfates with 6 EO groups; the actual concentrations were not stated.

Righton and Watts [195] used FAB-MS and detected in various effluents and water samples all six AES from C_{11} to C_{16}, containing, respectively, 3 to 9, 1 to 9, 1 to 8, 1 to 8, 5 to 7 and 1 to 3 EO groups; the concentrations were in the order of a few $\mu g\,L^{-1}$.

Environmental Fate

Primary Biodegradability

The data available, not as much as for LAS and AS, show that as a class AES are readily biodegradable (primary) both aerobically and anaerobically. From the still fewer results, it is clear that those AES available on the market as surfactants are also readily mineralized to CO_2 and H_2O, but AES tend to be slightly more resistant than AS. Many species of bacteria can degrade AES; Goodnow and

Harrison [277] found that of 45 strains tested 43 could readily degrade C_{12}-AE_3S.

Many reports give high removal values of MBAS from AES; 90–100% MBAS was removed within 1 to 5 d. Sekiguchi et al. [211] obtained 100% MBAS removal in 1 d from C_{12}-AE_3S and C_{12-13}-AE_3S and Hales et al. [278] reported 100% MBAS removal from C_{12}-AE_3S in 4 to 5 d in inoculated medium and in uninoculated river water. Yoshimura and Masuda [279] found that it took 3 d for the removal of 100% MBAS from C_{12}-AE_3S and 4 d for 98% MBAS to be removed from C_{12-14}-AE_3S.

Ultimate Biodegradability

In two of the above examples, ultimate biodegradation was also demonstrated. Sekiguchi et al. [211] reported 100% DOC removal from C_{12}-AE_3S in 24 d and 97% DOC was removed from C_{12-14}-AE_3S in river water, but only 69% in 9 d in an inoculated medium [279]. Other evidence for the ultimate biodegradation of AES has been given by Fischer [203] for C_{12-18}-AE_2S with 78% of ThOD taken up in 30 d in an inoculated medium, while Urano and Saito [122] reported 101% of ThOD and 72% of ThOD from 10 and 30 mg L^{-1}, respectively, of C_{12}-AE_5S.

Gerike [3] reported 98% MBAS removal in the OECD test and 100% of ThOD in 28 d in the Closed Bottle test from C_{12-14}-AE_2S. Schöberl et al. [4] found 96–100% DOC removal in the modified OECD test, 58–100% of ThOD in the Closed Bottle test and 65–83% of $ThCO_2$ in the Sturm test for C_{12-14}-AE_2S and C_{12-15}-oxo-alcohol-AE_3S. An unspecified AES, $C_{12}AE_xS$ was degraded by 100% DOC in 55 d in uninoculated sea water, which was faster than LAS and almost as fast as AOS [211].

In cultures of a *Pseudomonas* sp. isolated from sewage, Griffiths et al. [280] showed that 75% of ^{14}C from $Na[1-^{14}C]C_{12}$-AE_3S was evolved as $^{14}CO_2$, about 20% remained as water-soluble material and about 5% was present as ether-extractable residues.

Effect of Structure on Degradation

In a study of eleven AES, there were no significant differences in the ThOD removals after 5 or 20 d (50–68% and 70–100%, respectively,) for eight of the AES containing 10 to 16 C atoms and 2 to 6 EO groups [5, 166]. The other three AES tested contained C_{16-18} or 12 EO groups and had low $ThOD_5$ removals (35–44%) but were better degraded at 20 d.

There is evidence that branching of the alkyl chain hinders biodegradation. Berger [225] found that while AES based on natural linear primary alcohols degraded by 97% (MBAS) in 3 d and AES based on linear primary oxo alcohol degraded by 90% in 3 d, those based on tetra-propylene primary oxo-alcohol were removed by only 50% in the same time. Similarly, the length of the EO chain tended to increase the time needed for degradation; for a primary AE_3S the time needed to remove 90% was 4–6 d, and primary APE_4S took 4 d, but a primary APE_6S took 10 d [281].

The linear secondary AES are somewhat more poorly degraded than the linear primary; n-C_{12}-AE_3S took 1.6 d in river water for the removal of 90% MBAS, while sec-C_{12}-AE_3S took 4.5 d. The difference was more marked for ultimate biodegradation: 80% was removed in 7 d from the primary but only 39% DOC was removed from the secondary [279]. For the two secondary AES, 5-methylol-undecane C_{12}-AE_3S and 8-methylol-heptadecane C_{18}-AE_3S, the removals were only 12% and 0%, respectively.

The APES are less well degraded and there is a wide variation in biodegradability depending in the hydrophobe and a greater resistance with higher degrees of ethoxylation. The linear secondary APES are less degradable than the corresponding LAS (also linear secondary) and this seems to be due to the "internal" (5-phenyl) isomers of the sulfates being much less well degraded than the 2-phenyl isomers as compared with relatively small difference in degradability between the 2- and 5-phenyl isomers of LAS. Stead et al. [282] found 88% MBAS removal from a non-random linear C_{9-10}-APE_9S, having a preponderance of the 2-isomers, but only 45% from an ordinary linear C_{7-9}-APE_9S, and 20% from a tetrapropylene C_9-APE_8S, illustrating the better degradation of the 2-phenyl isomers.

Simulation and Field Tests

In the "coupled-units" mode of the Husmann activated sludge simulation test, 67–68% DOC was removed from C_{12-14}-AE_2S and 89–99% from C_{12-15}-AE_3S, while in the non-coupled mode 79–89% DOC was removed [4]. Gerike [3] reported $97 \pm 7\%$ DOC removal for a C_{12}-fatty alcohol-AE_xS, using the coupled units test.

The 1.3–1.4 mg L^{-1} combined AES + AS in two Japanese sewages was removed by 100% throughout the year [188].

In estuarine waters 96–99% of [1-^{14}C]-C_{16}-AE_9S was evolved exponentially as $^{14}CO_2$ in 30 d with a t_{50} of 2.4 d for the 1-C atom. When the ^{14}C label was put uniformly on the ethoxy chain of C_{16}-AE_3S, 70% of the ^{14}C was evolved exponentially as $^{14}CO_2$ in 20 d with t_{50} of 6.3 d [283].

Anaerobic Biodegradation

In a laboratory-scale septic tank system 72–81% MBAS was removed from $C_{14-16-18}$(14:32:54)-AE_3S over a period of 8 months, when dosed at 26 and 52 mg L^{-1}; when the effluent was poured through an aerobic seepage bed 98–99% MBAS was removed. In a 6-month study, more than 95% MBAS was removed from the same AES when dosed to an anaerobic digester [5, 166].

AES is not as rapidly degraded under anaerobic conditions as it is under aerobic conditions and, under anaerobic conditions it is not degraded as rapidly as AS. Oba [284] reported C_{12}-AE_3S to be removed by 64% MBAS in 28 d and Maurer et al. [197] found 70% removed in 17 d from C_{16}-AE_1S. An AE_xS was shown to produce CO_2 and methane anaerobically but was less rapidly degraded than AS and AE [231].

Other Alkoxylate Sulfates

Propoxylate and butoxylate sulfates appear to take longer to degrade than ethoxylates. A study of nine propoxylate and ethoxylate sulfates showed a tendency towards slower biodegradation with increasing P or E content. Little difference in the rate of primary biodegradation was observed, except for P_5 compounds. C_{12}-AE_3S was removed by 98% MBAS in 2 d and by 58% of ThOD in 6 d compared with only 67% MBAS removal and 36% of ThOD for C_{12}-AP_5E_3S. Only 86% MBAS and 31% of ThOD were recorded for C_{18}-AP_5E_3S in 6 d [279].

Metabolic Pathways and Intermediates

The compound C_{12}-AE_3S has been studied [278, 285] for degradation pathways used by a number of bacterial isolates. Of the three established routes

(a) hydrolytic desulfation to form an ether
(b) etherase cleavage of the ethoxylate moiety, and
(c) β-oxidation of the alkyl chain,

ether cleavage (b) predominates, and the pathways can operate simultaneously in one species. In cultures of the most studied organism, *Pseudomonas* DES 1, the products of reaction (b) are mono-, di- and tri-ethylene glycol monosulfates in the proportions 35%, 45% and 20%. These sulfates were identified as major metabolites by the use of ^{35}S-labelled C_{12}-AE_3S; minor metabolites were acetic acid-2-(ethoxy sulfate) and acetic acid-2-(diethoxysulfate) which are formed by oxidation of the free alcohol groups of di- and tri-ethylene glycol mono-sulfates, respectively. Inorganic ^{35}S-sulfate also appeared in significant quantities in culture liquids and was formed from the parent surfactant by reaction (a) but not from any of the five metabolites.

In mixed cultures, from river water, under dieaway conditions the tri-ethylene glycol ether monosulfate (E_3S) was the main metabolite with smaller amounts of the di- and mono-ethylene glycol ether mono-sulfates (E_2S and E_1S). Carboxylate derivatives of each glycol sulfate were identified, indicating oxidation of the terminal $-CH_2OH$ groups (ω-oxidation). Other metabolites identified indicated ω- and β-oxidation of the dodecyl from the AES, but the mixed culture could form sulfate from the sulfated glycols. Eventually these intermediates are converted to CO_2.

The same group [280] followed the oxidation, by *Pseudomonas* DES 1, of the hydrophobic portion of the AES by using $[1-^{14}C]-C_{12}-AE_3S$. The unsulfated ether $[1-^{14}C]-C_{12}-AE_3$, triethylene glycol dodecyl ether, was formed with release of sulfate (reaction (a)). Extended incubation of the ether or the surfactant produced about 20 radiolabelled ether-extractable metabolites, 8 in major amounts, but at no time did they amount to more than 20% of the total label. Among these were 3,6,9-trioxahene-eicosanoic acid, diethylene glycol dodecyl ether and the corresponding 3,6-dioxaoctadecanoic acid. None of the metabolites accumulated.

Effects on the Environment

Bacteria

AES seem not to be particularly inhibitory towards bacteria, except perhaps in seawater. Goodnow and Harrison [277] reported that, of 2 freshwater strains out of 45 which were unable to degrade AES, one strain was completely inhibited at 100 mg AES L^{-1}. The lowest concentrations which prevented growth of *Escherichia coli* on plates were as high as 18, 4 and 2 g L^{-1} for $C_{12}-AE_3S$, Ziegler $C_{12}-AE_3S$ and a $C_{12-14}-AE_{2.2}S$, respectively [202]. The EC_{50} for "lauryl ether sulfonate"—but presumably "sulfate"—in uninoculated sea water was 2.2 mg L^{-1} for thymidine incorporation and 1.5 mg L^{-1} for glucose metabolism.

Algae

The minimum concentrations, over a 5-d exposure, which inhibited the growth of *Selenastrum capricornutum* and *Navicula seminulum* were between 10 and 100 mg L^{-1} and of *Microcystis aeruginosa* between 100 and 1000 mg L^{-1} for an unstated AES [5, 166]. The EC_{50} for *S. capricornutum* was found to be 4 to 8 mg $C_x-AE_9S L^{-1}$, and for *Nizschia fonticula* and *M. aeruginosa* the values were 5–10 and 10–50 mg L^{-1} [286]. Nyberg [244] also reports greater toxicity than the Procter and Gamble Co [5, 166] study. *S. capricornutum* was inhibited (growth) by 68% at 10 mg $C_{12-14}-AE_nS L^{-1}$ and the same concentration of $C_{10-16}-AE_2S$ produced 40% inhibition. Little [5] reports an unexplained inverse relationship between toxicity and concentration of a coco-AE_xS. The marine red-tide dinoflagellate, *Gymnodinium breve*, was exposed for 48 h to the AE_xS and 2.5 $\mu g L^{-1}$ was said to cause 87% mortality, 12.5 $\mu g L^{-1}$ resulted in 63% deaths but only 44% perished at 50 $\mu g L^{-1}$ [287].

Gerike [3] reported a NOEC for the growth of algae (unnamed) as high as 65 mg L^{-1} of an unspecified AES; this value was repeated by Schöberl et al. [4], who specify C_{12-14}-fatty alcohol AE_2S and C_{12-15}-oxoalcohol-AE_3S.

Invertebrates

Both Gerike [3] and Schöberl et al. [4] quote the range for LC_{50} for the acute toxicity to Daphniae as 1 to 50 mg L^{-1} for the above quoted AES. Table 12 gives reported individual values for a variety of AES towards daphnids; the 24 to 96 h-LC_{50} values fall in the range 1 to 37 mg L^{-1}.

Maki [288] reported the NOEC for *D. magna* as 0.27 mg L^{-1} and the EC_{50} for reproduction over 21 d as 0.37 mg L^{-1} for $C_{13.67}$-$AE_{2.25}$S. There are insufficient data to decide whether there is a relationship between structure and toxicity.

Other invertebrates were found to have 24 h-LC_{50} of 11 mg AE_3S L^{-1} and LC_{100} of 12 mg L^{-1} (mosquito larvae—*Aedes aegypti*), and the LC_{50} for snails—*Biomphaloria glabrata*—was 12 mg L^{-1} [263]. In tests with the pink shrimp, *Penaeus duorarum*, it was found that the 96 h-LC_{50} was 350 mg L^{-1} for $C_{14-16}AE_{2.25}$-S and the NOEC for this acute exposure was less than 120 mg L^{-1} [6, 166].

The toxicity of C_{12ave}-AE_nS to *D. magna* decreased steadily with time as a result of biodegradation; after 30 h in static conditions the solution was almost non-toxic [252].

These few data suggest that these invertebrates are less susceptible to AES than fish are.

Fish

A selection of acute toxicity values for various AES and fish species is given in Table 13. The values confirm the range of 1.4 to 20 mg L^{-1} given by Gerike [3] and Schöberl et al. [4]. In all, some ten freshwater species and 1 marine species were reported. Some individual AES are considerably less toxic, for example,

Table 12. Toxicity of AES to *Daphnia magna*

AES type	Period of test	EC_{50} (mg L^{-1})	Ref.
C_{12-14}-AE_3S-Na	24 h	16.3	Continental Oil Co. [5]
,, ,, -NH_4	24 h	18.9	
C_{11-16}-AE_3S	24 h	19.6	Unilever Res. Lab. (unpub.) [5]
C_{12}-AE_3S	24 h	5.0	Lundahl et al. [202]
C_{12}-AE_3S-Ziegler	24 h	37.0	
C_{12-14}-$AE_{2.2}$S	24 h	21.0	
$C_{13.67}$-$AE_{2.25}$S	96 h	1.17	Maki [288]
	21 d	0.74	,, ,,
	(NOEC)	0.27	,, ,,
	21 d (reproduction)	0.37	,, ,,
C_x-AE_nS	48 h (*D. pulex*)	20.2	Moore et al. [251]

Table 13. Toxicity of AES to fish

Species	AES type	Hardness (mg CaCO$_3$ L^{-1})	Time h	LC$_{50}$ (mg L^{-1})	Ref.
Goldfish	C$_{12}$-AE$_{2.6}$S	100	6	55 (66.5)	Gafa [172]
	C$_{14}$-AE$_3$S			6 (8.1)	
	C$_{16}$-AE$_{3.4}$S			41	
	C$_{15}$-AE$_{3.2}$S†			37	
Goldfish	C$_{11-16}$-AE$_3$S	200	24	10–15	Unilever Research Lab. [5]
		100		10–15	
Fathead Minnow	AE$_2$S	100	24	C$_{14}$ 1.8 / C$_{16}$ 1.0 / C$_{18}$ 80	Monsanto Co. [5]
	AE$_4$S			4.0 / 0.9 / 15	
	AE$_6$S			9.3 / 0.8 / 2.1	
	C$_{11}$AE$_4$S			17.0	
	C$_{12}$AE$_2$S			1.5	
Guppy	C$_{11-16}$-AE$_3$S	150	24	4.7	Unilever Research Lab. [5]
Golden orfe	,,	150		4.3	
Harlequin	,,	20		4.4	
Minnow	,,	210		5.8	
Brown trout	,,	36		5.7	
	,,	20		2.4	
Rainbow trout	,,	300		3.0 — 4.6 (LC$_{0-100}$)	
Bluegill	C$_{14.7}$E$_1$S	35	48	2.1	Procter and Gamble [5, 166]

( ) values for branched isomers; † 5% branched

66.5 mg L^{-1} for a 5% branched C_{12}-$AE_{2.5}S$ (goldfish) [172] and 400–450 mg L^{-1} for a C_{9-10}-$AE_{2.5}S$ based on a Dobanol (rainbow trout) [5, 289]. The most toxic AES were $C_{15.9}$-$AE_{2.1}S$—LC_{50} 0.3 mg L^{-1} to bluegill—and $C_{14-16}AE_{2.25}S$—0.63 mg L^{-1} in a 45 d test [5, 166].

Reiff et al. [290] reported values of LC_{50} for C_{12-15} AE_3S determined in a number of laboratories on various fish species: brown trout, 1.0–2.5 and 1.5 mg L^{-1} (96 h); harlequin fish, 3.9 mg L^{-1} (96 h); golden orfe, 3.3–6.2 and 5.7 mg L^{-1} (48 h). The 96 h-LC_{50} for a marine species, sheepshead minnow (*Cyprindon variegatus*) was 0.39 mg $C_{14-16}AE_{2.25}S$ L^{-1} [5, 166].

The effects of C chain length and number of EO groups is complex and is not yet finally resolved. In general, changes in EO numbers affected toxicity more than changes in C chain length. In AES with alkyl chain of less then C_{16}, the toxicity tended to decrease with increasing numbers of EO but above 16 this was reversed. The toxicity of AES seems to peak at chain length C_{16}: for example, using AES of C chain length 8 to 19.6 and EO of 1 to 3, the LC_{50} for bluegill fell from >250 mg L^{-1} for C8 and 375 mg L^{-1} for C_{10} to 24 mg L^{-1} for C_{13}, 4–7 mg L^{-1} for C_{14}, 2 mg L^{-1} for C_{15} and 0.3 mg L^{-1} for C_{16}. For $C_{17.9}$ and $C_{19.6}$, LC_{50} again rose to 10.8 and 17 mg L^{-1} respectively [5, 166]. Data from the Shell Chemical Co. [6, 289] confirmed that toxicity increases with increasing alkyl chain length. The 96 h-LC_{50} for rainbow trout was 400–450 mg L^{-1} for C_{9-10}-$AE_{2.5}S$, 28 mg L^{-1} for C_{12-13}-AE_2S and down to 8.9 mg L^{-1} for $C_{12-15}AE_3S$. However, Gafa [172] found that $C_{16}AE_{3.4}S$ was one of the less toxic (to goldfish) AES which he tested, with a LC_{50} of 41 mg L^{-1}.

The LC_{50} over 45 d to fry of the fathead minnow was 0.63 mg L^{-1} and for juveniles it was 0.94 mg L^{-1} of $C_{14-16}AE_{2.25}S$ [6, 166]. Sub-lethal effects were reported by Maki [288]; the lowest level giving inhibition of growth of fathead minnow was 0.22 mg L^{-1} of $C_{13.7}$-$AE_{2.25}S$ over a period of 1 y and the NOEC was 0.1 mg L^{-1}. The ventilation rate of bluegill sunfish was increased by 0.39 mg C_{16}-AE_3S L^{-1} in 48 h flow-through tests [291].

Higher Plants

Plants, such as soybean, pea, onion, treated with 100–1000 mg AES L^{-1} gave increased germination rates and yields. For example, Dwarf *Coleus salicifolius* seeds treated before germination with AE_3S reduced the germination time from 10 to 6 days and yielded 29–33 healthy seedlings after 16 d compared with 17 for treatment with water alone [292].

Effects on Small Mammals and Man

The acute oral toxicity, LD_{50}, to rats of AES and APES is 1000–6000 mg kg^{-1} body weight, about the same as that of AS (see Gloxhuber, [181]). This range of values can be compared with estimates of total anionic surfactants taken in by humans from all sources which are put variously at 0.3–0.4 to 15–20 mg d^{-1}; the highest values are considered to be excessive. For adults, 15 mg d^{-1} would be about 0.2 mg kg^{-1} body weight per day and for 10 y-old children the higher intake would be equivalent to about 1.5 mg kg^{-1} per day, which are very much lower than the LC$_{50}$ values for rats.

In chronic tests lasting 90 d to 2 years, doses of C$_{12}$-AE$_3$S up to 1000 ppm in the diet had no effect on rats or mice; 5000 ppm sometimes resulted in increased kidney or liver weights [293]. Single doses were soon excreted, the main metabolite from C$_{12}$-AE$_3$S being acetic acid-2-triethylene glycol sulfate. From C$_{11}$-AE$_3$S the corresponding metabolite was propionic acid-3-triethylene glycol sulfate; these intermediates indicate that AES were being degraded by ω- and β-oxidation of the alkylchain.

Only small amounts of AES were absorbed through the skin and only little to slight irritation occurred when AES was applied at 1%. No evidence of reproductive, teratogenic or mutagenic effect of AES was seen.

Alkane Sulfonates (SAS)

Structure

Aliphatic alkane or alkyl paraffin sulfonates are normally, from the method of manufacture, mixtures of secondary alkane sulfonates, hence the abbreviation SAS. The alkyl chain, normally C$_{14}$ to C$_{18}$, is linear and the –SO$_3$ group is placed at random along the chain. Thus C$_{14}$ alkane sulfonate is a mixture of the six isomers 2-, 3-, 4-, 5-, 6- and 7-sulfotetradecane.

Primary alkane sulfonates (PAS) C$_n$H$_{2n+1}$SO$_3$Na, have been synthesized but seem not to have been produced commercially.

Nomenclature

The nature of the alkyl group is signified as n-C$_x$SAS or n-C$_x$avc SAS but n (=normal) could be omitted. No differentiation has so far been made between isomers having the sulfonic group placed at different points along the chain but this could be signified as n-C$_{14}$3-SAS, denoting that the hydrophilic group is on the C3 atom.

Production Methods

Because of their inertness towards sulfuric acid, paraffins cannot be directly sulfonated and a sulfo-oxidation process has to be used. SAS were first made in the 1930s from C_{14} to C_{18} paraffins produced by the Fischer-Tropsch hydrogenation process using sulfur dioxide and chlorine under gamma radiation or latterly UV-irradiation. The chloro-sulfonate was then saponified with caustic soda. More recently *n*-paraffins from the appropriate petroleum fractions, separated from *iso*-paraffins and aromatic hydrocarbons, were sulfoxidized with sulfur dioxide and oxygen added in the correct proportions, again under UV-irradiation. An intermediate, peroxysulfonate, appears to be formed and acts as initiator of the sulfonation reaction. In the latter process the sulfonated paraffin is separated from the by-product, sulfuric acid, by extraction with hexanol, followed by neutralization with sodium hydroxide and then bleaching.

Organic by-products include di- and poly-sulfonates but improvements in the process have reduced the proportions of these impurities. Other impurities are non-linear analogues derived from residues of iso-paraffins in the hydrocarbon stock. The product typically contains 85–87% mono-sulfonate, 7–9% di-sulfonate and 1% unreacted paraffins.

The well-known bisulfite addition to olefines has been used to produce alkane sulfonates but not on the commercial scale because the reaction did not go to completion. Recently, however, patents suggest that nearly complete reactions are possible in the presence of an initiator. If α-olefines are used, the normal rules of addition are not obeyed and terminal, or primary, alkane sulfonates are formed.

Amounts Produced

Matson [186] records that for the year 1976 50 000 tonnes of SAS were used in Western Europe; little was used in the U.S.A. Schöberl et al. [4] estimated the use of SAS in W. Germany for 1986 to be 16 000 tonnes.

Uses and Applications

SAS has not met with outstanding success compared with LAS presumably because of price and performance factors. Alkane sulfonates are used in liquid detergents (washing-up liquids etc.) often in conjunction with AES, and in concentrated shampoos. It is not much used in powder form since it is not easy to spray-dry, though incorporation of sodium toluene sulfonate and colloidal silica can overcome this.

SAS are good mercerising agents and are used as textile and leather auxilliaries. Because of their stability, SAS can be used in industrial alkaline preparations for cleaning metal, steam jets and pickling baths.

Physical and Chemical Properties

SAS are marketed as flakes (approx. 90% 'active' matter), soft yellowish pastes (approx. 65%) or slightly yellow or clear, colourless liquids (approx. 30%). As a class they have good detergency, foaming and wetting properties over a wide range of temperature, about the same as those of LAS. They have high stability in water and are more stable than LAS in hard waters and under alkaline conditions.

Analytical Methods

In common with other sulfonate and sulfate anionic surfactants, SAS react in the methylene blue test. Since, like other sulfonates, SAS are not hydrolysed by acids, the determination of MBAS before and after hydrolysis gives a measure of SAS plus (LAS + AOS).

TLC or paper chromatography has been applied for the direct determination of SAS [294] or indirectly after prior conversion to the sulfonate methyl esters; the separated spots were examined by NMR and GC [295, 296]. Matsutani et al. [295] also separated seven homologues from C_{12} to C_{18} of SAS products by GC by first converting the SAS to the corresponding olefins by alkaline fusion. Each peak indicated a set of isomers differing in the position of the double bond along the chain; on hydrogenation of the olefines seven 'clean' peaks were obtained representing the seven alkanes. An HPLC method [193] was also applied to the analysis of SAS (see p 45).

Environmental Concentrations

No record of monitoring of SAS in environmental samples was found, but MBAS values in such samples would include any SAS present. Environmental concentrations of SAS in the U.S.A. would presumably be very low since they have not been used extensively in that country.

Gilbert and Pettigrew [71] calculated the concentrations of SAS in various parts of the environment in the U.K. from SAS production, water usage, biodegradability, etc., to be approximately 2 mg L^{-1} in sewage, 1.4 mg L^{-1} in settled sewage, 0.04 mg L^{-1} in effluent, 0.02 mg L^{-1} in river water and 0.16% w/w in sewage solids and 3.2 mg kg^{-1} in soil. Righton and Watts [195] detected the five homologues from C_{13} to C_{17} SAS in effluents from the textile industry and printworks in the U.K.

Environmental Fate

Biodegradability

Both primary and secondary alkane sulfonates (SAS) are readily biodegradable, ranking between LPAS and LAS for speed of degradation in batch tests [297]. When pure, SAS reach virtually 100% MBAS removal, but commercial products, which are predominantly linear and secondary, may contain non-linear and other impurities which could reduce the degree of primary biodegradation to less than 100%. Di- and poly-sulfonates which are present as impurities degrade as readily as the mono-sulfonate [214].

There is also ample evidence that SAS are ultimately biodegraded in laboratory die-away tests; for example, 63–77% of $ThCO_2$ were evolved in 20 d from SAS ranging from C_{13} to C_{18} [5, 166], 100% DOC removal from C_{14}–C_{18}-SAS in 5 d [298] and 93% of ThOD from C_{12}–C_{18}-SAS in the 30-d Closed Bottle test [131]. Schöberl et al. [4] recorded 88–96% DOC removal, 63–95% of ThOD and 56–91% of $ThCO_2$ from C_{13}–C_{18}-SAS in laboratory tests, while Gerike [3] found 96% MBAS, 73% of ThOD and 80% DOC removal from C_{13}–C_{18}-SAS in the relevant OECD tests.

Little effect of chain length of biodegradation has been reported; for example, individual homologues from C_{13}–C_{18}-SAS gave similar values for %ThOD and for %$ThCO_2$ [5, 166], and there was little difference between the biodegradation of individual homologues from C_{13} to C_{19} as measured by COD removal after 21 d, although less degradation occurred with the C_{17} to C_{19} homologues in the first few days of incubation [299]. Lundahl et al. [249, 252] confirmed this for C_{19} and C_{21} and suggested that it was a reflection of the greater inhibitory effect of the higher homologues.

In activated sludge simulation tests, 20 mg C_{10}–C_{18}-SAS L^{-1} was removed by 96% (DOC) and at 100 mg L^{-1} 84% DOC was removed [221], while in small-scale trickling filters 97–99% MBAS was removed [300]. Also in activated sludge simulation tests, 83 to 96% DOC was removed from C_{13}–C_{18}-SAS [3,4]. In full-scale trickling filters SAS was found to be removed more efficiently than LAS [297] and SAS was removed efficiently in an oxidation ditch [301].

SAS is not degraded under anaerobic conditions [72], along with other sulfonates, and is inhibitory to anaerobic digestion at about 10 mg L^{-1} and above [232].

Metabolic Pathways

Primary alkane sulfonates, which are mineralized to the extent of 95–100%, are degraded to bisulfite and the corresponding aldehyde [2]. The bisulfite is presumably oxidized chemically to sulfate, while the aldehyde is oxidized to the carboxylic acid and thence mineralized by β-oxidation, starting at the carboxyl group.

The pathway for SAS is less well established. Thysse and Wanders [302] isolated an alkane sulfonate-α-hydroxylase from a pseudomonad grown on 1-C_8-PAS and showed that it desulfonated n-C_{12}-SAS (n-dodecyl-2-sulfonate) to 2-dodecanone. Hence, Swisher [2] suggested that the first step is the formation of a keto-bisulfite,

$$C_{10}H_{21}.CH(CH_3)SO_3^- \longrightarrow \underset{\text{keto-bisulfite}}{C_{10}H_{21}.CHOH(CH_3)SO_3^-}$$

$$\underset{\text{2-dodecanone}}{C_{10}H_{21}.CO.CH_3} \qquad HSO_3^-$$

then the formation of bisulfite and 2-dodecanone, which is further oxidized by sub-terminal oxidation to an ester, undecyl acetate. The ester is then converted to the corresponding alcohol and acetic acid and a similar process is repeated until mineralization is complete.

Microbial Growth

Two species of pseudomonads, *P. rathonis* and *P. alcaligenes* were able to degrade SAS completely using the sulfonate as sole source of carbon and sulfur, producing sulfite and carboxylic acids [303,304].

Using mixed cultures derived from river water and sewage, Hrsak et al. [131] calculated the specific growth rate (μ) on SAS and LAS to be about $0.1 \, h^{-1}$ at $20 \, mg \, L^{-1}$; at $95 \, mg \, L^{-1}$, μ for SAS was $0.12 \, h^{-1}$ but only $0.01 \, h^{-1}$ for LAS, while at $465 \, mg \, L^{-1}$, μ for SAS was $0.1 \, h^{-1}$ but LAS was not degraded at this concentration.

Effects on the Environment

Only a small number of studies on the toxicity of SAS to aquatic organisms were found.

Bacteria

Lundahl et al. [202] reported the lowest concentrations of various SAS which did not allow development of cultures of *Escherichia coli* on gelatin plates to be as high as 20 g L^{-1} for C_{14}–C_{15}-SAS, 200 g L^{-1} for C_{14}-SAS and more than 200 g L^{-1} for C_{13}–C_{17}-SAS. The significance of these results for the aquatic environment is difficult to assess.

Algae

Chlamydomonas variabilis showed increased effects with increasing chain length, the 24 h-EC_{50} falling from 125 mg L^{-1} for $C_{10.3}$-SAS to 32.4 mg L^{-1} for C_{14}-SAS, 9.4 mg L^{-1} for C_{16}-SAS and 3.7 mg L^{-1} for $C_{18.9}$-SAS. The highest chain length tested, $C_{20.7}$, was somewhat less toxic than $C_{18.9}$-SAS at 8.4 mg L^{-1} [248]. On the other hand, Guhl and Gode [305] reported an LC_0 as high as 183 mg SAS L^{-1} (unstated composition) for an unstated alga.

Daphnia

Lundahl and Cabridenc [248] similarly found SAS was increasingly toxic to *Daphnia magna* as the chain length increased. The 24 h-EC_{50} for $C_{10.3}$-SAS was 319 mg L^{-1}, 111 mg L^{-1} for C_{14}-SAS, 30 mg L^{-1} for C_{16}-SAS and 3.3 mg L^{-1} for $C_{18.9}$-SAS. Again, the highest chain length tested, $C_{20.7}$, was slightly less toxic, with EC_{50} of 6.3 mg L^{-1}. A pure C_{14}-SAS had a 24 h-EC_{50} of 60 mg L^{-1}.

Schöberl et al. [4] reported a range of 8.7–13.5 mg L^{-1} for the EC_{50} for C_{13} to C_{18}-SAS, while Gerike [3] found a much wider range of values, namely, 4 to as high as 250 mg L^{-1} for C_{13-15} to $C_{16.3}$-SAS and 0.7–6 mg L^{-1} for C_{15-18}-SAS. Guhl and Gode [305] reported an LC_0 value of 1.8 mg L^{-1} for an unstated SAS.

Fish

The longer chain length SAS are also more toxic to fish than are the shorter homologues. This has been shown for the bluegill [5, 166], guppy [306] and minnow [248].

The 96 h-LC_{50} for bluegill was 1.3 mg L^{-1} for C_{18}-SAS, 4.6 mg L^{-1} for C_{16}-SAS and 144 mg L^{-1} for C_{13}-SAS. For *Phoxinus phoxinus*, the 24 h-LC_{50} rose from 3.1 mg C_{16}-SAS L^{-1} to 8.5 mg C_{15}-SAS L^{-1} and finally to 34.5 mg C_{14}-SAS L^{-1}. A commercial product $C_{14.5}(C_{12}$–$C_{17})$ gave a 6 h-LC_{50} for *Carassius auratus* of 9.5 mg L^{-1} [172]. Schöberl et al. [4] presented a range of LC_{50} values of 3–24 mg L^{-1} for C_{13}–C_{18}-SAS against golden orfe, trout, carp and zebra fish, while in a similar presentation the LC_{50} for largely the same species was 2 to 10 mg L^{-1} for C_{13-15} and $C_{16.3}$-SAS and 1 to 2 mg L^{-1} for C_{15-18} and C_{18}-SAS

[3]. Guhl and Gode [305] reported the LC_0 for "fish" to be 0.5 mg L^{-1} of an unnamed SAS and for a laboratory microcosm it was 0.3 mg L^{-1}.

Effects on Small Mammals and Man

A few studies on rats and mice gave the acute oral toxicity (LC_{50}) of SAS to be 1440 to 4200 mg kg^{-1} body-weight, making SAS about as toxic as sodium chloride and sodium carbonate. Insufficient data are available to discern any trends with chain length, but generally for other surfactants the middle members of the range of chain lengths found in commercial products are usually more toxic than higher and lower homologues [181]. Chronic tests on rats receiving 300 mg SAS kg^{-1} daily for 45 and 90 d revealed no adverse effects. Similarly, rats fed 0.5% in their diets for 91 d developed no adverse symptoms [293]. Single doses to rats of *prim.* C_{12} alkane sulfonate were fairly rapidly excreted and metabolism occurred by ω- and then β-oxidation. The final product was 4-sulfobutyric acid; no sulfite was found.

The above LC_{50} doses are to be compared with the estimated daily intake by man of anionic surfactants from all sources of 0.3–20 mg d^{-1} equivalent to 0.2 mg kg^{-1} body-weight for an adult and about 1.5 mg kg^{-1} for a 10 year-old child.

No adverse reactions have been reported in skin tests and no reports were seen of results of studies on teratogenic, carcinogenic or reproductive effects.

α-Olefine Sulfonates (AOS)

Structure and Nomenclature

Present day olefine sulfonates are manufactured from linear α-olefines and because the process is complex (and not completely understood) the product is not homogeneous but consists of two main types, plus varying amounts of other impurities. The major (about 60 to 65%) portion consists of alkenyl sulfonates, followed by hydroxy-alkane sulfonates (35 to 40%), together with minor components such as 1,2-, 1,3- and 1,4-sultones as well as disulfonates and sulfone sulfonic acid.

The alkene 1-sulfonates have the structure

$RCH=CH(CH_2)_nSO_3Na$

where $n = 1, 2$ or 3; that is, the double bond is in the 2,3; 3,4; or 4,5 positions and they are primary sulfonates. The C chain length can be from 11 to 20, but is more usually 14 to 18 C atoms and is normally linear.

The hydroxy-alkane 1-sulfonates have the structure

$RCH_2 . CHOH . CH_2 . CH_2 . SO_3Na$

with the hydroxy group on the C_3 or C_4 atoms and contain the same total number of C atoms as the alkene sulfonates.

The sultones have the structure

$$\begin{array}{cc} RCH.CH_2.CH_2 & RCH.CH_2.CH_2.CH_2 \\ |\quad\quad\quad\quad| & |\quad\quad\quad\quad\quad\quad/ \\ O\text{———}SO_2 & O\text{———}SO_2 \\ \text{1,3 sultone} & \text{1,4 sultone} \end{array}$$

The α-olefine sulfonates are expressed as, for example, C_{18}-AOS or C_x-AOS if the number of C atoms is not known.

The hydroxy alkane-sulfonates are expressed as

C_{18}–3OH–AOS or C_{16}–4OH–AOS

the numbers 3 and 4 indicating the position of the –OH group on the C chain.

It seems that little has been reported on individual alkane sulfonates; if needed, the double bond is designated as

C_{18}–Δ3–AOS

meaning that the double bond is at C3 to C4.

Production Methods

Early attempts based on reacting olefines, having internal double bonds, with bisulfite were unsuccessful; the products were not good detergents. When α-olefines, that is the, double bond in the 1,2 positions, became available they were reacted with an SO_3-organo-compound. Better results are now obtained by sulfonating with diluted sulfur trioxide in a film reactor.

The hydrocarbon source was first formed by the thermal cracking of linear paraffins yielding linear α-olefines, but it can also be produced by the Ziegler process involving the use of aluminium or by the Shell SHOP process employing a nickel catalyst.

The method of sulfonation is important since some methods lead to the need to bleach coloured product with hypochlorite. The use of hypochlorite appears to cause traces of sultones to remain after subsequent alkaline treatment; sultones have been implicated in causing effects on human skin. By sulfonating in film reactors the time of contact between olefines and sulfur trioxide can be adjusted to the relatively short time needed for mono-sulfonation. The products under these conditions are light in colour, needing no bleaching, and do not have harmful effects on the skin. Even so, the reaction is complex and large amounts of various sultones are formed as intermediates.

The acid reaction mixture has to be converted to water-soluble surfactants by neutralization with alkali and saponified with caustic soda to convert the sultones to hydroxy-alkane sulfonates. The final product is a mixture of alkene sulfonates, largely the 2,3-, 3,4- and 4,5-alkene sulfonates, and hydroxy-alkane sulfonates, mostly the 3- and 4-isomers. The 2- (or β) sultones are unstable and isomerize to the 3-sultone, so that after saponification little of the poorly water-soluble 2-hydroxy-alkane sulfonate is formed. The 3-sultone is less readily converted to the 4-sultone, which is, in turn, more resistant to hydrolysis.

Thus, more perhaps than other anionic sufactants, the properties and effects of commercial AOS may be due not to "AOS" itself but to the complex mixture produced during the manufacturing processes.

Amounts Produced

Matson [186] reported the production in 1976 of 10 000 tonnes evenly divided between U.S.A. and W. Europe; the usage in the U.S.A. in 1980 was about 6800 tonnes.

Schöberl et al. [4] reported the use in 1986 of C_{14}–C_{18}-AOS in W. Germany of less than 1000 tonnes.

Uses and Applications

AOS have not made great strides in the field of heavy-duty laundry formulations but they have been successful in light-duty detergents, hand dishwashing, shampoos, bubble baths and synthetic soap bars. They are also used for cleaning upholstery and carpets, and in the textile, leather and cosmetic industries.

Physical Properties

Commercially produced AOS are available as the sodium salt in the form of clear to pale yellow liquids (25–40% active matter) or free-flowing slightly pale yellow powder (90–100% active matter). They are also available as blends with lauryl ether sulfate.

AOS are effective cleaning agents, imparting a "soft feeling" to washed fabrics, equal to LAS in detergency and foaming. They have a high degree of stability, high solubility in water, having a slight advantage over LAS in hard water and are not hydrolysed by acids. They also become rancid at a slower rate than LAS.

Analysis

Like other sulfonates AOS (both main components) are not hydrolysed in acid solution and form complexes with methylene blue. It has been shown [307] that the disulfonates do not react stoichiometrically in the overall process since the MB-complexes formed tend to remain in the aqueous phase and are not completely extracted by chloroform.

A range of chromatographic methods can be applied as for the other anionic surfactants. TLC methods are available for the sultones [308] and also for separating and quantifying the alkene and hydroxyalkane-monosulfonates [309]. Alternatively, the methyl sulfonate ester can be formed before application of TLC and identification of the individuals by NMR/GC [295, 296].

GC has been used by the same author [295] to separate the trimethysilyl esters with good separations of the even homologues of both the sulfonates and hydroxyalkane sulfonates. Also, the sulfonyl chlorides of the hydrogenated alkene sulfonates have been separated by GC [310].

Reversed phase HPLC was applied by Johannessen [311] using 0.4 M $NaNO_3$ in the 75:25 methanol:water mobile phase to improve the separation. Besides separation of C_{14} from C_{16}, there was partial separation of the 2- and 3- alkene sulfonates and the 3- and 4-hydroxyalkane sulfonates.

Far infra red absorption was used to determine AOS as a class. The MB-complex was hydrolysed by acid to remove organic sulfates and the sulfonates were hydrogenated to convert alkene to alkane sulfonates. After separation of released methylene blue by ion-exchange, the sulfonates were converted to the sulfonyl chlorides and the amount of AOS was determined by far infra red absorbance [188].

Environmental Concentrations

Because of the low tonnage produced and its ease of biodegradability, the concentration of AOS in the environment is likely to be low, and since it reacts with methylene blue any AOS present will be included in MBAS.

The only values found for AOS as a group were for two Japanese sewages and treated effluents [188]. The annual average concentrations in the sewages were 0.164 and 0.160 mg L^{-1} out of 8.2 and 8.4 mg L^{-1} total MBAS, respectively; AOS was not detected in the treatment plant effluents.

Environmental Fate

There are limited data for the biodegradation of AOS, but available results show that AOS are rapidly biodegraded and mineralized at rates slightly lower than AS and slightly faster than LAS. Many authors report that MBAS was removed by 95–100% in 2 to 8 d in river water and in inoculated media, for example, C_{15}–C_{18}-AOS [211]; C_x-AOS [130]; C_{12}–C_{18}-AOS [312]. In sea water 5 mg AOS L^{-1} was removed by 100% in 5 d [211]. The time for removal in river water of MBAS from 20 mg L^{-1} was 2 d for C_x-AOS compared with 6 d for LAS and 0.5 d for AS [210]; this was confirmed by Itoh et al. [313] and by soil perfusion tests [220].

Many laboratory tests showed that AOS were readily mineralized; for example, in the Bunch-Chambers tests C_{15}–C_{18}-AOS lost 99% MBAS and 90% DOC in 1 d, and 100% in 5 d [298] and with an activated sludge inoculum 100 mg C_x-AOS L^{-1} lost 90% DOC in 8 d and 98% in 15 d [199]. Kawasaki et al. [210] reported C_x-AOS to be "completely" degraded to CO_2 in 12–14 d in river water, but Kravetz et al. [314] were more precise in reporting 57–71% of $ThCO_2$ for a series of AOS (C_{14}–C_{18}). Earlier it had been reported an average of 65% of $ThCO_2$ and 52% of ThOD (in 5 d) for the range C_{12} to C_{18}-AOS containing about 40% hydroxyalkane sulfonate [5, 166]. Urano and Saito [122], applying the 14 d-MITI test found the % of ThOD fell from about 100% at 3 mg C_{12}-AOS L^{-1} to 72% at 100 mg L^{-1}. The first order decay rate was $2 \times 10^{-2} h^{-1}$ at 100 mg L^{-1}, which was lower than the rates for both AES and AS (namely, 5.1×10^{-2} and $8.8 \times 10^{-2} h^{-1}$, respectively).

In activated sludge simulation tests, C_x-AOS was removed by 100% MBAS and 88% DOC [130] and in field trials C_{18ave}-AOS was removed by 100% MBAS [315]. Oba et al. [188] carried out year-long tests at two sewage treatment works in Japan and found 100% removal of about 0.16 mg AOS L^{-1} present in the sewage.

Whether or not chain length plays a significant part in degradation of AOS is unclear; some reports indicate that C_{16} and C_{18}-AOS are less degradable than C_{12}- and C_{14}-AOS, for example, 58% $ThCO_2$ in 28 d compared with 69% [314]; while others indicate no trend, e.g. Pitter [316] found 91, 88, 91 and 92% ThCOD removal after 8 d for C_{12}, C_{14}, C_{16} and C_{18} respectively. Also, Tuvell [317] reported 77, 74 and 73% of $ThCO_2$ produced from C_{14}-, C_{16}- and C_{18}-AOS respectively. It seems that initially degradation may start at different times after inoculation for each homologue, but that by the end of the test the degrees of degradation are virtually the same.

The various components of commercial AOS have been tested separately and found to be well biodegraded. Tuvell [317] found that C_{14}-, C_{16}- and C_{18}-alkene sulfonates each degraded by about 84% $ThCO_2$ within 27 d and the corresponding 3-hydroxy-alkane sulfonates were degraded by about 86% under the same conditions. However, Tomiyama et al. [318] showed that the hydroxy-alkane sulfonates were more slowly degraded. They used GC analysis of the sulfonyl chlorides on samples from biodegradability tests and found that, when about 50% total MBAS had been removed, 65–75% of the alkene sulfonates had disappeared compared with only 28–32% of the hydroxy-alkane sulfonate. The same author found that the disulfonates were slightly more slowly degraded than the monosulfonates, but ultimate biodegradation of the disulfonates levelled off at about 40% of $ThCO_2$ compared with 85% for the monosulfonate [317].

Anaerobic Biodegradability

From the studies of Wagener and Schink [232] it could be expected that AOS were not degraded anaerobically, but Oba et al. [229] gave a value of 31% to 43% removal of MBAS under anaerobic conditions in 28 d. Itoh et al. [231] reported that of a number of surfactants tested, LAS and AOS were the least degraded anaerobically.

Metabolic Pathways

The metabolic pathways by which AOS are metabolized are not known. On the basis of enzyme studies by Thysse and Wanders [302] involving C_9-AOS, Swisher [2] speculated that the first step is the formation of an unsaturated carboxylic acid, for example, nonenoic acid from C_9-AOS. For an even numbered AOS he suggested the first product might be an unsaturated aldehyde bisulfite followed by release of bisulfite and formation of a ketene and then a saturated carboxylate.

Effects on the Environment

Although many studies have been made on the toxicity of AOS to fish, few studies have been made on other organisms. The range of LC_{50} values in acute toxicity tests was 0.5 to 50 mg AOS L^{-1}.

Algae

Schöberl et al. [4] reported a range of 10–100 mg L^{-1} for C_{14}–C_{18}-AOS as being toxic against the growth of algae; it is not clear whether EC_{50} or some other function is meant and neither are the algal species named.

Invertebrates

The same authors [4] gave the EC_{50} for *Daphnia magna* as 5–50 mg C_{14}–C_{18}-AOS L^{-1}. Little [5] quotes unpublished work; the 24 h-EC_{50} was 15.6 mg L^{-1} for an AOS of MW 299 [319] and 9.3 mg L^{-1} for C_x-AOS [289]. In a third study [5] C_{14}–C_{16}-AOS had an EC_{50} for *D. magna* of 16.6 mg L^{-1} while a C_{16}–C_{18}-AOS was more toxic at 7.7 mg L^{-1}. Lundahl and Cabridenc [252] reported that the toxicity of AOS decreased with time; thus, inoculated media taken from biodegradability tests showed negligible toxicity to *D. magna* after incubation for 48 h or longer.

Fish

It is quite clear that the higher homologues are more toxic than the lower ones (Table 14). The C_{16}–C_{18}-AOS were on average just over four times as toxic as the C_{14}–C_{16}-AOS for several species of fish, that is LC_{50} of 0.5–2.0 mg L^{-1} and 2.5–10 mg L^{-1}, respectively. The overall range for LC_{50} is 0.5 to 20 mg L^{-1}, which includes the single homologues C_{14}, C_{16} and C_{18} and which has a lower minimum than the range of 2 to 20 mg L^{-1} quoted by Schöberl et al. [4] for C_{14}–C_{18}-AOS.

In a rather limited test, using only two concentrations of a C_{15-18} "Technical"-AOS, the 24 h-LC_{50} for white tilapia (*Tilapia melanopleura*) was 2.0 mg L^{-1} [322]. In the discussion to the presented paper the authors gave further data on the same species, namely, that over a three-month period 14 out of 20 fish died in 0.3 mg C_{15-18}-AOS (Technical) L^{-1}. In another study 3.2 mg C_{14-16}-AOS L^{-1} and higher allowed no survival of hatched eggs of fathead minnow and no effect was observed at 1.8 mg L^{-1} [5, 320]. The same authors reported serious effects on midges (*Chironomid tentans*) when exposed to 9 mg C_{14-16}-AOS L^{-1} continuously through two generations (0–66% survival) but no effect was seen at 4.5 mg L^{-1}.

Table 14. Toxicity of AOS to fish

Fish species	24 h-LC_{50} C_{14}–C_{16}	(mg L^{-1}) C_{16}–C_{18}	Ref.
Goldfish	7.1	1–3	Unilever Research Lab. [5]
Golden orfe	4.7	1.2	Unilever Research Lab. [5]
	4.9*	1.0*	Reiff et al. [290]
	3.4*	0.9*	,, ,,
	5.7*	1.9*	,, ,,
Fathead minnow	8.6	1.8	Colgate-Palmolive Co. [320]
Guppy	10.1	1–2 (LC_0–LC_{100})	Unilever Research Lab. [5]
Harlequin	6.2	1.3	Unilever Research Lab. [5]
	3.3*	0.5*	Reiff et al. [290]
Brown Trout	2.5–5.0*	0.5*	Reiff et al. [290]
	3.5†	<0.3–0.5†	Unilever Research Lab. [5]
Minnow	5.3	1.4	Unilever Research Lab. [5]
Rainbow Trout	5.1	0.8	Unilever Research Lab. [5]
	3.5†	0.6†	,, ,,
Fathead Minnow	8.2	—	Monsanto Co. [321]
	C_{14} 15–21	C_{16} 3.2–6.9	C_{18} 0.5–0.8

* 96 h; † 48 h

Higher Plants

No effects were observed on germination or growth of tomato, barley and bean plants in pots or plots of land watered daily with concentrations of C_{15-18}-technical AOS from 0 to 40 mg L^{-1} [322]. However, Takita [323] indicated that the threshold concentration for effects on the early stages of growth of hydroponic cultures of radish, China cabbage and rice was 10 mg L^{-1} and that LAS was more toxic than AOS.

Small Mammals and Man

Tests with rats and mice have shown that AOS have moderately low oral toxicity; the acute LD_{50} for rats was 1300–2400 mg kg^{-1} and for mice it was 2500–4300 mg kg^{-1}. In longer term tests no adverse effects were observed when 1000 mg kg^{-1} d^{-1} was fed over 90 d nor when 1000–5000 mg AOS L^{-1} was fed daily over a 2 year period. In the latter case the dose was equivalent to 195 mg kg^{-1} d^{-1} for males and 295 mg kg^{-1} d^{-1} for females.

References

1. Davidsohn AS, Milwidsky B (1987) Synthetic detergents. Longman, London
2. Swisher RD (1987) Surfactant biodegradation, Marcel Dekker, New York
3. Gerike P (1987) In: Falbe J (ed) Surfactants in consumer products, Springer, Berlin Heidelberg New York, p 450
4. Schöberl P, Bock KJ, Huber L (1988) Tenside Surfactants Detergents 25:2
5. Little AD (1977) Human safety and environmental aspects of major surfactants. Prepared for Soap and Detergent Association New York. AD Little, Cambridge MA
6. Little AD (1981) As 5
7. Sisley PJ (1964) Encyclopaedia of surface-active agents. Chemical Publishing Co., New York
8. Cain RB (pers. comm.)
9. Berth P, Jeschke P (1989) Tenside Surfactants Detergents 26:75
10. FRG Bundesgetzblatt Pt 1 Nr 128 (1976) pp 3017–3032
11. Control of Pollution Act (1974) Part 2 Pollution of Water HMSO, London
12. FRG Bundesgesetzblatt Part 1. No. 49 Dec. 12 1962. 698–706
13. Soap and detergent association (1965) JAOCS 42:986
14. Organisation for economic co-operation and development Pollution by detergents: Determination of the biodegradability of anionic synthetic surface active agents, OECD, Paris, 1971
15. OECD proposed method for the determination of the biodegradability of surfactants used in synthetic detergents, Paris, 1976
16. Council directives 73/404 and 73/405 Off J European Comm L347/53 7th December 1973
17. Council directive, amending 73/405 Off J European Comm C.112/4 14th May 1981
18. Council directive, amending 73/404 Off J European Comm L109/1 22nd April 1982
19. Council directive, 67/548, 6th amendment Off J European Comm L259/10 15th October 1979
20. Sweeney WA, Foote JK (1964) JWPCF 36:14
21. Sweeney WA (1966) JWPCF 38:1023
22. Berna JL, Ferrer J, Moreno A, Prats D, Ruiz Bevia F (1989) Tenside Surfactants Deterg 26:101

23. Giger W, Brunner PH, Ahel M, McEvoy J, Marcomini A, Schaffner C (1987) Gas Wass-Abwass 67:3
24. Matthijs E, De Henau H (1985) Tenside Deterg 22:299
25. Urano K, Saito M, Murata C (1984) Chemosphere 13:293
26. Swanwick JD, Shurben DG, Jackson S (1969) Wat Pollut Control 68:639
27. Janicke W, Niemitz W (1973) Vom Wasser 40:369
28. Urano K, Saito M (1984) Chemosphere 13:285
29. Games LM (1982) In: Dickson KL, Maki AW, Cairns J (eds) 4th meeting Modelling the fate of chemicals in the aquatic environment. Ann Arbor Sci Ann Arbor Michigan U.S.A. 1981
30. Hand VC, Williams GK (1987) Env Sci Technol 21:370
31. Llenado RA, Neubecker TA (1983) Anal Chem 55:93R
32. Llenado RA, Jamieson RA (1981) Anal Chem 53:174R
33. International organisation for standardization (1984) ISO 7875-1. Determination of anionic surfactants by the methylene blue spectrometric method
34. Abbott DC (1962) Analyst 87:266
35. Longwell J, Maniece WD (1955) Analyst 80:167
36. Standing committee of analysts (1982) Analysis of surfactants in waters, waste waters and sludges. Methods for the examination of waters and associated materials, HMSO, London
37. Uchijama M, Takamura M (1980) Jap J Wat Pollut Res 3:133
38. Waters J, Garrigan JT (1983) Wat Res 17:1549
39. Osburn QW (1986) JAOCS 63:257
40. Hon-Nami H, Hanya T (1978) J Chromatog 161:205
41. Hon-Nami H, Hanya T (1980) Wat Res 14:1251
42. McEvoy J, Giger W (1986) Env Sci Technol 20:376
43. Tsukioka T, Murakami T (1983) Bunseki Kagaku 32:723
44. Matsutani S, Shige T, Nagai T (1979) Yukagaku 28:847
45. Eganhouse RP, Ruth EC, Kaplan IR (1983) Anal Chem 55:2120
46. Tanaka Y, Ikebe K, Yoshida M, Tanaka R, Kanita N (1976) Ken Kyusho Kenku Hokuku, Shokunin Eisei Hen 7:21
47. Gloor R, Johnson EL (1977) J Chrom Sci 15:413
48. Jandera P, Churacek J (1980) J Chromatog 197:181
49. Linder DE, Allen MC (1982) JAOCS 59:152
50. Nakae A, Tsuji K, Yamanaka N (1980) Anal Chem 52:2275
51. Nakae A, Tsuki K, Yamanaka M (1981) Anal Chem 53:1818
52. De Henau H, Matthijs E, Hopping WD (1986) Int J Environ Anal Chem 26:279
53. Matthijs E, De Henau H (1987) Tenside Deterg 24:193
54. Castles MA, Ward SR (1987) Analysis of LAS in environmental matrices by reverse phase HPLC with fluorescence detection. Presented at 8th annual meeting. Soc Env Tox Chem Nov 9–12, Pensacola, Florida, U.S.A.
55. Kikuchi M, Tokai A, Yoshida T (1986) Wat Res 20:643
56. Hellmann H (1982) Zf Wass-u Abwass-Forsch 15:15
57. Swisher RD (1981) J Environ Qual 10:243
58. Schaumberg GD, Le Vesque-Madore CS, Sposito G, Lund LJ (1981) J Environ Qual 10:244
59. Uchiyama M (1977) Wat Res 11:205
60. Mizuno M, Kawasaki T, Kojima I (1984) Mizu Shori Gijutsu 25:207
61. Sedlak RI, Booman KA (1986) Soap Cosmet Chem Spec 62:44
62. Rapaport RA, Hopping WD, Eckhoff WS (1987) Monitoring LAS in the environment. Presented at 8th annual meeting. Soc Env Tox Chem Nov 9–12, Pensacola, Florida, U.S.A.
63. Wagner R (1978) Gas v. Wasserfach (Wasser Abwasser) 119:235
64. Painter HA, King EF (1979) Water Research Centre, Rep. 709-S Stevenage U.K.
65. Painter HA, King EF, Moorhouse P (1980) Water Research Centre, Rep. 4-M Stevenage U.K.
66. Stiff MJ (1987) (pers. comm.)
67. 18th Annual Report, Standing Technical Committee on Synthetic Detergents, HMSO, London 1978.
68. Nara T, Oyamada K, Toanaka M, Hayakari T, Noda M, Yamasaki Y, Imis T, Ishida H (1983) Aomori-ken Kogai Senta Shoho 6:39
69. Kondoh H, Aoi T (1982) Hokkaido Kogai Boshi Kenkyushoho 9:69
70. Painter HA (1977) Water Research Centre Report 626-S Stevenage U.K.
71. Gilbert PA, Pettigrew R (1984) Int J Cosmet Sci 6:149
72. Bruce AM, Swanwick JD, Ownsworth RA (1966) J Proc Inst Sew Purif Pt 5:427

73. McEvoy J, Giger W (1985) Naturwiss 72:429
74. Yoshimura K (1984) XV Jornados Com Espanol Deterg p 103
75. Waters J, Holt MS, Matthijs E (1989) Tenside Surfactants Deterg 26:129
76. Kobuke Y (1985) Rikusuigaku Zasshi 46:279
77. Hennes EC, Rapaport RA (1989) Tenside Surfactants Deterg 26:141
78. Fischer WK (1980) Tenside Deterg 17:250
79. Gerike P, Winkler K, Schneider W, Jakob W (1989) Tenside Deterg 26:136
80. Morena Danvila A (1987) Private presentation, Petroquiaicu Espanola SA, Madrid, Spain
81. Institut d'Hygiene et d'Epidemiologie (1985) Reseau de mesure de la qualité' des eaux superficielles Belges en 1985. 14 rue Juliette Wytsman, Bruxelles
82. Motoyama S, Mukai H (1981) Niigata Rikagaku 7:21
83. Okamoto T, Shirane Y (1982) Hiroshima-Ken Kankyo Senta Kenkyu Hokoku 3:60
84. Takada H, Ishiwatari R (1987) Env Sci Tech 21:875
85. Katsuno T, Tsukuoka T, Kono Y (1983) Nagano-Ken Eisei Kogai Kenkyusho Kenkyu Hokoku 6:23
86. De Henau H, Matthijs E, Hopping WD (1986) Int J Environ Anal Chem 26:279
87. Uchiyama M (1979) Wat Res 13:847
88. Siron R, Giusti G (1985) CR Acad Sci Ser 3 301:313
89. Mochalov OS, Autonova NM, Nesterova MP (1984) Geol Istov Geokhim Balt Morja p 165
90. Utsunomiya A, Ito S, Setsuda S, Naato S, Shimozata T (1980) Eisei Kaguku 26:159
91. Waters J (1976) Vom Wasser 47:131
92. Anon (1978) Yushi 31:30
93. Imaida M, Sumimoto T, Miyano K, Yoshida A (1979) Osaka-furitsa Koshu Eisei Kenkyusho Kenkyu Hokoku, Shokuhin Esei Hen 10:63
94. Mancini P, Bellandi S, Lan Conca P, Pantani F (1984) Riv Mercoed 23:143
95. Yoshimura K, Hayashi K, Kawase J, Tsuji K (1984) Jap J Limnol 45:51
96. Painter HA, Zabel TF (1988) Review of the environmental safety of LAS Con 1652 Water Research Centre, August 1988, Medmenham, U.K.
97. Cain RB, (1987) Biochem Soc Trans 15 (Suppl) :7S
98. Schöberl P (1989) Tenside Surfactants Deterg 26:86
99. Willetts AJ (1973) Envir and Change 2:110
100. Divo C, Cardini G (1980) Tenside Deterg 17:30
101. Phillips WK (1978) Symbiotic degradation of linear alkyl benzene sulfonate. Thesis. In: Diss Abstr Int B 39:3158
102. Huddleston RL, Allred RC (1963) Dev Ind Microbiol 4:24
103. Bock KJ, Wickbold R (1966) Vom Wasser 33:242
104. Swisher RD (1972) Yukagaku 21:130
105. Gledhill WE (1975) Appl Microbiol 30:922
106. Simko JP, Emery EM, Blank EW (1965) JAOCS 42:627
107. Abe S, Seno M (1985) Nippon Kagaku Kaishi 5:814
108. Utsunomiya A, Naito S, Tomita I (1986) Eisei Kagaku 32:258
109. Larson RJ, De Henau H (1988) In: Proceedings of 2nd World surfactants congress, Paris, France 24–27 May
110. Ward TE (1986) In: 7th ann meeting of Soc Environ Toxicol Chem. November, Arlington, VA, U.S.A.
111. OECD (1981) Chemicals Group. Guidelines for testing chemicals ISBN 92-64-1222-4, Paris
112. International Organization for Standardization (1984) ISO 7827 Evaluation of the ultimate biodegradability of organic compounds—Method by analysis of dissolved organic carbon
113. Painter HA, King EF (1985) In: Hutzinger O (ed) The handbook of environmental chemistry, vol 2, Part C, Springer, Berlin Heidelberg New York
114. Gerike P, Fischer WK (1979) Ecotoxicol Env Safety 3:159
115. Gerike P, Fischer WK (1981) Ecotoxicol Env Safety 5:45
116. Painter HA, King EF (1983) Reg Toxicol Pharmacol 3:144
117. Leidner H, Gloor R, Wuhrmann K (1976) Tenside Deterg 13:122
118. Painter HA (1978) Water Research Centre, Rep 663-S Stevenage, U.K.
119. Larson RJ (1979) Appl Env Microbiol 38:1153
120. Gilbert PA, Kleiser HA (1986) Presentation at 32nd Referate-Tagung Waschereiforschung, Krefeld, FRG 23 April 1986
121. Lotzsch K, Neufahrt A, Tauber G (1979) Tenside Deterg 16:150
122. Urano K, Saito M (1985) Chemosphere 14:1333

123. Larson RJ, Wentler GE (1982) Soap Cosmet Chem Spec (May):33
124. Painter HA, Durrant K (1976) Water Research Centre, Tech Rep TR19, Stevenage, U.K.
125. De Oude NT (1977) Tenside Deterg 14:189
126. Yoshimura K, Ara K, Hayashi K, Kawase J, Tsuji K (1984) Jap J Limnol 45:204
127. Swisher RD (1967) JAOCS 44:717
128. Nielsen AM, Huddleston RL (1981) Dev Ind Microbiol 22:415
129. Commission Directive of 18th November 1987 adapting Council Directive 67/548/EEC (1988) Off J European Comm L133 10th May 1988
130. Brown D (1976) 7th Int Surfactant Congress, 4 vols U.S.S.R. Nat Comm on Surface Active Substances. Moscow 4:44
131. Hrsak D, Bosnjak M, Johanides V (1981) Tenside Deterg 18:137
132. Painter HA (1979) Water Research Centre, Rep 699-S, Stevenage, U.K.
133. Brown VM, King EF, Painter HA (1981) Water Research Centre. Rep 68-M. Stevenage, U.K.
134. Brown VM, King EF, Painter HA (1981) Water Research Centre. Rep 222-M. Stevenage, U.K.
135. Petresa (1987) Private Comm
136. Gerike P, Jasiak W (1985) Tenside Deterg 22:305
137. Gerike P, Fischer WK, Holtmann W (1980) Wat Res 14:753
138. Painter HA, King EF (1978) Wat Res 12:909
139. Steber J (1979) Tenside Deterg 16:140
140. Nielsen AM, Meyers JD, Bleckmann CA, Huddleston RL (1980) Soap Cosmet Chem Spec 56:48
141. Shimp RJ, Procter and Gamble Co (to be published)
142. Stiff MJ, Rootham RC (1973) Wat Res 7:1407
143. Gledhill WE (1974) Adv Appl Microbiol 17:265
144. Larson RJ, Perry RL (1981) Wat Res 15:697
145. Larson RJ, Payne AG (1981) Appl Env Microbiol 41:621
146. Larson RJ (1983) Residue Rev 85:159
147. Ventullo RM, Larson RJ, Procter and Gamble Co (to be published)
148. Ventullo RM, Ward TE, Larson RJ, Procter and Gamble Co (to be published)
149. Larson RJ, Federle TW, Shimp RJ, Ventullo RM (1989) Tenside Surfactants Deterg 26:116
150. Ward TE, Larson RJ (1988) Ecotoxicol Environ Saf 17:119
151. Larson RJ, Bishop WE (1988) Soap Cosmet Chem Spec
152. Giger W, Brunner PH, Ahel M, McEvoy J, Marcomini A, Schaffner C (1987) Gas-Wasser-Abwasser 67:111
153. Haberl R, Konig M, Pollhammer P (1980) Osterreichische Abwasser-Rundschau 25:53
154. Pitter P, Veselsky J, Ciperova M (1971) Sb Vvys SK, Chem-Technol Praze, Technol Vody 16:37
155. Larson RJ, Ventullo RM (to be published)
156. Vandoni MV, Goldberg-Federico L (1973) Riv Ital Sostanze Grasse 50:185
157. Litz N, Doering HW, Thiele M, Blume H-P (1987) Ecotoxicol Environ Saf 14:103
158. Kimerle RA (1989) Tenside Surfactant Deterg 26:169
159. Lewis MA (1986) Environ Toxicol Chem 5:319
160. Bishop WE, Perry RL (1981) In: Branson DR, Dickson KL (eds) Aquatic toxicology and hazard assessment. 4th Conf ASTM STP 737 pp 421–435
161. Hitchcock WS, Martin DF (1977) Bull Environ Contam Toxicol 18:291
162. Lal H, Misra V, Viswanathan PN, Krishna Murti CR (1983) Ecotoxicol Environ Saf 7:538
163. Lal H, Misra V, Viswanathan PN, Krishna Murti CR (1984) Ecotoxicol Environ Saf 8:447
164. Lewis MA, Perry RL (1981) In: Branson DR, Dickson KL (eds) 4th Conf Aquatic Toxicity and Hazard Assessment. ASTM STP 737 pp 402–418
165. Moreno-Danvila A (1983) 14th Jorn Com Espanol Deterg pp 23–36
166. Proctor and Gamble Co (unpublished)
167. Kimerle RA, Swisher RD (1977) Wat Res 11:31
168. Monsanto Co (unpublished)
169. Lewis MA, Supprenant D (1983) Ecotoxicol Environ Saf 7:313
170. Calabrese DM, Davis HC (1967) Proc Nat Shellfisheries Assoc 57:11
171. Ladle M, House WA, Armitage PD, Farr IS (1989) Tenside Surfactant Deterg 26:159
172. Gafa S (1974) Riv Ital Sostanze Grasse 51:183
173. Macek KJ, Sleight BH (1977) In: Mayer FL, Hamelink GL (eds) Aquatic toxicity and hazard assessment, ASTM STP 634 pp 137–146
174. Swisher RD, Gledhill WE, Kimerle RA, Taulli TA (1976) 7th Surf Cong U.S.S.R. Nat Comm Surface active substances, Moscow, 1978, 4:218

175. Pickering QH, Thatcher TO (1970) JWPCF 42:243
176. Vailati G, Calamari D, Marchetti R (1975) Naovi Ann Ig Microbiol 26:69
177. Pittager CA, Woltering DM, Masters JA, Procter and Gamble (to be published)
178. Bressan M, Brunetti R, Casellato S, Fara GC, Giro P, Marin M, Negrisolo P, Tallandini L, Thomann S, Tosoni L, Turchetto M (1989) Tenside Surfactants Deterg 26:148
179. Windeatt AJ (1987) Unpublished study (Brixham Study No Q334/A) Unilever Research Laboratories, 1987
180. Mieure JP Unpublished study, Monsanto Co St Louis, MO
181. Gloxhuber C (ed) (1980) Anionic surfactants. Biochemistry toxicology, dermatology. Marcel Dekker Inc New York
182. Hill JT In: 181 pp 405–410
183. Swisher RD (1968) Arch Environ Health 17:232
184. Oba K (1980) In: Ref 181. "Carcinogenic, mutagenic and teratogenic properties" pp 327–404
185. Krienke EG, Ritter S, v Muhlendahl KE (1980) In: Ref 181 "Observations on tolerance in cases of ingestion" pp 127–138
186. Matson TP (1978) JAOCS 55:66
187. Longman GF (1975) The analysis of detergents and detergent products. J Wiley and Sons, London, New York
188. Oba K, Miura K, Sekiguchi R, Yagi R, Mori A (1976) Wat Res 10:149
189. Arisue J, Inoue T, Konishi K (1985) Hokkaido Kogai Boshi Kenkyushko 12:52
190. Smedes F, Kraak JC, Werkhoven-Goewie CF, Brinkman UAT, Rei RW (1982) J Chromatogr 247:123
191. Nakamura K, Morikawa Y, Matsumoto I (1981) JAOCS 58:72
192. Nakamura K, Morikawa Y (1982) JAOCS 59:64
193. Irgolic KJ, Hobill JE (1987) Spectrochim Acta 42B:269
194. Lyon PA, Stebbings WL, Crow FW, Tomer KB, Lippstreu DL, Gross ML (1984) Anal Chem 56:8
195. Righton MJG, Watts CD (1986) Water Research Centre, Rep ER1194-M, Medmenham, U.K.
196. Rivera J, Fraisse D, Ventura F, Caixach J, Figueras A (1987) Fresenius Z Anal Chem 328:577
197. Maurer EW, Cordon TC, Stirton AJ (1971) JAOCS 48:163
198. Borstlap C, Kooijman PL (1963) JAOCS 40:78
199. Miura K, Yamanaka K, Sangai T, Yoshimura K, Hayashi N (1979) Yukagaku 28:351
200. Itoh S, Setsuda S, Utsunomiya A, Naito S (1976) Eisei Kagaku 22:254
201. De Fulvio S, Marconi A, Pierini N (1980) Inquinamento 22:31
202. Lundahl P, Cabridenc R, Xuereff R (1972) 6th Surf Cong Zurich, Chemie and Anwendungstechnik der Grenzflachenacktiven Stoffe, Carl Hauser, Munich 1973, 3:689
203. Fischer WK (1981) In: Fettalkohole—Rohstoffe, Verfahren und Verwendung, Henkel, Dusseldlorf, English edn pp 181–222
204. Sales Marquez D, Galan Vallejo M, Munoz Cueto MJ, Flores Luque V (1981) Comun Journ Com Esp Deterg 12th 193–225 Asoc Invest Detergentes, Tensioactivos, Afines: Barcelona, Spain
205. Degens PN, Vander Zee H, Commer JD, Kamphuis HA (1950) J Proc Inst Sew Purif pp 63–68
206. Hammerton C (1955) J Appl Chem 5:517
207. Crauland M, Courtier A, Bolle J (1964) 4th Surf Cong Brussels, Chemistry physics and application of surface active substances 1:93 Gordon and Breach, London 1967
208. Hammerton C (1956) Proc Soc Water Treat Exam 5:145
209. Bosari GB, Buosi F, Fuochi EP (1974) Riv Ital Sost Grasse 51:193
210. Kawaski T, Mizuno M, Sato M, Kojima I (1983) Jap J Water Waste 25:1251
211. Sekiguchi H, Miura K, Oba K (1975) Yakagaku 24:451
212. Steber J, Gode P, Guhl W (1988) Fett Wiss Technol 90:32
213. Livingston JR, Drogin R, Kelly RJ (1965) Ind Eng Chem Prod Res Dev 4:28
214. Winter W (1962) Wasserwirtsch-Wassertech 12:265
215. Gebril BA, Abou-Zeid H (1966) Tenside 3:150
216. Flores Luque V, Sales Marquez D, Torregrosa RE (1980) Inq Quim (Madrid) 12:123
217. Cook TM, Goldman CK (1974) Chesapeake Sci 15:52
218. Sales D, Quiroga JM, Gomez-Parra A (1987) Bull Environ Contam Toxicol 39:385
219. Vives-Rego J, Vaque MD, Leal JS, Parra J (1987) Tenside, Surfactants, Deterg 24:20
220. Abe S (1984) Nippon Kagaku Kaishi 9:1465
221. Janicke W (1971) Wat Res 5:917
222. Fischer WK, Gerike P (1975) Wat Res 9:1131

223. Gerike P, Jasiak W (1984) Welt-Tensid-Kongress, Munich, Kongressberichte, Kürle Druck Geinhausen, 1:195
224. Moreno Danvila A (1979) X Jorn Com Esp Deterg pp 59–80
225. Berger B (1964) Ind Chim (Paris) 51:421
226. Mann AH, Reid VW (1971) JAOCS 48:798
227. Pitter P (1964) 4th Surf Cong Brussels, Chemistry, physics and application of surface active substances, Gordon and Breach, London (1967) 3:861
228. Maurer EW, Cordon TC, Weil JK, Nunez-Ponzoa MV, Ault WC, Stirton AJ (1965) JAOCS 42:189
229. Oba K, Yoshida Y, Tomiyama S (1967) Yukagaku 16:517
230. Itoh S, Naito S, Unemoto T (1986) Eisei Kagaku 32:101
231. Itoh S, Naito S, Unemoto T (1987) Eisei Kagaku 33:415
232. Wagener S, Schink B (1987) Wat Res 21:615
233. Steber J, Wierich P (1987) Wat Res 21:661
234. Birch RR, Biver C, Campagna R, Gledhill WE, Pagga U, Steber J, Reust H, Bentinck WJ (1989) Chemosphere 19:1527
235. Dodgson KS, White GF, Massey JA, Shapleigh J, Payne WJ (1984) FEMS Microbiol Lett 24:53
236. White GF, Dodgson KS, Davies I, Matts PJ, Shapleigh JP, Payne WJ (1987) FEMS Microbiol Lett 40:173
237. Cain RB (1987) Biochem Soc Trans 15 (Suppl):7S
238. Elnabarawy MT, Robideau RR, Beach SA (1988) Tox Assess 3:361
239. Dutka BJ, Nyholm N, Petersen J (1983) Wat Res 17:1363
240. King EF (1984) In: Toxicity screening procedures using bacterial systems. Liu D, Dutka BJ (eds) Marcel Dekker, Inc. New York, Basel, 175–194
241. Xu H, Dutka BJ (1987) Tox Assess 2:149
242. Vives-Rego J, Vaque D, Martinez J (1986) Wat Res 20:1411
243. Beaubien A, Bouchard A, Jolicoeur C (1986) Tox Assess 1:187
244. Nyberg H (1988) Wat Res 22:217
245. Lenova LI, Stavskaya SS, Ratushnaya MY (1980) Gidrobiol ZH 16:83
246. Roderer G (1987) Arch Environ Contam Toxicol 16:291
247. Ukeles R (1965) J Phycol 1:102
248. Roberts MH, Warrinner JE, Tsai CF, Wright D, Cronin LE (1982) Arch Environ Contam Toxicol 11:681
249. Lundahl P, Cabridenc R (1978) Wat Res 12:25
250. LeBlanc GA (1982) Environ Pollut (Ser A) 27:309
251. Moore SB, Diehl RA, Barnhardt JM, Avery GB (1987) World Textile Abstracts 19:290
252. Lundahl P, Cabridenc R (1976) J Fran Hydrol 7:143
253. Bode H, Ernst R, Arditti J (1978) Environ Pollut 17:175
254. Patzner AM, Adam H (1979) Zool Anz Jena 202:199
255. Conti E (1987) Aquatic Toxicol 10:325
256. Price KS, Waggy GT, Conway RA (1974) JWPCF 46:63
257. Tatem HE, Anderson JW, Neff JM (1976) Bull Environ Contam Toxicol 16:368
258. Piper WD, Maxwell KE (1971) J Econ Entomol 64:601
259. Graney RL, Giesy JP (1988) Environ Toxicol Chem 7:301
260. Cardwell RD, Woelke CE, Carr MI, Sanborn E (1978) Bull Environ Contam Toxicol 20:128
261. Hidu H (1965) JWPCF 37:262
262. Verna SR, Mohan D, Dalela RC (1978) Acta Hydrochim Hydrobiol 6:121
263. Van Embden HM, Kroon CCM, Schoeman EN, Van Seventer HA (1974) Environ Pollut 6:297
264. Kikuchi K, Wakabayashi M, Nakamura T, Inoue W, Takahashi K, Kawana T, Kawahara H, Koido Y (1976) Ann Rep Tokyo Metrop Res Inst Environ Prot 57–69
265. Tovell PWA, Newsome CS, Howes D (1974) Wat Res 8:291
266. Fogels A, Sprague JB (1977) Wat Res 11:811
267. Hara TJ, Thompson BE (1978) Wat Res 12:893
268. Dalela RC, Tyagi AK, Pal N, Verna SR (1981) Wat Air Soil Pollut 15:3
269. Wakabayashi M, Kikuchi M, Kojima H, Yoshida T (1978) Chemosphere 11:917
270. Kikuchi M, Wakabayashi M (1979) Seitai Kagaku 1:195
271. Nadasy M, Dobozy OK, Bartha B, Pafli D, Kolesei M (1972) 6th Int Cong Surface Activity. Zurich 1972, Chemie, physikalische Chemie und Anwendungstechnik der grenzflachenaktiven Stoffe, 4 vols, Carl Hanser, Munich 1973

272. Antonielli M, Lapattelli M (1977) Agrochimica 21:15
273. Taniyana T, Nomura T (1978) Mie Daigaku Kankyo Kagaku Kenkyo Kiyo 3:93
274. Muramoto S, Oki Y (1988) J Environ Sci Health A23:603
275. Yagami K, Yokota Y, Iwashita S, Nakayama M (1984) Yukagaku 33:284
276. Neubecker TA, Brag JP (1983) ACS Env Chem Div Ext Abstr 23:112
277. Goodnow RA, Harrison AP (1972) Appl Microbiol 24:555
278. Hales SG, Dodgson KS, White GF, Jones N, Watson GK (1982) Appl Env Microbiol 44:790
279. Yoshimura K, Masuda F (1982) JAOCS 59:328
280. Griffiths ET, Hales SG, Russell NJ, White GF (1987) Biotechnol Appl Biochem 9:217
281. Steinle EC, Myerly RC, Vath CA (1984) JAOCS 41:804
282. Stead JB, Pugh AT, Koduji II, Morland RA (1972) 6th Int Surf Cong Zurich Chemie, physikalische Chemie und Anwendungstechnik der grenzflachenaktiven Stoffe, Carl Hanser, Munich, 1973, 3:721
283. Vashon RD, Schwab BS (1982) Env Sci Technol 16:433
284. Oba K (1971) Nippon Eiseigaku Zasshi 25:494
285. Hales SG, Watson GK, Dodgson KS, White GF (1986) J gen Microbiol 132:953
286. Yamane AN, Okada M, Sudo R (1984) Wat Res 18:1101
287. Kutt EC, Martin DF (1974) Marine Biology 28:253
288. Maki AW (1979) J Fish Res Bd Can 36:411
289. Shell Chemical Company, unpublished
290. Reiff B, Lloyd R, How MJ, Brown D, Alabaster JS (1979) Wat Res 13:20
291. Maki AW (1979) Spec Tech Publ 667, Am Soc Testing Materials pp 77–95
292. Farone WA (1979) U.S. Pat App 4171968 23 Oct 1979, pp 7
293. Scailteur V, Maurer JK, Walker AP, Calvin G (1986) Fd Chem Toxic 24:175
294. Takeshita R, Jinnai N, Yoshida H (1976) J Chromatog 123:301
295. Matsutani S, Shige T, Nagai T (1980) Yukagaku 29:189
296. Matsutani S, Shige T, Nagai T (1980) Yukagaku 29:926
297. Kroner M, Schneider G (1968) Fette-Seifen-Anstrichmittel 70:753
298. Sekiguchi H, Miura K, Oba K, Mori A (1975) Yukagaku 24:145
299. McAteer JH, Kinnard LM (1964) 4th Int Surf Cong Burssels, Chemistry, physics and application of surface active substances Gordon and Breach, London, 1967 (3 vols) 1:127
300. Eden GE, Truesdale GA, Stennett GV (1968) Wat Pollut Control 67:107
301. Walther H, Knaack D (1970) Wasserwirt-Wassertech 20:306
302. Thysse GJE, Wanders TH (1974) A van Leeuwenhoek 40:25
303. Stavskaya SS, Taranova LA, Grigor'eva TY, Degtyarev VA, Pisarev VT, Rotmistrov MN (1984) Mikrobiologiya 53:218
304. Rotmistrov MN, Taranova LA, Radchenko OS, Stavskaya SS, Dumanskii AV (1986) Dokl Akad Nouk SSSR 288:246
305. Guhl W, Gode P (1989) Tenside Surfactants Detergents 26:383
306. Knauf W (1973) Tenside 10:251
307. Oba K, Mori A, Tomiyama S (1968) Yukagaku 17:517
308. Wolf T, McPherson BP (1976) JAOCS 54:347
309. Allen MC, Martin TT (1971) JAOCS 48:790
310. Nagai T, Hashimoto S, Yamane I, Mori A (1970) JAOCS 47:505
311. Johannessen RO, DeWitt WJ, Smith RS, Tuvell ME (1983) JAOCS 60:858
312. Kravetz L, Chung H, Guin KF, Shebs WT (1979) Presented at AOCS Am Oil Chemists Society 70th Ann Meeting, San Francisco, May 1979
313. Itoh S, Setsuda S, Utsunomiya A, Naito S (1979) Yukagaku 28:199
314. Kravetz L, Chung H, Rapean LC (1982) JAOCS 59:206
315. Miura K, Nishizawa H (1982) Yukagaku 31:367
316. Pitter P, Horakova M, Fuka T, Struharik L (1980) Chem Prum 30:188
317. Tuvell ME, Kuehnhanss GO, Heidebrecht GD, Hu PC, Zielinski AD (1978) JAOCS 55:70
318. Tomiyama S, Takao M, Mori A, Sekiguchi H, Oba K (1968) Ann Meeting Am Oil Chemists Society, New York, October 1968
319. Continental Oil Company, unpublished
320. Colgate Palmolive Company, unpublished
321. Monsanto Company, unpublished
322. Lopez-Zavala A, de Alujz AS, Elias BL, Manjarrez L, Buchmann A, Mercado L, Caltenco S (1975) Prog Water Tech 7:73
323. Takita Y (1982) Yukagaku 31:507

The Environmental Chemistry, Fate and Effects of Nonionic Surfactants

M. S. Holt, G. C. Mitchell and R. J. Watkinson

Shell Research Ltd., Sittingbourne Research Centre, Sittingbourne, Kent, U.K.

List of Abbreviations	90
Introduction	91
Chemical Nature of Surfactants	92
Hydrophobe Structures	93
Hydrophile Structures	95
Manufacture of Nonionic Surfactants	95
Chemical Stability	97
Physico-chemical Properties of Nonionic Surfactants	98
Analytical Methods	101
Sampling	102
Extraction and Purification	103
Non-specific Methods	106
Specific Chromatographic Methods	107
Environmental Concentrations	110
Pathways of Nonionics into the Environment	110
Surface Waters	115
Marine Waters	118
Soils and Sediments	119
Biodegradability	119
Laboratory Tests	119
Alcohol Ethoxylates	120
Biodegradation Pathways	122
Polyglycol Degradation	127
Alkylphenol Ethoxylates	128
Other Nonionics	130
Ecotoxicology	131
References	139

Summary

This review of nonionic surfactants could not hope to cover all the extensive knowledge of these materials. Its aim is to give sufficient information on the nature, production and properties of these compounds to support the major part of the review on environmental analysis, biodegradability and effects on the flora and fauna of several environmental niches. From the flavour of the topics covered the reader is guided to the major literature on the subjects where one can gain the more in-depth scientific information.

The Handbook of Environmental Chemistry,
Volume 3, Part F, Ed. O. Hutzinger
© Springer-Verlag Berlin Heidelberg 1992
© Shell Research Ltd.

List of Abbreviations

AA	Atomic absorption
AE	Alcohol ethoxylate
APE	Alkylphenol ethoxylate
APE_n	Alkylphenol ethoxylate averaging n mols EO per mol
APEC	Alkylphenol ethoxycarboxylate
AS	Activated sludge
BAE	Branched alcohol ethoxylate
BiAS	Bismuth iodide active substances
BO	Butylene oxide
BOD	Biological oxygen demand
C	Centigrade or Carbon
C_n	Alkyl chain with n carbons
CI	Chemical ionisation
CTAS	Cobalt thiocyanate active substances
CU	Coupled unit test
DCI	Desorption chemical ionisation
DOC	Dissolved organic carbon
E_n	Polyethoxylate moiety averaging n mols of EO
ECD	Electron capture detector
EG	Ethylene glycol
EI	Electron impact
ELS	Evaporative light scattering
EO	Ethylene oxide
FA	Fatty alcohol
FAB	Fast atom bombardment
FDMS	Field desorption mass spectrometry
GC	Gas chromatography
HBr	Hydrogen bromide
HPLC	High performance liquid chromatography
HRGC	High resolution gas chromatography
IR	Infra red
LAE	Linear primary alcohol ethoxylate
LSAE	Linear secondary alcohol ethoxylate
MIKE	Mass analysed ion kinetic energy spectroscopy
MOST	Modified OECD screening test
MS	Mass spectrometry
NMR	Nuclear magnetic resonance
NP	Nonylphenol
NP1EO	Nonylphenol monoethoxylate replacement of the O by C indicates the ethoxylate has been carboxylated
NP2EO	Nonylphenol diethoxylate
NP3EO	Nonylphenol triethoxylate

NPE	Nonylphenol ethoxylate
NPE_n	Nonylphenol ethoxylate averaging n mols EO
OP	Octylphenol—other variations as for nonylphenols
Oxo	Oxo alcohol
PIC	Phenylisocyanate
PEG	Polyethylene glycol
PG	Polypropylene glycol
PO	Propylene oxide
PPAS	Potassium picrate active substances
ppb	Parts per billion
ppm	Parts per million
R	Alkyl group
SDA	American Soap and Detergents Association
SFC	Supercritical fluid chromatography
SFE	Supercritical fluid extraction
SPE	Solid phase extraction
TF	Trickling filter
TLC	Thin layer chromatography
TOC	Total organic carbon
UV	Ultraviolet
UVF	Ultraviolet fluorescence

Introduction

Surface active agents that have the molecular characteristics of distinct hydrophilic and hydrophobic regions and have no electronic charge are referred to as nonionic surfactants. Such molecules find application within a range of chemical products, with a major usage in detergents. The term detergent means a cleaning agent that is usually a formulated product that contains other additives to enhance the specific performance required. Products, other than detergents, that contain nonionic surfactants include cosmetics, pharmaceuticals, pesticides, paints, fibres and fibre processing aids. They may be used wherever their interfacial effects of wetting, detergency, foaming and defoaming, emulsification and demulsification, dispersion and solubilisation can enhance product or process performance. With such a potentially wide range of uses it suggests that there are multiple routes for these materials to reach the environment. Since they are currently produced at a rate approaching 750 000 tonnes per annum within Western Europe and U.S.A., they constitute a significant chemical input to the environment. Thus it is necessary that there is an adequate understanding of their chemistry as well as the knowledge and techniques available for studying their fate, effects and behaviour in the environment.

The routes to the environment will be dictated to a large extent by their pattern of usage. By far the greatest use will be in domestic detergents, household cleaners, personal care products such as shampoos and commercial cleaning processes. The major routes for their disposal will be to sewage systems. Ultimately, small traces will be discharged to surface waters and the terrestrial environment where their fate will be determined by their chemistry and the nature of the receiving environment. For those locations with little or no biological treatment, the discharge will be direct to surface waters. However modern domestic detergent components are designed to be readily biodegradable and will be extensively removed in sewage treatment systems using biological treatments such as activated sludge plants or trickling filters. The wider usage in small volume applications via products other than detergents and surfactant use in several manufacturing and treatment processes, will probably lead to a wider distribution to the general environment. Their ready biodegradability will also mean that those materials directly released to the environment will be degraded by the natural population of microorganisms of surface waters and soil to inorganic end products. It is clear from the large tonnages of products that have been used and the lack of any indication of accumulation in the environment that they are indeed effectively degraded in the various environmental compartments.

It is possible to calculate on a per capita basis for certain populations the daily use of surfactants and to compare this with the concentrations in sewage plant influents. Frequently there is good agreement between the two values which indicates that this route captures the majority of materials used. These mass balance assessments further support the observed removal of these materials in the environment as well as serving as a predictive model for expected environmental concentrations.

Chemical Nature of Nonionic Surfactants

It is clear that the surfactant properties of these materials arise from the separate polar (hydrophilic) and nonpolar (hydrophobic) regions of their molecules. The nonpolar regions are mainly derived from hydrocarbon, alcohol or fatty acid (or amine) sources with carbon skeletons in the range of C8–C20. The polar region is usually provided by a polyoxyethylene glycol, perhaps with some propylene glycol residues. Alternative hydrophilic regions may be provided by the sugar glycols. Sugar residues (aldoses) themselves may be substituted with fatty acids or alkyl groups to provide surfactants that are similar to some naturally produced lipids. Fatty acid ethanolamines and fatty amines allow two hydrophilic chains to be added to one structural centre thus changing the molecular morphology and the nature of the interactions at surfaces. Figure 1 shows the structures of these molecules and the likely ranges of hydrophobe chain length

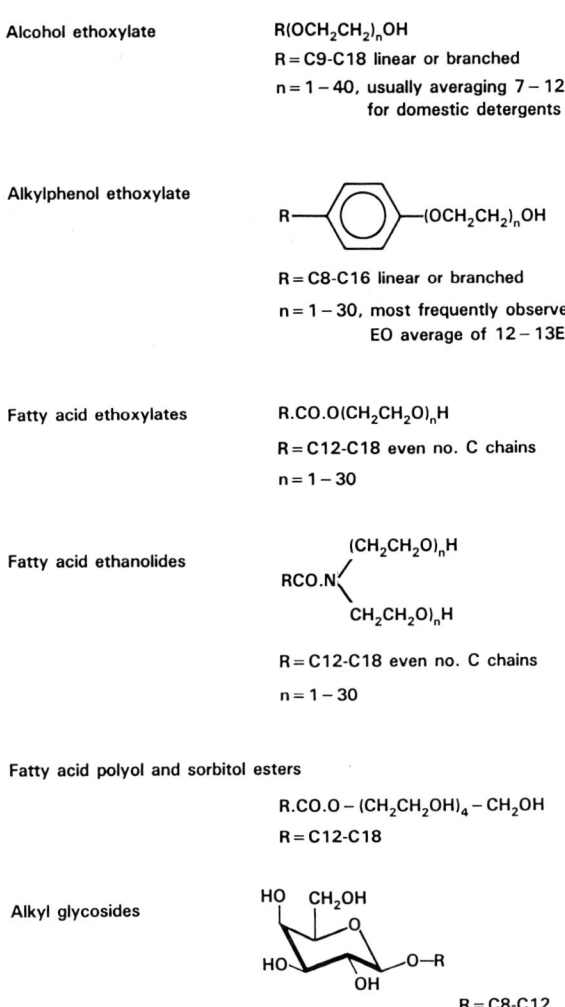

Fig. 1. Main classes of nonionic surfactants

and sizes of hydrophilic group that may be met. The routes to the manufacture of these components will be covered later.

Hydrophobe Structures

The major routes to the manufacture of nonionic surfactant hydrophobes are shown in Fig. 2. The base materials are crude oil and vegetable oil. The crude oil on distillation will yield a paraffin fraction that can be purified into specific

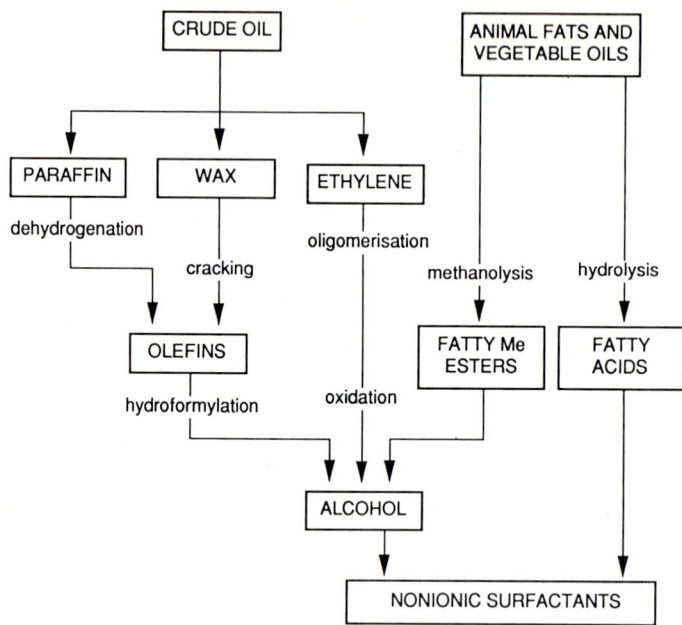

Fig. 2. Routes to surfactant hydrophobes

n-paraffin fractions by molecular sieve technology. Crude oil will also provide wax which when catalytically cracked gives rise to a range of olefins which can be distilled into various ranges. The naphtha fraction of crude oil is used to produce ethylene, one of the basic chemical building blocks in petrochemistry. Natural fats and oils offer an alternative source of hydrocarbon molecules that are not reliant on fossil fuels and in the modern 'green' movements are being popularised. The fats and oils are hydrolysed to yield fatty acids which can be reduced with hydrogen to provide alcohols. By far the greatest production of hydrophobes for surfactants comes via the olefin route. Based on ethylene, propylene and butylene monomers, the polymerisation reactions used can give rise to branched carbon chains dependent upon the nature of the starting feedstock. Early materials were based on butene and isobutene polymerisations with trimers providing the major quantity of alkyl moiety for the detergent applications. The materials were thus highly branched but also gave rise to a wide number of isomers. A significant aspect of the branching with respect to its impact on the biodegradability of the materials is the abundance of quaternary carbon atoms. For this reason the di- and triisobutylene derived materials, with a high proportion of quaternary carbons in their structure are relatively poorly biodegradable.

Such molecules were utilised for alkylating phenol in the synthesis of the alkylphenol ethoxylates. More recent hydrophobes are derived from propylene oligomers, which provide more regular and less severe branching in the carbon

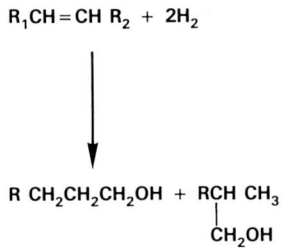

Fig. 3. 'Oxo' route to detergent alcohol

skeleton. But it is only the oligomerisation of ethylene that provides linear olefins (and directly purified linear materials) and these now provide the majority of hydrophobes for domestic detergent products. The petrochemical industry has optimised its ethylene production and utilisation to meet the demand for these more desirable surfactant intermediates. Alcohols are derived from the olefins by carbon extension via hydroformylation reactions using carbon monoxide and hydrogen chemistry. One such process in common use is the 'OXO' process (Fig. 3).

Hydrophile Structures

Ethylene oxide oligomers are the most prevalent of the hydrophilic groups found in commercial surfactants. The reagent ethylene oxide (EO) is derived from the feedstock ethylene making this grouping very cost efficient. The number of EO units within a surfactant molecule can range from one to forty but in domestic detergents the mean value is usually between 7–11 EO residues. Propylene oxide units may also be incorporated, frequently to depress foaming properties, but as we shall discuss later this may reduce the biodegradability of the molecules significantly and thus materials derived in this way tend to be used as specialist chemicals. Alternative hydrophiles arise from sugar chemistry and examples of polyols (e.g. sorbitol) and glycosides are beginning to receive renewed attention for specific applications.

Manufacture of Nonionic Surfactants

The synthetic routes to the final surfactant products are shown in Fig. 4. The production and properties of the most important surfactants have been reported by Maag [1]. The general outline for the manufacture of surfactants has been presented in the last section. It is the intention of this section to address some of

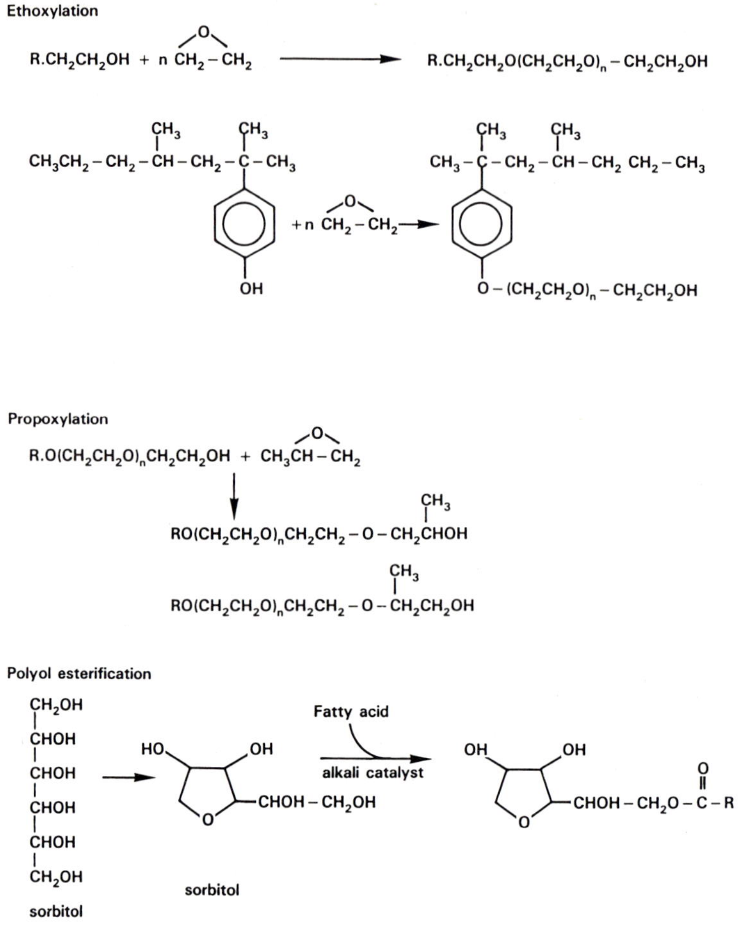

Fig. 4. Reactions leading to the manufacture of nonionic surfactants

the details of some processes and also to indicate likely by-products. In the area of environmental analysis and monitoring the identification of unreacted starting materials, intermediates and by-products may act as a 'fingerprint' of the parent products. However it should also be recognised that many of the surfactant feedstocks are used for petrochemical products other than surfactants (e.g. alkylbenzene for cable oils). Therefore the detection in the environment of these materials in the environment may not always reflect an input of surfactants.

Alcohol ethoxylate surfactants may be produced from linear or branched alcohols although their behaviour in the environment, and biodegradation in particular, is very different. The branched materials are usually based on the propylene trimer or tetramer with their carbon chains extended by an "OXO" process; the resulting branched decanol, or tridecanol, is ethoxylated.

Linear alcohols are produced by an "OXO" reaction with α-olefin, a reaction which yields 80–85% primary alcohol with decreasing amounts of α, β or γ branched alcohol. This is reacted with EO to yield the final product with an EO range of 2–20 EO per surfactant molecules.

Ethoxylations are normally carried out using KOH/EtOH as the catalyst and supplying a known molar ratio of EO to alcohol. Ethoxylation takes place with a kinetic preference of primary alcohol > secondary > tertiary alcohol. It is possible to generate polyoxyethylene glycols as well as cyclisation products such as dioxane but these can be minimised through process parameter optimisation. These by-products are more frequent impurities of acid catalysed ethoxylation procedures.

Alkylphenol ethoxylates are produced from the same olefin starting materials as described for the alcohol ethoxylates. Earlier processes used alkenes derived from butylene and propylene which gave rise to p-alkylphenol whereas the more modern linear alkenes based on ethylene provide o-alkylphenol as the major product. These acid catalysed alkylations give rise to a number of cyclised side products in small proportions, for example tetralins and indanes. The alkylphenol intermediates are ethoxylated in an analogous reaction to the alcohol ethoxylates.

The syntheses of sugar-based nonionic surfactants have been based on the protection of the sugar alcohols with acetate groups and reaction with an alcohol [2]. The acetate protecting groups are subsequently released to yield alkyl glycosides. Alternative reactions are the interesterification of the sugar glycols to give fatty esters similar to the biosurfactants produced by some microorganisms. The commercialisation of these products is still in its infancy but developments continue to be made.

Chemical Stability

The major nonionic surfactants composed of saturated alkyl, alkylphenyl and polyoxyalkylene groupings have been generally considered to be stable under normal conditions of use and exposure. However their chemical stability is analogous to that of those polyoxyalkenes which serve as models for the surfactant types. The polyoxyalkenes are subject to the normal autoxidation reactions of ethers and, although not to the same extent as the dialkyl ethers, they are capable of forming hydroperoxides and breakdown products derived from these by free radical, metal ion and radiation-induced autoxidation. This reaction could be important in the environmental chemistry and fate of these materials. The first stage in the autoxidation process is a free radical initiation reaction that can generate a polyoxyalkyl radical. In the presence of oxygen this radical can generate hydroperoxides. These unstable compounds are subject to further chemical degradation reactions via several potential routes. Common products found as a result of these reactions include formic acid and formates, aldehydes and carboxylic acids.

It is not within the scope of this chapter to review the complex chemistry of the stability of these products, since it has been well covered by Donbrow [3]. It is sufficient to alert the environmental chemist to the possibility that these autoxidative reactions occur both in the environment itself and during the extraction and purification procedures relevant to further analysis. It is good practice to minimise these reactions with the use of nitrogen screens and by avoiding high temperatures and light.

Physico-Chemical Properties of Nonionic Surfactants

It is not the intention to document here those properties which provide the basis for the use of nonionic surfactants in their various product roles as these have been adequately covered in many books and reviews [4, 5]. However, those properties which give these molecules their specialist activities have implications for their fate and behaviour in the environment and how they may be handled in environmental chemistry. We shall therefore concentrate on their performance in aqueous solution and their interfacial properties, particularly those involving liquid/air interfaces and solid supports. This choice is made because we are primarily concerned in this review with the fate and behaviour of these surfactants in surface waters, sediments and soils.

The behaviour of surfactants in water is driven by the hydrophobicity of the molecule, that is the degree to which the polar part of the molecule can be solvated and accommodated by water. The measurement of surface tension of aqueous solutions of nonionics with varying concentrations of surfactant provides data similar to that shown in Fig. 5. This behaviour is interpreted as the formation of a surface film because of the rejection of the water phase by the non-polar region of the molecule which then forms a surface monolayer of hydrocarbon tails. A point of inflection is reached at some concentration and it is the concentration of surfactant at this point of inflection we term the critical micelle concentration (CMC). Above this concentration the molecules form aggregates and thus the addition of more surfactant only increases the concentration in the aggregates and has little impact upon the surface properties at boundary layers. CMC values for alcohol and alkylphenol ethoxylates in water are in the range of 10^{-3} to 10^{-5} M. A characteristic for anionic surfactants is the Krafft point which may be defined as the temperature where the solubility of the surfactant equals the CMC. The Krafft point is usually characterised by a rapidly increasing solubility curve over a small temperature range. With the very low CMC values shown above for the nonionic detergents this would drive the Krafft point to temperatures close to and below the freezing point of water. Thus Krafft points are not usually observed for many nonionic surfactants [5]. Increasing the

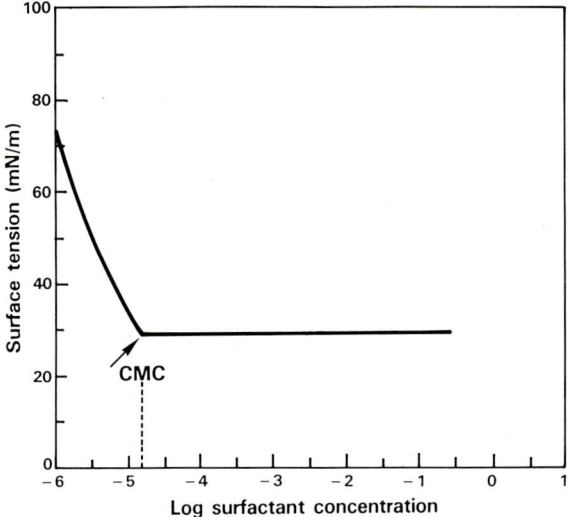

Fig. 5. Surface tension of surfactant solutions

temperature of a surfactant solution with a concentration above its CMC results in a phase separation. This temperature is referred to as its cloud point. It is normally observed as an appearance of opacity in a clear solution resulting from the production of two liquid phases. On cooling these again become one clear solution. These phase changes are indicated in the binary phase diagram for water and surfactant schematically shown in Fig. 6. The cloud points are determined by the molecular composition and Fig. 7 shows the trend in cloud points for a range of alcohol ethoxylates.

The property of surfactant accumulation at air/solution interfaces is utilised in environmental analysis for the concentration and extraction of aqueous solutions by the technique of sublation. This technique involves the dispersion of small nitrogen or air bubbles through a water column containing the surfactant and their collection in an overlying organic phase, usually ethyl acetate.

The adsorption of nonionic surfactants at solid-liquid interfaces would appear to be less complex than that of ionic surfactants. For nonionic materials the adsorption mechanisms are obviously not directly subject to ion exchange and ion pairing chemistry, although by forming complexes with ionic materials some secondary mechanisms could be invoked. The types of adsorption which these materials undergo involve the polarisation of electrons where the molecules have pi electron rich centres, which may be attracted by the positive ion centres of the adsorbent. Another mechanism can involve adsorption by London-Van der Waals dispersion forces acting between the surfactant and adsorbent. Hydrogen bonding can occur between polyoxyalkylene groups and suitable hydroxyl or amino groups on the adsorbent. For example hydrogen

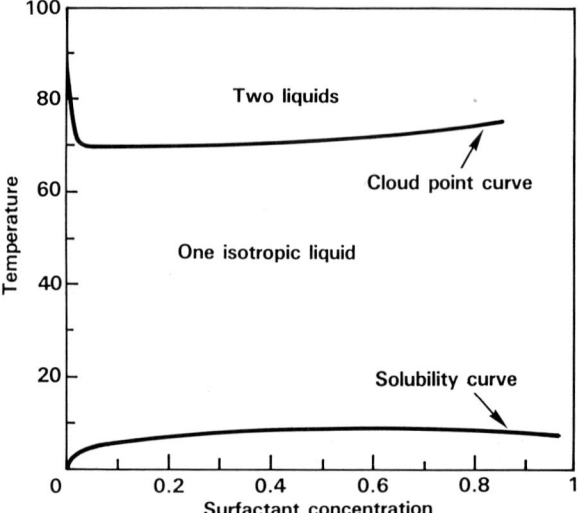

Fig. 6. Schematic surfactant—water phase diagram

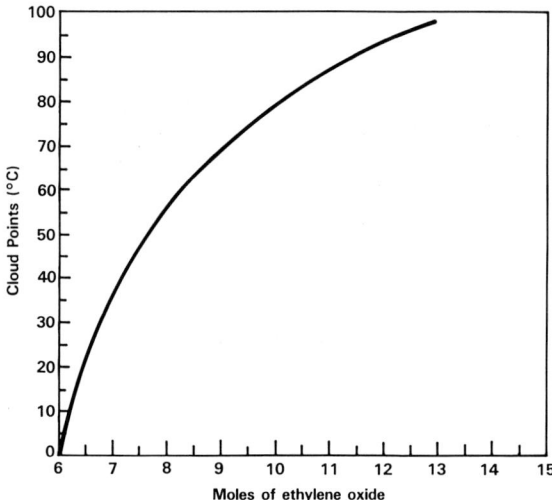

Fig. 7. Cloud points for alcohol etholxylates (1%)

bonds may form between SiOH groups and the oxygen of the oxyalkylene groups.

For nonionic surfactants the adsorption isotherms are frequently of the Langmuir type that follow the general form:

$C_s = Cm \cdot C_1/(C_1 + a)$

where

C_s = surface concentration of surfactant (mol/cm^2)
C_m = surface concentration at monolayer adsorption
C_l = concentration of surfactant in the liquid (mol/l)
a = a constant derived from the free energy of adsorption

This equation provides the idealised asymptotic graph shown in Fig. 8 when plotting the quantity of material adsorbed against equilibrium liquid concentration. There are many influencing factors that provide deviations from this ideal behaviour resulting in varying isotherms. In the environment the adsorbents are frequently non-homogeneous, the phases involved do not exhibit ideal behaviour and the adsorption film is probably not monomolecular, all of which invalidates the theoretical conditions for this type of adsorption. That the isotherm seems to describe the observed adsorption probably arises from the mutual compensation of some of these factors.

The property of surfactants to bind and accumulate at surfaces provides one of the major problems in their handling and analysis. It requires special care in the preparation of glassware and equipment. In environmental analysis of trace concentrations the possibilities for contamination are many. Practised laboratories have usually spent much time in developing their own techniques and preparation of apparatus to overcome these problems.

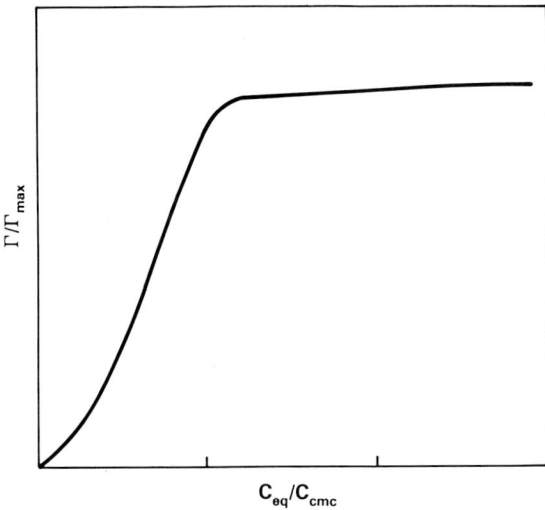

Fig. 8. Schematic of a normalised adsorption isotherm

Analytical Methods

The analysis of nonionic surfactants is complex, the majority of methods being applicable only to the analysis of formulations and for product quality control,

Table 1. Techniques suitable for nonionic analysis

Technique	References General	Environmental
Atomic Absorption		59–63
GC and GC/MS	64–68	26, 27, 29, 43, 44, 49, 50, 51, 52, 56, 58, 69, 70–76
GC/ECD or GC/MS		57
HPLC	77, 78	10
HPLC/ELS	79	
HPLC/UV	10, 46, 77, 80–85	17, 27, 29, 45, 53, 54, 56
HPLC/UVF		12, 15, 22, 47, 48, 55, 86,
NMR	87, 88	
Spectroscopic/Potentiometric	89–92	24, 30, 31, 33, 34, 36, 37–42, 76, 93
SFC	87, 94–97	
Reviews		6–9

since they are neither selective or sensitive enough to determine environmental concentrations.

Swischer [6] has described in considerable detail the wide variety of techniques and analytical procedures available for the determination of nonionic surfactants and has also outlined their associated problems and limitations. Llenado et al. [7, 8] have also summarised the more recent developments in nonionic surfactant analysis and Thomas [9] reviewed the methods available for water and effluent analysis. A list of the techniques used for the analysis of nonionics is given in Table 1. The increasing interest in environmental acceptability of surfactants has resulted in the development of highly specific methods of analysis although some physical and non specific methods are still routinely used. The methods that have been applied to environmental analysis are appropriately listed. Spectrophotometric methods, including IR, NMR and MS, have been used mainly for structural and identification purposes and not primarily for quantification. GC and HPLC in particular however have proved to be valuable techniques for quantitative analysis. For a review of the use of HPLC in the analysis of nonionics, including polyoxyethylene alkylamides, the reader is referred to Garti et al. [10]. In this review the emphasis will be placed on reviewing the most recent developments in sample preparation, extraction and clean up and in specific end analyses of environmental samples.

Sampling

Nonionics will be adsorbed, because of their physico-chemical properties, onto any suspended solids as well as the walls of the containing vessel and, as a result, glass containers are recommended by Sedlak and Booman [11]. It must be clearly stated what is to be analysed, i.e. total nonionic or only that in solution. If the former is required it must first be ascertained whether the level of solids in the sample will allow a satisfactory estimation of the total nonionics. If the suspended solids concentrations are very high then centrifugation or filtration of the sample may be necessary. Sample bottles should be completely filled and preserved with 10 ml of 40% v/v formaldehyde solution per litre and stored at 0–5 °C, although Kubeck and Naylor [12] concluded that refrigeration alone was sufficient to preserve river water samples for up to 4 weeks. Other methods of sample preservation include acidification to pH 2 with hydrochloric acid (Sheldon and Hites [13]) or by the addition of mercuric chloride. To obtain a representative sample the bottle should not be shaken but slowly inverted to avoid the formation of foam. Ideally, because of the daily and indeed hourly variations in the nonionics concentrations in sewage influents, when monitoring sewage treatment plants time proportional samples should be collected and then mixed to give 24 hour flow proportional composite samples for each sampling point.

The increased awareness of the fate of organics adsorbed to sludge that is subsequently disposed of to land has resulted in the need to monitor concentrations of detergents in sludge amended soils and landfills. Composite representative samples should be collected with a screw auger or corer to a depth of between 5 and 20 cm, depending on the mode of sludge application. There are many sampling patterns available, one of which is that at least 25 samples/acre are collected at regular intervals along a W shaped path as suggested by HMSO [14], however Marcomini et al. [15] found 5 random samples to be sufficient. No one sampling pattern will be suitable for all occasions and in general the method chosen will depend on the specific features of the sampling site but should be designed to yield statistically relevant data.

Extraction and Purification

A consequence of the biodegradability of nonionic surfactants is that their concentrations in surface waters are close to the limits of the standard methods of detection. It is therefore essential to incorporate concentration and purification steps into the analytical procedure to improve sensitivity and selectivity. Many of the methods are outlined by Swischer [6]. The most commonly used extraction technique for nonionics in aqueous samples is that of sublation, however there have been considerable advances in the use of solid phase extraction in recent years.

Sublation

Solvent sublation may be used to selectively concentrate intact surfactants free from non-surface active materials, including PEG and carboxylated nonionic surfactant intermediates. With this technique all surface active materials are extracted from the aqueous phase into an overlying layer of organic solvent (ethyl acetate) by passing finely dispersed bubbles of nitrogen through the solution.

Solid Phase Extraction (SPE)

Amberlite XAD-4 resin was evaluated by Jones and Nickless [16] and used in a study on the fate of nonionics in various water systems. The nonionics from large volumes of river water (3.5–50 l) were passed through a narrow glass column containing 5 g resin at 100 ml/min and eluted from the column by four solvent systems. The XAD resin technique is very convenient for handling up to several hundred litres of water but the presence of impurities can result in high blank values. Clean up of the XAD is therefore an important prerequisite for surfactant analysis and can be accomplished by Soxhlet extraction of the resin with methanol for 8 hours prior to use. The resin can then be used indefinitely if it is washed successively with 0.05 M NaOH in methanol, HCl/methanol/chloroform (1:5:5) and methanol between samples (Schmitt et al. [17]).

Neufahrt et al. [18] developed an automated system in which 25 l aqueous samples were passed through a cation exchanger (Dowex 50WX2) and an anion exchanger (Dowex 1X2) before the nonionics were adsorbed on to XAD-2. The adsorption capacity of the XAD-2 was said to be 7 mg fatty alcohol ethoxylate per g resin. The recovery of cationics was significantly improved (from 34 to 93%) by lowering the pH of the sample to 3. For anionics the recovery improved from 89 to 93% and for nonionics in river water from 79 to 86% $\pm$ 9%. For effluents the recovery on nonionics was 82%.

Water samples (500–1000 ml) were shaken with 1 g XAD-2 for 4 hours and the nonionics eluted with 100 ml methanol with recoveries of 90–100% by Saito and Hagiwara [19].

Thurman et al. [20] used XAD-8 resin to extract nonionics from groundwater. 300 l of water was pumped through a 1.2 l column at 300 ml/min. The nonionics adsorbed on the XAD-8 resin were eluted in a Soxhlet extraction with methanol. Recoveries using 1 l of water spiked with APE were 100 $\pm$ 5%. Interfering anionics were removed by trapping on an A-7 column which did not adsorb the APE. Recovery through this step of the procedure was 84 $\pm$ 5%. Valls et al. [21] studied the sequential recoveries of NP and NPnE from seawater by XAD-2 and liquid-liquid extraction and found that recovery of NPnE decreased according to ethoxy chain length (94.5% for NP and 62.6% for total NPnE). In contrast, 100% of both NP and NPnE were recovered by liquid-liquid extraction. For large volumes (up to 500 l) of seawater however, the authors recommend adsorption on to polyurethane foam plugs packed into a teflon column and elution with acetone and hexane.

A simple and rapid dual column technique has recently been reported by Kubeck and Naylor [12]. Water samples (1 l) were passed through 400 ml of a mixed bed ion exchange resin (Biorad 501X8 D, 20–50 mesh) to remove all ionic species, adding 150 ml methanol to the water improved the removal. The nonionics were trapped on a small bed (0.7 g) of octadecylsilica and were eluted with warm (55 °C) methanol. Recoveries of 84% for NPE from river water were obtained. The authors also challenged the widely held view that AP1EO and AP2EO are 'hard' biodegradation intermediates and conclude that reports of high levels of low EO oligomers of alkylphenol and loss of the higher EO oligomers in environmental samples are probably due to improper work up procedures rather than biodegradation.

Concentration of NP from aqueous samples using octadecyl silica columns was also reported by Marcomini et al. [22]. 50 ml raw sewage or 250 ml waste water were filtered through 1 μm glass fiber filters. The filter cake was rinsed with 5 ml methanol which was added to the filtrate. The filtered solution was passed through an octadecylsilica cartridge under vacuum. The cartridge was preconditioned with 3 ml acetonitrile, 3 ml methanol and 5 ml double distilled water before use. Flow rate was 10–15 ml/min and the nonionics were eluted with 3 ml acetone. Recoveries from water were only 66% but this was significantly improved by the addition of 8% sodium chloride to the sample prior to trapping on the C18 column.

In a second series of experiments C18 minicolumns (400 mg) were used (Marcomini et al. [23]) for marine samples. Seawater (100–300 ml), previously centrifuged, and fortified with 3% sodium chloride was passed through the column at <20 ml/min. Salts were removed from the column by washing with 1 ml double distilled water and the nonionics eluted with acetone. Recoveries for 0.1–3.1 μg/l spikes of NPE were 99 ± 4%.

As outlined in some of the methods above, the extracts obtained in the concentration procedure are generally ion exchanged under non-aqueous conditions to remove anionic and cationic surface active materials.

Gorenc et al. [24] evaluated the adsorption efficiency of Amberlite IRA 900 and Dowex 1 of various degrees of cross linking and granulation as a means of cleaning up water samples prior to analysis by the polarographic method of Kozarac et al. [25] and concluded that Dowex 1X4 (50–100 mesh) and Dowex 1X8 (50–100 mesh) were the most efficient at removing the anionics which normally interfere with the end determination.

Other Methods

Continuous liquid-liquid extraction in a system designed for organic solvents heavier than water was used by Stephanou and Giger [26] to extract AP, AP1EO and AP2EO from water samples with recoveries of 101 ± 4, $87 + 3$, and 93 ± 5 respectively. This method was also reported to be applicable to parent APE. A modified continuous liquid-liquid extractor was used by Shiraishi et al. [27] in which 200 l of effluent was extracted with 150 ml hexane at a flow rate of 68 ml/min however the analysis of NP, NP1EO and NP2EO was only semi-quantitative.

An exhaustive steam-distillation unit for simultaneous distillation and solvent extraction (Veith and Kiwus [28]) was used by Giger's group for the analysis of NP, NP1EO, NP2EO and NP3EO. After adding 30 g of sodium chloride 2 l water samples were distilled and the condensed distillate extracted by passing through a small layer (1–2 ml cyclohexane) of a lower density solvent. Continuous reflux ensured exhaustive extraction of the steam distillable organic constituents. The same extraction procedure was used for sludges and soils after suspending 10–50 g in 1.5 l water. Extraction efficiency with this technique decreased with increase in ethoxylate chain length but was remarkably selective for the low EO oligomers.

Comparison of the efficiencies of extraction of NP, NP1EO and NP2EO from sludges by Soxhlet with methanol and steam distillation/solvent extraction was made by Ahel and Giger [29]. Soxhlet recovery was only 50%, however addition of solid sodium hydroxide (20%) to the dried sludge improved the recovery to 99% which made the procedure as efficient as the steam distillation method.

In a later paper Marcomini et al. [23] reported extraction efficiencies of 86–100% for soils and sludges which had been Soxhlet extracted with methanol for 8 hours after the addition of sodium hydroxide (10% w/w).

A method for the future and one that was not been reported to date for surfactants but which nevertheless has considerable potential is the use of supercritical fluid extraction (SFE).

Non-Specific Methods

Several spectrophotometric and titration methods are routinely used in nonionic analysis. These techniques are relatively simple to perform and require little specialised equipment. They can be applied to environmental samples but they are not particularly suitable for the estimation of the low concentrations of nonionics found in most surface waters. Results are expressed arbitrarily in terms of a suitable reference standard.

Bismuth Active Substances (BiAS)

The bismuth active substances (BiAS) procedure of Wickbold [30, 31] has been adopted in the OECD Guidelines for laboratory investigations on the biodegradability of nonionic surfactants. The orange precipitate formed by the reaction of nonionic surfactant with tetraiodobismuthate (modified Dragendroff reagent) is dissolved with ammonium tartrate solution and the liberated bismuth is determined by titration (Wickbold [31]), atomic absorption (HMSO) [32] or UV, (Waters and Longman) [33]. The surfactants analysed (in the range 250–800 μg) must have 6–50 ethylene oxide groups although propoxylated surfactants also react. Short chain nonionics containing an average of 4 or less ethylene oxide groups provide little or no response. In the absence of suitable specific techniques for environmental analysis this method has been used for the analysis of nonionics in sewage influents, effluents and receiving waters. Waters et al. [34] have shown that an optimised procedure involving unfiltered samples, four 10-minute sublation steps and a cation/anion exchange clean-up of the sublation extracts is required for the analysis of environmental samples. The authors suggest that BiAS procedures which include non-optimal variables can result in BiAS concentrations in raw sewage samples being more seriously underestimated than in effluents and consequently underestimating removal during sewage treatment. This conclusion was also reached by Gledhill et al. [35] when comparing CTAS values with specific AE determinations.

Cobalt Thiocyanate Active Substances (CATS)

This procedure has been recommended by the American SDA on the grounds that it is simpler and faster than the BiAS procedure. The nonionic is reacted with ammonium cobaltothiocyanate in aqueous solution to form a coloured ion association complex which is readily extractable into dichloromethane. The intensity of the DCM extract is measured at 620 nm (Boyer et al. [36]). The CTAS procedure has been critically reviewed by Nozawa et al. [37].

Potassium Picrate Active Substances (PPAS)

The potassium picrate method was developed by Favretto et al. [38] to analyse nonionic surfactants after extraction with 1,2-dichloroethane. The cation formed by complexing the EO chain with aqueous K^+ is paired with picrate anion and extracted into methylene chloride, the picrate content of the organic phase being measured at 378 nm. After progressively refined versions (Favretto et al.) [39, 40], which were still sensitive to cationic interferences, a method was published (Favretto et al.) [41] which overcame the cationic problem and could be used to detect nonionic at concentrations of 2–200 μg/l without the use of ion exchange resins. An interlaboratory comparison carried out in 14 laboratories (Assanelli et al.) [42] concluded that the method could be satisfactorily used for the analysis of surface waters but was unsuitable for untreated urban liquid wastes.

Specific Chromatographic Methods

Alcohol Ethoxylates

Wee [43] described a method based on GC of the hydrogen bromide cleavage products of AE in environmental samples. The efficiency of the cleavage based on

4 samples of C14–15/EO-7 was 77 + 1.5%. And the sensitivity was in the ppb range for river waters. Tobin et al. [44] had previously used a similar method for biodegradation studies.

The HPLC procedure currently receiving the most attention is based on sublation, ion exchange and alumina chromatography (optional) to prepare a nonionic residue which is derivatised with phenyl isocyanate essentially as outlined by Allen and Linder [45]. An aqueous methanol-methanol gradient solvent system is used to separate the derivatised AE's according to their hydrophobe chain length on a C18 reverse phase column. Separation according to the degree of ethoxylation can be carried out by normal phase HPLC on a μBondapak NH_2 column with a hexane:dichloroethane (35:15) to acetonitrile:isopropanol (185:65) gradient (Schmitt et al. [17]) with the minimum level of detection said to be 0.1 ppm.

An alternative procedure was proposed by Nozawa and Ohnuma [46] who converted AE to the 3,5-dinitrobenzoate derivatives, a method that was used for environmental samples by Shiraishi et al. [28]. A second method (Kudoh et al. [47]) involves derivatisation of AE with 1-anthroyl nitrile followed by reverse phase chromatography. This technique was used successfully for the analysis of AE in river waters and sediments by Yoshimura [48].

Alkylphenol Ethoxylates

There have been some rapid developments in recent years for detecting APE at low concentrations. The use of GC/MS using electron impact (EI) (Giger et al. [49]) or chemical ionisation (CI) (Stephanou) [50] has proved useful for the identification of nonionics with a short ethoxylate chain. There has also been increased use of new ionisation techniques such as desorption chemical ionisation (DCI), field desorption (FD), fast atom bombardment (FAB) and the combination of FD or FAB with collision induced decomposition and mass analysed ion kinetic energy spectroscopy (CID-MIKE). Recently Rivera et al. [51] demonstrated the presence of nonylphenols with up to seven ethoxylate residues in river water by FAB and FAB-CID-MIKE in raw and drinking water after HPLC diode array fractionation of organic extracts. Ventura et al. [52] used GC/MS and FAB to demonstrate the presence of AP, APBr and acidic derivatives in water and drinking water.

The simplest and most suitable technique however for the quantitative analysis of APE is HPLC. The sensitivity of the procedure is dependent upon the method of detection. Ahel and Giger [53, 54] have described normal phase chromatography on aminosilica columns with gradient elution for the separation of APE_n coupled to UV detection and quote a detection limit of 1 μg/l. This system separated the ethoxymers without resolving the hydrophobes whilst reversed phase chromatography on 10 μm octylsilica resolved the hydrophobes only eluting the ethoxymers as a single peak. The use of a 5 μm LiChrosorb Si 60 column coupled to a fluorescence detector was reported by Kudoh et al. [47] but the most sensitive method reported to date is that of Holt et al. [55]. Using gradient elution with MTBE, acetonitrile and methanol they were able to use excitation wavelengths of 230 nm and report a minimum level of detection of 0.2 ng for individual homologues of the APE_n. Alkylphenol ethoxylates from AP1EO up to AP19EO were separated on a Zorbax NH_2 column and the biodegradation intermediates AP, AP1EO and AP2EO were separated on a Partisil 5 PAC column.

Other reports of the use of fluorescence detection include Marcomini et al. [22] and Kubeck and Naylor [12]. Marcomini described the simultaneous analysis of linear alkylbenzene sulphonates (LAS) and APE_n by reverse phase HPLC on 10 μm octylsilica columns with water/2-propanol/acetonitrile (containing 0.02 M sodium perchlorate) elution. Concentrations of 1–3 μg/l could be detected in river waters in the method reported by Kubeck and Naylor who used a 5 μm CN column with gradient elution and detection at 229 and 310 nm excitation and emission respectively.

Directly coupled HRGC/MS provided the basis for identification and quantitation of NP, NP1EO and NP2EO in sewage influents, effluents and sludge (Stephanou and Giger [26], Giger et al. [56]) however, this technique is limited to these lower EO oligomers.

Wahlberg et al. [57] have developed a method for the quantitation of NP, NP1EO, NP2EO and NP3EO based on GC/ECD or GC/MS of the pentafluorobenzoyl derivatives and have used the method for the analysis of these compounds in sewage sludges and the blue mussel (*Mytilus edulis*).

Acidic biodegradation intermediates of APE have been identified by GC/MS. The aqueous samples were adjusted to pH 2 with sulfuric acid and extracted into dichloromethane. Gas chromatography of the silylated derivatives was performed on a fused silica (SE 54 25 m × 0.25 mm) capillary column (Stephanou [58]).

AP1EC and AP2EC have been analysed in sewage samples by normal phase HPLC coupled to UV detection at 277 nm (Ahel and Giger [29]). In that paper the authors compare the relative merits of GC/MS and HPLC for the analysis of AP1EC and AP2EC. With GC it was possible to determine the distribution of nonyl isomers and also distinguish between the octyl and nonyl homologues. In contrast all the isomers and alkyl homologues are coeluted using HPLC. In practice this loss of resolution is not significant if rapid quantitation of APEC is required. The authors also indicated that the relatively selective analysis at 277 nm allowed better quantitative determination in non-treated sewage and primary effluents, which because of their high organic loading resulted in too many interfering compounds for GC/MS analysis.

Environmental Concentrations

Pathways of Nonionics into the Environment

The absence of a specific method for the analysis of alcohol ethoxylates has precluded the quantitative measurement of the concentration of nonionics in the environment. HPLC of the phenyl isocyanate derivatives of nonionics has recently been applied to environmental extracts but the technique has some limitations. In contrast there are a number of specific HPLC and HRGC methods available for the analysis of APE and their breakdown products and there is therefore a good understanding of the fate of APE in both sewage treatment plants and receiving waters.

Sewage Treatment

There have been a limited number of studies on the concentration and behaviour of alcohol ethoxylates in sewage treatment plants and these are summarised in Table 2 together with non specific nonionic data.

Concentrations of AE were determined in the influent and effluent at the activated sludge sewage treatment plant at Enid in the U.S.A. (Sedlak and Booman [11]) with the authors concluding that the AE specific and CTAS analyses were similar except for effluent samples. Average concentrations of 0.86 mg/l (1.62 CTAS) in raw influent and 0.01 mg/l (0.18 CTAS) in final effluent were found indicating that greater than 99% removal of AE and 91% removal of CTAS had occurred. No sludge analysis was possible. This study was reported in greater detail by Gledhill et al. [35] who list the individual analyses for the three sampling periods and for all the sampling points. The ratio of AE:CTAS was 0.53 in the raw influent and primary effluent but had fallen to 0.06 in the final effluent. Thus the non-specific analysis suffers from interfering substances to differing extents from the various samples.

Schmitt et al. [17] reported the results of two collaborative studies on the levels of nonionics and specifically alcohol ethoxylates in the influent and effluent of an activated sludge plant in the U.S.A. In the first study the mean of the 4 laboratories for AE in the influent was 0.91 mg/l (range 0.67–1.34) and for the effluents was 0.02 (range <0.01–0.022). CTAS values were 3.22 mg/l and 0.32 mg/l for the influent and effluent respectively. In the second study the influent AE concentration was 3.4 mg/l and the effluent 0.04 mg/l indicating >98% removal. The corresponding CTAS determinations were 4.7 mg/l and 0.44 mg/l which corresponds to 91% removal.

An optimised BiAS procedure was reported by Brown et al. [98, 99] for a monitoring exercise at both an activated sludge plant and a trickling filter plant in W. Germany. At the activated sludge plant removals were greater than 90%.

Table 2. Concentrations of nonionics in sewage treatment plants

Compound	Concentration (mg/l)	Compartment	Plant	Method	Ref.
BiAS	2.3	Influent	Marl	BiAS	Bock [101]
	0.5	Effluent	Germany		
AE	0.19–0.47	Influent	U.S.A.	GC/HBR	Wee [43]
(C14-15E07)	0.22	Activated sludge			
	0.006–0.012	Effluent			
nonionics	3.12	Influent		CTAS	
	5.32	Activated sludge			
	0.37	Effluent			
nonionics	3–4	effluent	Israel	CTAS	Zoller [102]
nonionics	3.1–8.5	Raw settled sewage	Hochdahl	BiAS	Brown [98]
	0.19–0.23	Final effluent	A.S.		
AE	0.03–0.06	Final effluent	plant	HPLC/PIC	
propoxylates	0.11–0.16	Raw settled sewage		GC/HBR	
	0.03	Final effluent			
nonionics	3.7–4.2	Raw settled sewage	Hosel	BiAS	Brown [99]
	0.7–0.9	Final effluent	T.F. plant		
	(March)				
	3.6–4.7	Raw settled sewage			
	0.4–0.6	Final effluent			
	(September)				
nonionics	0.03–3.6	Influent	35 plants	BiAS	Klopp [103]
	0.26	Effluent	Germany		
nonionics	3.22	Influent	U.S.A.	CTAS	Schmitt [17]
	0.52	Effluent			
AE	0.91	Influent		PIC/HPLC	
	0.02	Effluent			
nonionics	3.4–10.0	Total raw sewage Israel		BiAS	Narkis [104]
	0.8–6.2	Total effluent			
	1.2–5.9	Dissolved raw sewage			
	0.7–2.1	Dissolved effluent			
nonionics	1.62	Raw influent	Enid	CTAS	Gledhill [35]
	1.28	Primary influent	U.S.A.		
	0.13–0.18	Final effluent			
AE	0.86	Raw influent		PIC/HPLC	
	0.62	Primary influent			
	0.01	Final effluent			
	0.01	Final effluent			
nonionics	4.4	raw sewage-total	Haifi	BiAS	Narkis [105]
	2.5	raw sewage-dissolved			
	1.9	trickling filter-total			
	0.7	t.f-dissolved			
	0.8	Activated sludge-total			
	0.4	A.S. dissolved			

Concentrations of APE found in the various stages of sewage treatment are listed in Table 3

The corresponding removals for BiAS in a trickling filter plant were 88% in summer and 81% in winter. Through the application of specific HPLC procedures it was shown that 20% of the sewage BiAS at both plants was comprised of APE's. In the trickling filter plant the removals of APE were 70% and 75% in the winter and summer respectively and at the activated sludge plant the removal was 89%. When the contribution of APE's to the BiAS concentrations were taken into account the non-APE BiAS (largely thought to be alcohol ethoxylates in domestic sewage) removal was 91% for the trickling filter and 98% for the activated sludge plant.

The concentrations of 4-nonylphenol in 30 anaerobically stabilised sludges sampled in Switzerland ranged from 0.45–2.53 g/kg (mean 1.01 g/kg) whereas concentrations in 8 aerobically stabilised sludge samples were significantly lower and ranged from 0.08–0.5 g/kg (mean 0.28). Activated sludge and mixed primary and secondary sludges were also analysed and found to contain 0.09–0.15 and 0.04–0.14 g/kg respectively (Giger et al. [56]). The authors suggest that the formation of 4-nonylphenol is enhanced under mesophilic anaerobic conditions. Sludges from the same plants were examined again in 1986 and showed similar results (Giger et al. [100]). During this programme a more comprehensive study of the behaviour of NPE during sewage treatment at 13 plants was investigated. The raw sewage contained on average between 0.8–1.0 mg/l but ranged from 0.5–2.2 mg/l NPE. No significant differences were found between 6 raw sludges and 21 effluents from the primary sedimentation tanks. The ethoxymer distribution in the raw and presedimented sludges contained higher proportions of the lower EO components than the commercially available NPE and support the hypothesis that aerobic and anaerobic degradation was taking place in the sewers and during primary sedimentation. Approximately 50% (molar basis) of NPnE in sewage was transformed to NP and accumulated in the digested sludge. Giger [110, 100, 117] also reported that biological treatment produces carboxylated intermediates, particularly NP1EC and NP2EC (Fig. 11). Details of elimination of the nonylphenol breakdown products in 4 activated sludge plants were given and indicate 76–93% removal of NPnEO, 29–79% removal of NP1EO+NP2EO, 37–93% removal of NP but an accumulation of 230–450% of NP1EC+NP2EC.

Concentrations of NP, NP1EO and NP2EO from 29 of the anaerobically digested sludges analysed in the 1986 study ranged from 1.0–2.2, 0.1–0.7 and 0.02–0.22 mg/kg respectively (Brunner et al. [106]). Values for the quantity of materials adsorbed on particulate matter and dissolved are also given and confirm that >99.9% of the NP is adsorbed to the particulate matter in anaerobic sludge.

Table 3. Concentration of APEs in sewage treatment plants

Compound	Concentration (μg/l)	Sample	Location	Ref.
NP	0.45–2.53 g/kg	anaerobic	30 plants	Giger [56]
	0.08–0.5 g/kg	aerobic stabilised	8 plants	
	0.09–0.15 g/kg	activated sludge		
	0.04–0.14 g/kg	mixed primary and secondary sludge		
NP	5.2–13.6	effluent	Zurich-Glatt	Ahel [116]
NP1EO	14–63	effluent		
NP2EO	23–72	effluent		

Table 3. (*Continued*)

Compound	Concentration ($\mu g/l$)	Sample	Location	Ref.
NP	2.1–8.9	effluent	Niederglatt	
NP1EO	13–42	effluent		
NP2EO	7.7–50	effluent		
NP	14	raw wastewater		Ahel [53]
NP	8	effluent		
NP	128 mg/kg	activated sludge		
NP	1000 mg/kg	anaerobic digest		
NP	467	effluent from anaerobic digester		
NP1EO	18	raw wastewater		
NP1EO	49	effluent		
NP1EO	76 mg/kg	activated sludge		
NP1EO	79 mg/kg	anaerobic digest		
NP1EO	53	effluent from anaerobic digester		
NP2EO	18	raw wastewater		
NP2EO	44	effluent		
NP2EO	61 mg/kg	activated sludge		
NP2EO	0 mg/kg	anaerobic digest		
NP2EO	6	effluent from anaerobic digester		
NPE	400–2200	influent		Giger [110, 100]
NPE	12–120	effluent		
NP	20–47	influent		
NP	1–13	effluent		
NP	0.15–2.2 mg/g	sludge		
APE	126–410	influent	U.K.	Holt [55]
	40–228	effluent		
NP	280	Raw waste water	Zurich-Glatt	Marcomini [22]
	30	Effluent	Switzerland	
	180	Filtrate from activated sludge		
	190	Filtrate from digested sludge		
NPE	1920	Raw waste water		
	230	Effluent		
	120	Filtrate from activated sludge		
	100	Filtrate from digested sludge		
APE	900–1600	Raw settled sewage	Hochdahl	Brown [99]
	120–140	Final effluent	A.S. plant	
	680–800	Raw settled sewage	Hosel	
	190–250 (March)	Final effluent	T.F. plant	
	640–780	Raw settled sewage		
	150–200 (September)	Final effluent		
NP	10–15	Influent	W. Germany	Kunkel [107]
NP	1–2	Effluent		
NP	20–50 $\mu g/g$	Activated sludge		
NP	130–400 $\mu g/g$	Anaerobic sludge		
NP1EO	2–10	Influent		
NP1EO	1–7	Effluent		

Table 3. (*Continued*)

Compound	Concentration (μg/l)	Sample	Location	Ref.
NP1EO	5–30 μg/g	Activated sludge		
NP1EO	5–40 μg/g	Anaerobic sludge		
NP2EO	2–10	Influent		
NP2EO	<10	Effluent		
NP2EO	5–10 μg/g	Activated sludge		
NP2EO	<3 μg/g	Anaerobic sludge		
NP	1–2.2 g/kg	Anaerobic sludge	29 plants	Brunner [106]
NP1EO	0.1–0.7 g/kg	,, ,,	Switzerland	
NP2EO	0.02–0.22 g/kg	,, ,,		
NPE	1600–2500	Influent	High Point	Kubeck [12]
	50–100	Effluent	U.S.A.	
NP	1–2.5	Effluent		
NPE	11	Effluent	Govalle U.S.A.	
NPE	37–123	Effluent	Barcelona. Spain	Valls [21]
NP	400–1100 mg/kg	Anaerobic sludge	4 plants in	Wahlberg [57]
NP1EO	20–190 mg/kg		Sweden	
NP2EO	1–50 mg/kg			

Concentrations are in μg/l unless otherwise stated

Kunkel [107] reported on the activities of the analytical work group from the German Committee on Detergents and indicated concentrations of NP, NP1EO and NP2EO similar to those found by Giger [100] in Switzerland.

Samples of raw and treated wastewater from the High Point plant in N. Carolina were analysed by Kubeck and Naylor [12]. The Eastside plant treats primarily industrial waste (mainly from textile processing and furniture manufacture) whereas the Westside plant receives primarily domestic waste. Influent concentrations ranged from 1.6–1.98 mg/l in the industrial waste and from 2.27–2.52 in the domestic waste. Effluent concentrations were 0.05–0.1 mg/l indicating removals of 93–98%. They found no increase of NP or the EO oligomers and quote NP1EO removal of 89–95%. No sludges were analysed and therefore a complete balance was not obtained.

The concentration of nonionic surfactants in raw sewage in Haifa sewage treatment plant in Israel were up to 10 mg/l in 1976 and are currently 4–8 mg/l. Concentrations in the effluents following trickling filter treatment however have dropped from 4 mg/l in 1976 to about 1 mg/l (Narkis and Weinberg [105]) and reflect a change in the nonionic surfactants used in washing powder formulations in Israel. The nonionic identified in the effluents was primarily dinonylphenol ethoxylate with up to 50% adsorbed on to the suspended solids in the effluent.

In addition to the studies on the concentration of APE biodegradation intermediates chlorinated OPE were found in river water samples (Sheldon and Hites [13]), while chlorination of wastewaters resulted in the formation of both chlorinated and brominated OPE and OPEC (Ball and Reinhard [108], Reinhard et al. [109]) with halogenated OP2EC the major product. A wide range of halogenated OPE were determined in the municipal secondary sewage treatment plant effluent at concentrations up to 51 μg/l while nonhalogenated OPE residues were present at concentrations of 36–112 μg/l. Ventura et al. [52] also reported that chlorination at a drinking water treatment plant treating river water containing NPE and NPEC residues resulted in the formation of chlorinated NPE and NPEC compounds that were detected in tap water.

In general removal of nonionics (BiAS or CTAS) in sewage treatment is >90% for activated sludge plants and 80–90% for trickling filter plants. Specific analysis for alcohol ethoxylates indicates >98% removal in activated sludge and around 90% in the trickling filter plants. The removal of APE is less consistent and is more related to the plant efficiency and ambient temperature. During sewage treatment, AP1EO and AP2EO are formed and are partly adsorbed to solids and partly discharged in the effluent [110]. During anaerobic sludge treatment the degradation of these intermediates results in the formation of AP which does not appear to be significantly degraded under anaerobic operating conditions of many digestors and as a result accumulates on the digested sewage sludge.

Septic Tanks and Agricultural Run-off

Although septic tanks are a potential source of entry of nonionics into the environment and large tonnages of nonionics are used in agrochemical formulations the only report published concerned with nonionics in the soil is referred to the section on sludge amended soils.

Surface Waters

River Water

Specific analyses of river waters for alcohol ethoxylates are limited. A summary of nonionic concentrations in river waters is given in Table 4. In general the concentrations fall in the range 0.02–0.1 mg/l. Noticeably higher concentrations were reported following analysis by the TLC method of Patterson et al. [111]. The CTAS results for two streams in Israel (Zoller [112]) are also significantly higher (1.6–2.6 mg/l) than for other surface waters. These values are said by the authors to be typical of the concentrations found in Israeli surface waters in 1985–1986 and reflect the large usage and poor biodegradability of branched alkylphenol ethoxylates in Israel. These concentrations have serious implications since these levels are of the order of acute toxicity effect concentrations to sensitive fish and aquatic species.

Table 4. Concentrations of nonionics found in surface waters

Compound	River	Country	Method	Concentration (mg/l)	Ref.
nonionics	2 rivers	U.K.	TLC	0.2–1.0	Patterson 1967 [111]
nonionics	2 rivers	Germany	BiAS	0.01–0.03	Wickbold 1972 [31]
nonionics	15 rivers	Italy	BiAS	0–0.08	Arpino 1973 [121] 1974 [122]
nonionics	Lippe	Germany	BiAS	0.02–0.1	Bock 1973 [101]
nonionics	Rhine	Germany	BiAS	0.02–0.1	Bock 1973 [101]
nonionics	Vltava	Chechosl	BiAS	0.1	Pitter 1974 [123]
nonionics	6 rivers	Germany	BiAS	0.01–0.1	Fischer 1976 [124] 1980 [125]
nonionics	Main	Germany	BiAS	0.02–0.03	Kupfer 1979 [126]
NPE	Sanno	Japan	HPLC/FDMS	0.024	Otsuki 1979 [127]
nonionics	6 rivers	France	BiAS	0.05–2	Cadridenc 1980 [128]
nonionics	Avon	U.K.	TLC	0.008	Jones 1978 [16]
C14-15 EO	Ohio	U.S.A.	GC/HBR	0.004	Wee 1981 [43]
LAE		Japan	HPLC	0.2	Shiraishi 1982 [27]
APE	2 rivers	Japan	CTAS	0.05–0.07	Saito 1982 [19]
nonionics	3 rivers	Japan	PPAS	0.05–0.3	Saito 1983 [129]
nonionics	Meuse	France	A.A.	<0.001–0.5	Van Hoof 1985 [63]
nonionics	Rhine	Germany	BiAS	0.02	Berth 1988 [130]
	Ruhr	,,	,,	0.01	
	Neckar	,,	,,	0.02	
	Main	,,	,,	0.02	
nonionics	2 streams	Israel	CTAS	1.6–2.6	Zoller 1989 [112]
nonionics	Rhine	Germany	BiAS	0.02–0.08 0.004–0.03	Gerike 1989b [113] 1989a [131]
NPE	Colorado	U.S.A.	HPLC	1.1–1.9 µg/l	Kubeck 1989 [12]
Metabolites				µg/l	
OP1EO	Delaware	U.S.A.	GC/MS	5	Sheldon 1979 [132]
OP2EO				0.6–5	
OP3EO				0.4	
NP	Glatt	Switzer	HPLC	1–2	Ahel 1984 [116]
NP1EO	,,	,,	,,	0.5–12	
NP2EO	,,	,,	,,	0.5–15	
NP	Glatt	Switzer	HPLC/HGRC	0.5–2	Ahel 1985 [53]
NP1EO	,,	,,	,,	0.5–18	
NP2EO	,,	,,	,,	0.5–16	
NP	Marl	Germany	HPLC/HRGC	2–5	Kunkel 1987 [107]
NP1EO	,,	,,	,,	1–2	

Gerike [113] has reported BiAS concentrations of 0.01 mg/l in the river Rhine which indicated a drop in concentration from 1951 to 1987 of 67%. In contrast, trend analysis by Klopp [103] showed a constant level of nonionics in the lower Ruhr between 1979 and 1985 and the author suggests that the degradation of nonionics in treatment plants and receiving waters is worse than that of anionic surfactants. Average winter concentrations of BiAS were 0.12 mg/l in comparison to summer concentrations of 0.05 mg/l. No details of the analytical procedure are given. However, the application of BiAS to the analysis of surfactants in river waters is highly problematical since the concentrations found are very close to the limits of detection of the method and little is known about interfering substances. This again highlights the need for the development of specific analytical techniques in order to make categorical statements on environmental concentrations.

The Japanese Environmental Agency (JEA) has investigated the existence of nonionic surfactants in river waters and sediments [114, 115]. Data for 1978 showed APE concentrations ranged from 0 (80 samples) to 0.13–0.93 ppm (25 samples) for river water and from 0 (19 samples) to 2.1–50 ppm (69 samples) for sediments. Lower concentrations were found in both compartments in the 1982 survey ranging from 0 (29 samples) to 0.09 (1 sample) for river water and from 0 (22 samples) to 2.6–4.9 ppm (8 samples) for sediments. AE in 1982 on the other hand were below the 0.005 ppm detection limit in river waters (30 samples) and ranged from 0 (11 samples) to 0.22–1.0 ppm (19 samples) for river sediments.

Kubeck and Naylor [12] analysed surface water from the Delaware River in the U.S.A. for NPE. Total NPE (including NP) concentration upstream of a sewage treatment plant was 1.1 μg/l with an EO distribution close to that found in commercial NPE cleaning products. Downstream of the sewage discharge point the total NPE concentration was 1.9 μg/l and showed a higher percentage of the lower EO (NP to NP5EO) oligomers with the NP2EO+NP3EO especially high. Analysis of water from a second sampling point downstream of two other large treatment plants resulted in similar concentrations and oligomer distributions. The authors also point out that the distribution profile was similar to that of the treated sewage.

Ahel et al. [116] monitored NP, NP1EO and NP2EO concentrations at 4 locations on the Glatt River in Switzerland. The concentrations of NP1EO and NP2EO showed increases from 0.5 to 12–15 μg/l in areas close to treated sewage discharge points, however, the concentration of NP showed only slight increases to 1–2 μg/l near the sewage outfalls. Additional analyses by the Ahel and Giger [53] compared the detection of the degradation products of NPEs by HPLC, HGRC and HGRC/MS; the authors concluded that all three techniques gave remarkably similar quantitative data. HPLC was the easiest and most rapid technique and therefore the most suitable for routine analysis of a large number of samples. Details of the concentrations of nonylphenol carboxylic acids in the Glatt river in relation to the proximity of the sewage discharge plant are given by Ahel et al. [117]. As in secondary effluents there is a predominance of NP2EC over NP1EC. Concentrations of 2–116 μg/l NPEC were found which were, therefore, higher than the concentration of NP, NP1EO and NP2EO.

In a water quality survey of the river Meuse along its 900 km course through France, Belgium and the Netherlands, Van Hoof et al. [63] analysed 70 samples from 38 sites by atomic absorption. In general the concentration of nonionics was 0.01–0.1 mg/l. High values of 0.2 mg/l were found where some tributaries discharged partially treated waste waters into the river basin. Immediately downstream of the waste water discharge from a detergent factory near Liege the concentration was 0.5 mg/l. PEG were measured in the lower Mississipi river by ^{1}H and ^{13}C NMR and ranged from undetectable to 145 µg/l (Leenheer et al. [231]).

Groundwater

There are limited data available on the concentrations of nonionics in groundwater as a result of either point or non-point sources. Thurman [20] analysed groundwater from wells 500 m and 3000 m downfield of sewage infiltration ponds in the U.S.A. and found 0.01 mg/l CTAS in the water from the 500 m well but nothing in the 3000 m well. Zoller [112] reported concentrations of nonionics, determined as CTAS, ranging from 0.12–0.78 mg/l which corresponded to approximately 23% of the concentrations of nonionic found in the adjacent streams. NP, NP1EO and NP2EO were found at concentrations of 2, 2 and 4 µg/l respectively in a sample of groundwater analysed by GC/MS however no NP3EO was detected (Schaffner et al. [118]).

Marine Waters

Seawater samples from the Adriatic (Trieste harbour) over a five year period were analysed by Favretto et al. [38, 41, 119] using the potassium picrate method and were shown to contain 0.039–0.187 ppm PPAS.

Marcomini et al. [15] presented a methodology for the determination of NPE in marine waters. Extraction of NPE from centrifuged surface water was carried out by adsorption on to C18 mini-column cartridges at a flow rate of 20 ml/min and desorbed from the column with acetone. Surface water contained 19.6 µg/l NP1EO-NP13EO and 0.2 µg/l NP. The same authors, in a second paper, [120] reported the concentrations of NPE and NP in sediment, resuspended sediment and water from five locations in the Venice lagoon at various times of the year. NPE concentrations in water in April, July, and October were 2–5, 1–5 and 1–3 µg/l respectively. Average concentrations of NP, NP1EO and NP2EO were 8.3–10.1 mg/kg in resuspended sediment and 0.25 mg/kg in macroalgae.

Valls et al. [21] reported concentrations of NPE on dissolved and particulate phases of sewage, seawater and sediments. The oligomeric distribution was shown to depend on the environmental compartment. The dissolved phase in waters contained predominantly NP2EO and longer ethoxylates whereas the sediments contain NP to NP3EO. Coastal water was reported to contain 845 ng/l NPE dissolved and 0.5 ng/l associated with the particulate matter. Concentrations of 0.1–6.6 µg NPE/g were detected in marine sediments.

Soils and Sediments

Sludge Amended Soils

Concentrations of the major degradation products of NPE (i.e.: NP, NP1EO and NP2EO) in the top 5 cm of soil from a sludge treated experimental plot have been monitored on 19 occasions during a 1 year period by Marcomini et al. [120]. The plot had been treated with 5 tonnes of digested sludge per hectare per year for 10 years. The initial concentrations of NP, NP1EO and NP2EO immediately after sludge application were 4.7, 1.1 and 0.1 mg/kg respectively which had fallen to approximately 20% of their initial value in the first 3 weeks. Further slow degradation was reported but residual concentrations of 0.5, 0.1 and 0.01 mg/kg remained after 320 days and probably reflect the availability of the compounds for microbial attack.

Sediments

The only reports found relating to concentrations of nonionics in sediments were published by Marcomini and Giger [86] who analysed surface sediment from the river Rhine and found 0.9, 0.8 and 0.7 mg/kg of NP, NP1EO and NP2EO respectively and the Japanese Environmental Agency values reported earlier [114, 115].

Biodegradability

Laboratory Tests

The biodegradation of nonionic surfactants and in particular, AE and APE, has been followed as BiAS removal, CO_2 evolution in the Sturm Test, by biochemical oxygen demand (BOD) in the Closed Bottle Test and by organic carbon analysis in the OECD Screening Test and Coupled Units Test. This data has been comprehensively reviewed by Swisher [6].

The structure-biodegradability relationship for nonionic surfactants with the various routine laboratory testing methodologies in papers published since 1984 are detailed in Tables 5 and 6. It should be remembered however that, in general, simulation tests give lower removals than occur in full scale treatment plants which is probably a feature of the 'limitations' of the synthetic sewage feed and the differences in microbial populations. For example Gilbert and Klaiser [133] reported that removal of AE was increased from $53 \pm 8\%$ to $88 \pm 9\%$ DOC

Table 5. Biodegradation in simulation tests

Compound	Confirmatory test % (BiAS) removal	Coupled units test % removal	Metabolite test	Ref.
C8 FA+6EO	91			[175]
C8 FA+6EO			95.5	[175]
C16–18 FA+10EO	98	95		[130]
C16–18 FA+10EO		62	98.8	[176]
C12–14 FA+30EO	98	93		[130]
		59	87.5	[176]
C16–18 FA+14EO		96		[175]
C12–14 FA+5EO/4PO		60		[175]
		60	97.4	[176]
C12–14 FA+4EO/5PO		49	82.1	,,
C12–18 FA+9EO-nC4H9		88	102.0	,,
C16–18 Alkylamino+10EO		33	95.4	,,
EO/PO block polymer	7	2–4		[130]
NP+10 EO		90		[175]
iNP+9EO	97	90		[130]

in units operated at 6 d sludge retention time (SRT) when natural sewage was used instead of artificial sewage.

Alcohol Ethoxylates

All alcohol ethoxylate surfactants derived from straight chain primary or secondary alcohols undergo rapid and ultimate biodegradation. The three features which have a significant influence on the biodegradation pathway, on the rate of biodegradation and the extent of biodegradation are:

1) The structure of the hydrophobic moiety. Swisher [6] states that ultimate biodegradation is most rapid with linear hydrophobes and that single methyl branches exert little or no retarding effect. A secondary alcohol linkage to the EO chain retards ultimate biodegradation slightly even when the hydrophobe is linear and that highly branched hydrophobes are much more slowly mineralised. The influence of the structure of the hydrophobe has also recently been summarised by Kravetz [134].
2) Chain length of the hydrophile. Gledhill [135] reported that modifications in either the length of the alkyl chain (C8–20) or in the number of EO units (EO3–11) have little effect on the rate and extent of CO_2 evolution. Sturm [186] also found no effect up to 11 EO units but that the presence of 20 EO units markedly reduced the rate of ultimate biodegradation.
3) Incorporation of other glycols into the hydrophile. Insertion of oxypropylene groups into the polyoxyethylene hydrophile of oxo alcohol ethoxylates were shown by Naylor et al. [136] to decrease the rate and extent of biodegradation in 12 week SCAS inherent biodegradability tests. The surfactants chosen

Table 6. Summary of laboratory studies on primary and ultimate biodegradation of nonionic surfactants

Compound	% Primary degradation Screen	Confirm	Ultimate degradation Test	Parameter	%	Ref.
C16–18 FA+5EO	96		CB	oxygen	67–75	[177]
+14EO	99		MOST	TOC	80	[130]
C12–18 FA+10–14EO	98–99	93–98	CB	oxygen	69–86	[177]
			MOST	TOC	94	
			CU	TOC	95	
			CU	TOC	96	
C12–14 FA+5EO			CU	TOC	84	[177]
+30EO	99		—	—	—	[130]
+50EO	98		—	—	—	[130]
C16–18 FA+30EO	99	98	CB	O_2	27	[177]
+25EO			CU	TOC	75	[177]
C9–11 Oxo+7EO		86	CU	TOC	36	[177]
C13–15 Oxo+3–12EO		95	MOST	TOC	75	[177]
C14–15 Oxo+7EO	93					[130]
C14–15 Oxo+9–20EO		83–93	Sturm	CO_2	65–75	[177]
NP+5–6EO	85	91–93	CU	TOC	91	[177]
NP+9–10EO	80	87–97	MOST	TOC	8–17	[177]
			CB	O_2	5–10	
			Sturm	CO_2	40	
			CU	TOC	77–90	
NP+20EO	80	85–90	CU	TOC	70	[177]
NP+25EO		85	CU	TOC	50	[177]
nC8–10 Ap+9EO	84					[130]
C16–18 fatty amine +10EO	16	97	CU	TOC	33	[177]
		98	CU	TOC	6	
C12–18 fatty amine +12EO	88		CB	O_2	33	[177]
+20EO			CU	TOC	70	[177]
+2EO			Sturm	CO_2	85	[177]
C12–18 fatty acid ester + 5–29EO	95–99	92–96	CB	O_2	60–80	[177]
			MOST	TOC	100	
			CU	TOC	71–92	
FA +2–5EO/4PO	90	93–97	MOST	TOC	52	[177]
			CB	O_2	52	
			Sturm	CO_2	32	
			CU	TOC	60	
C12–18 FA +2.5EO/6PO	79	87	MOST	TOC	43	[177]
			CB	O_2	36	
			CU	TOC	37	
+5EO/8PO	70		—	—	—	[130]
C12–18 FA +6EO/2PO	95		MOST	TOC	69	[177]
			CB	O_2	83	
C13–15 Oxo +6PO/4EO		76	CU	TOC	17	[177]
C12–14 FA +10PO	50–63		MOST	TOC	11	[130]
C12–14 FA +9EO n-butyl ether	98		CB	O_2	80	[177]
			CU	TOC	88	
C12–14 FS ethanolamide+4EO	no BiAS		CB	O_2	47	[177]
+10EO	no BiAS		CB	O_2	35	

for these studies were linear and branched alcohol-EO-PO adducts in which either the PO block was located between two EO blocks of equal length or the EO-PO was completely random. They conclude that biodegradation is inversely proportional to the amount of PO incorporated into the surfactant. For linear alcohol (fatty or oxo) EO-PO adducts the maximum PO block which undergoes complete biodegradation is about four equivalents if located in the middle of the ethoxylate chain. The 2.8 PO adduct gave >90% DOC removal while the 4.2 PO adduct gave 50–55% removal. The PO capped fatty alcohol ethoxylate (9EO-8PO) gave 40% DOC removal. Primary biodegradation studies also suggest that less PO can be accommodated if the hydrophobe is branched. Extensive data on EO-PO surfactants has also been reported by Henkel [137].

Biodegradation Pathways

A comprehensive review of the metabolic pathways involved in AE and APE biodegradation is beyond the scope of this chapter. Some generalisations can be made but for a more detailed account the reader is referred to Swisher [6]. Table 7 lists a number of the studies reported on the mechanism of ultimate biodegradation of nonionic surfactants.

The initial point of attack for the biodegradation of AE can be at any one of three sites.

1) A central fission mechanism between the hydrophilic and hydrophobic moieties resulting in an alcohol or alkyl carboxylic acid which then undergo β-oxidation. β-Oxidation proceeds very rapidly and does not produce any measurable intermediates. The hydrophile is released as either PEG or carboxylated PEG. The mechanisms for the PEG degradation would seem to involve oxyethylene carboxylic acid intermediates or a diol cleavage step producing an ethoxymer with one less EO unit.
2) Oxidation of the terminal carbon of the alkyl chain to produce a carboxylic acid. β-Oxidation then shortens the alkyl chain by two carbon units at a time.
3) Oxidation of the terminal EO to give a carboxylic acid.

The proposed pathways are illustrated in Fig. 9 and some of the evidence supporting these pathways is outlined below. Essentially the biodegradation of LAE occurs in two stages, rapid mineralisation of the hydrophobe in conjunction with a slower but nevertheless extensive biodegradation of the PEG or carboxylated PEG.

Detailed mechanistic studies with ^{14}C labelled nonionics have been conducted by Kravetz et al. [138]. The results were consistent with a biodegradation mechanism in which the initial step is cleavage of the molecule at the hydrophobe—hydrophile ether bond to form hydrophobic and hydrophilic products followed by alkyl chain degradation to CO_2 beginning at the functional end group rather than at the terminal methyl group. In contrast Nooi et al. [139]

Table 7. Summary of microbial metabolism studies with nonionic surfactants and polyglycols

Compound	Microorganism	Ref.
APE and LAE	Activated sludge	Patterson 1970 [178]
C18 +nEO	Activated sludge	Nooi 1970 [139]
C12–15 +9EO	Secondary effluent	Tobin 1976 [44]
C12+5EO	*Pseudomonas*	Ichikawa 1978 [179]
s-butyl EEG	*Nocardia*	Baggi (1977) [180]
C14/15-7EO	Activated sludge	Cook 1979 [181]
C12–15 +9EO and NPE	Activated sludge	Kravetz 1982, 1984 [169], [168]
NPE and C12–13 oxo alcohol 12EO	Activated sludge	Schöberl 1981 [162]
C12+9EO and C16+3EO	River water	Larson 1981, 1982, 1984a, b [140] [142, 141, 182]
C12+9EO and C16+3EO	Esturaine water	Vashon 1982 [143]
C18 +nEO	Activated sludge	Neufahrt 1982 [144]
APE	Activated sludge	Brueschweiler 1983 [164]
C11–15+7EO	*Alcaligenes faecalis*	Grant 1983 [150]
Tetronics	Bacillus	Plucinski 1983 [183]
C18 +7EO	Activated sludge	Steber 1983, 1985 [145, 146]
APE	River water	Hellman 1985 [170]
LAE	*Ps.* sp DES1	Griffiths 1987 [147]
LAE	Anaerobic sludge	Wagener 1987 [151]
LAE	*Pelobacter propionicus* and *Acetobacterium*	Wagener 1988 [152]
C18+7EO	Anaerobic sludge	Steber 1987 [153]
NPE	Activated sludge	Neufahrt 1987 [171]
diEG-PEG 400	Soil bacterium	Fincher 1962 [184]
PEG 400 and 1000	Activated sludge	Borstlap 1967 [185]
<PEG 400	,, ,,	Patterson 1970 [178]
<PEG 1000	Acclimated sludge	Sturm 1973 [186]
EG-PEG 3500	Sludge and pure culture bacterium	Pitter 1973 [187]
triEG-PEG 400	*Pseudomonas*?	Ohmata 1974 [188]
triEG-PEG 400	Sludge and *Alcaligenes*	Harada 1975 [189]
PEG up to 20 000	Unidentified pure cultures	Ogata 1975 [190]
PEG up to 20 000	*Ps. aeruginosa*	Haines 1975 [191]
PEG 6000	Flavobacterium and *Ps.* sp	Kawai 1977 [192] 1978, 1986 [193, 194, 195]
PEG up to 4000	Acclimated sludge	Cox 1978 [155]
EG-PEG 400	Possibly *Acinetobacter*	Jones 1976 [196]
EG-PEG 1500	*Acinetobacter*, *Ps*, and *Flavobacter*	Watson 1977 [157]
PEG 400 and 6000	Soil bacteria	Hosoya 1978 [197]
EG	*Flavobacterium*	Childs 1978, [198] Willetts 1981 [199]
EG-PEG 4000	Estuarine mud or activated sludge	Jenkins 1979 [158]
EG	*Mycobacterium* E44	Wiegant 1980 [200]
diEG-PEG 400	*Pseudomonas*	Thelu 1980 [159]
EG-PEG 1500	*Acinotobacter, Pseudomonas Aeromonas*	Pearce 1980 [201]
EG	Halophile	Caskey 1981 [202]
diEG and triEG	*Ps.* fluorescens	Schöberl 1983 [203]
PEG 20 000	*Pelobacter venetianus*	Schink 1983 [161] Strass 1986 [204]
EG-PEG 20 000	Methanogenic consortia	Dwyer 1983 [160]

Table 7. (*Continued*)

Compound	Microorganism	Ref.
EG-PEG 20 000	*Desulfovibrio desulfuricans* and a *Bacteroides*	Dwyer 1986 [205]
PEG 400	Activated sludge	Steber 1985 [146]
PPG 2000–4000	Soil and sludge possibly *Corynebacterium*	Kawai 1977 [192, 193]
PG	*Propionibacterium freudenreichii*	Hosoi 1978 [206]
PG and EG	Enterobacteriaceae	Toraya 1979 [207]
PG	*Flavobacterium*	Willetts 1979 [208]
PG	*Nocardia* A60	De Bont 1982 [209]
EG and PG	*Acetobacterium* sp.	Tanaka 1988 [210]

indicated that the initial oxidation is at the terminal methyl group prior to hydrolytic cleavage of the hydrophobe from the hydrophile. These findings confirm that different bacterial strains exist with the potential for a number of different pathways for AE biodegradation.

Larson studied the kinetics and affect of temperature on the biodegradation of $C_{16}E_3$ and $C_{12}E_9$ in river water [140] and groundwater [141] in the U.S.A. These pure chain compounds were labelled either in the alkyl carbon or uniformly in the ethoxylate chain. The rate of biodegradation was proportional to the concentration of surfactant over the range 1–1000 µg/l. Conversion to $^{14}CO_2$ was more rapid for the hydrophobe than the hydrophile but in both cases was extensive. The effect of temperature was described by the Arrhenius equation. Data on the fate of ^{14}C incorporated into biomass and remaining in solution are reported in Larson and Wentler [142]. The half-life for LAE in Ohio river water was 36 hours. Not surprisingly V_{max} in groundwater was 121 ng/l/h compared to 8081 ng/l/h in river water and probably reflects the difference in organism concentration in these two environmental compartments.

The rate and extent of mineralisation of the same compounds in estuarine water were reported by Vashon and Schwab [143]. Half-lives for the alkyl chain carbon in water was 2–3 days and for the ethoxylate carbon was 5.8 days. First order kinetics were demonstrated over a concentration range 580 ng/l to 140 µg/l for the alkyl carbon and at 1 µg/l or less for the ethoxylate carbon. These authors argue that, from an environmental exposure standpoint, the key characteristic controlling the exposure of a chemical is not the rate and extent of biodegradation over an arbitrary time period per se, but the kinetics or rate of biodegradation (half-life) compared to its residence time in a particular environmental compartment.

Further confirmation of the extensive ultimate biodegradation of LAE and the more rapid mineralisation of the alkyl chain was given by Neufahrt et al. [144]. They also demonstrated that a significant percentage of the added label was adsorbed onto the biomass at the start of the experiment.

Steber and Wierich [145, 146] using a model activated sludge plant fed with alkyl or glycol labelled stearyl alcohol ethoxylate followed the distribution of

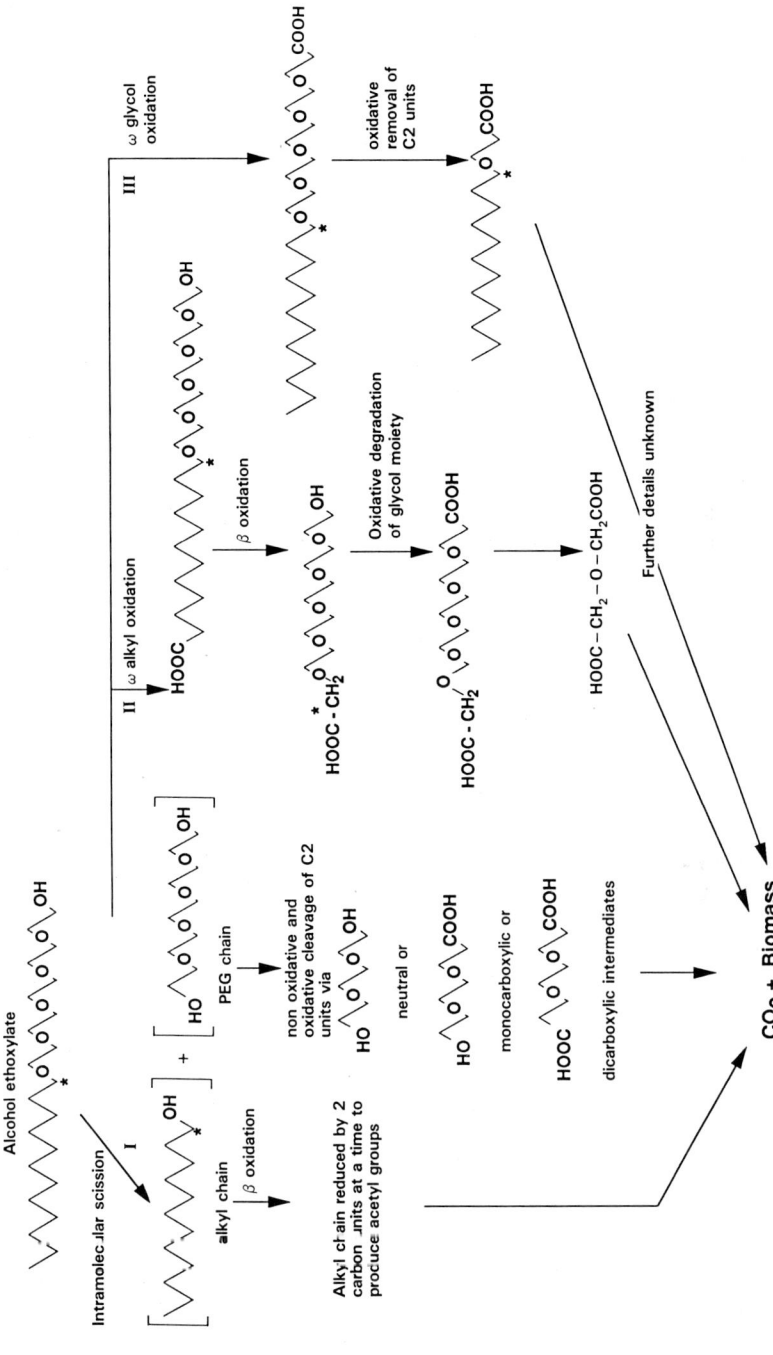

Fig. 9. Pathways of degradation of linear alcohol ethoxylates

radiolabel as CO_2, material adsorbed onto the sludge, carbon incorporated into biomass and in the plant effluent. Removal was 99% with 1% in the effluent. Their results confirmed the more rapid degradation of the alkyl chain but indicated that there were two distinct primary degradation pathways acting simultaneously—intramolecular scission coupled to either ω- or β-oxidation. They also identified a homologous series of neutral and acidic polyglycols as intermediates in the breakdown of the hydrophilic moiety (see Fig. 9).

Pure culture and cell lysate studies with a sewage isolate *Pseudomonas* sp. DES 1 were described by Griffiths et al. [147] in an attempt to identify hydrophobic metabolites produced during the biodegradation of $C_{12}E_3$. They concluded that the alkyl chain was metabolised to CO_2 via a large number of intermediates, all of which occurred in small amounts. The absence of significant accumulations or precursor-product relationships made it unclear which class of compound was the product of the primary degradation step. A study (Federle and Schwab [148]) has been published relating the mineralisation rates of $C_{15}E_7$ by the microbiota associated with duckweed and cattail from either a pristine pond or a pond which received wastewater from a laundromat. The microbiota mineralised the LAE without a lag phase with half-lives of 5 and 2–3 days for cattail and duckweed respectively regardless of their source. In a second study (Federle and Ventullo [149]), half-lives of 6.4–8.4 days were found with oak leaf detritus from the pond receiving the laundromat wastewater.

Information on the anaerobic degradation of nonionics is limited. Preliminary experiments on the anaerobic degradation of AE were described by Grant and Payne [150] who found that the linear C12–15 alcohols with 3, 5 or 7 EO units could support the growth of *Alcaligenes faecalis*. Wagener and Schink [151, 152] studied the anaerobic biodegradation and inhibitory effects of AE and APE on methanogens in sediments and sewage sludges. Nonionics enhanced methane production by anoxic sediments with 80 and 70% of the C12–23 EO and C10/12–7.5 EO respectively converted to methane after 37 days incubation. In contrast the APE exhibited only 45–50% conversion to methane which the authors say reflects the fact that only the polyglcol chain was degraded. In their second paper the biodegradation of nonionics was studied with two pure cultures isolated from anaerobic sewage sludge. A metabolic pathway was proposed (see Fig. 10) in which the glycol chain is degraded, by a stepwise release of C_2 units as acetaldehyde, to a fatty acid which then undergoes further degradation by other bacterial strains in the sludge. These two strains could not metabolise nonionics containing polypropylene glycols.

Stearyl alcohol ethoxylate radiolabelled with ^{14}C in either the alkyl or heptaglycol chain was used in anaerobic degradation studies in a model sludge system by Steber and Wierich [153]. After 4 weeks incubation at 35 °C more than 80% of the radioactivity was found as methane or carbon dioxide and a further 10% was incorporated into biomass. In contrast to the conclusions of Wagener these authors suggest that the first step is the intramolecular scission of the nonionic into alkyl and PEG moieties followed by β-oxidation of the alkyl chain and stepwise removal of C_2 units from the PEG. Biodegradation of linear AE and secondary AE have been shown to result in the release of significant amounts

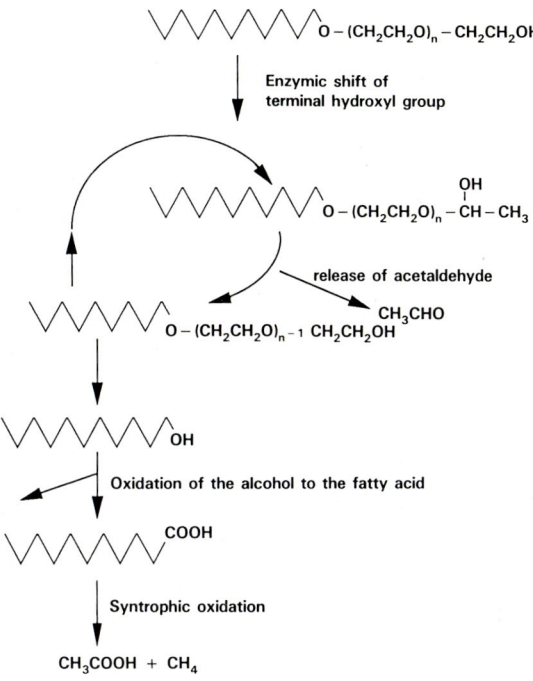

Fig. 10. Proposed pathway for anaerobic degradation of AE (Wagener and Schink 1988) [152]

of PEG, (for example Birch [154]), indicating the central fission pathway. However no PEG is produced in the biodegradation of BAE indicating an ethoxylate oxidation mechanism to give BAE with shorter EO chains. As will be seen later this parallels one of the suggested mechanisms for alkylphenol metabolism.

Polyglycol Degradation

Table 7 summarises the studies reported on PEG biodegradation. For a review of the recent advances in the biochemistry of polyglycol degradation the reader is referred to Cox [155] and Kawai [156] in which details of the biodegradation pathways for PPG are also given. In the last 10 years studies have shown that PEGs up to M.Wt. 20 000 are degraded not only aerobically but also anaerobically. The pathway for aerobic biodegradation involves oxidation of the terminal primary alcohol followed by depolymerisation and acidic intermediates have been identified in culture filtrates (Watson and Jones [157], Jenkins et al. [158], Thelu et al. [159] and Steber and Wierich [146]). Although other routes have been suggested conclusive evidence has not been reported. Several authors have reported anaerobic degradation of PEGs however the metabolic pathway remains unclear. Dwyer and Tiedje [160] suggest that PEG is first transformed to EG and then fermented to acetate and alcohol whereas Schink and Stieb [161] propose a pathway in which PEG undergoes depolymerisation from either end but does not involve ethylene glycol. The B12 mediated ethanediol dehydratase that takes ethylene glycol to acetaldehyde may have a broader specificity that encompasses EO polymers or a similar mechanism by a specific enzyme may be invoked to explain the observed reactions.

Alkylphenol Ethoxylates

In contrast to alcohol ethoxylates, the surfactants derived from straight or branched chain alkylphenols are not as rapidly or completely biodegraded. APE have been and still are the focus of discussion on environmental acceptability. They undergo substantial primary biodegradation however the pathway of ultimate biodegradation remains unproven (see Fig. 11). Biodegradation of APE is initiated by hydrophile attack at the ethoxylate end in which the EO_n chain is sequentially shortened by one EO group at a time. In some instances the glycol retains its terminal hydroxyl group as it is shortened (Schöberl et al. [162]). These authors were unable to clarify the pathway for the catabolism of the

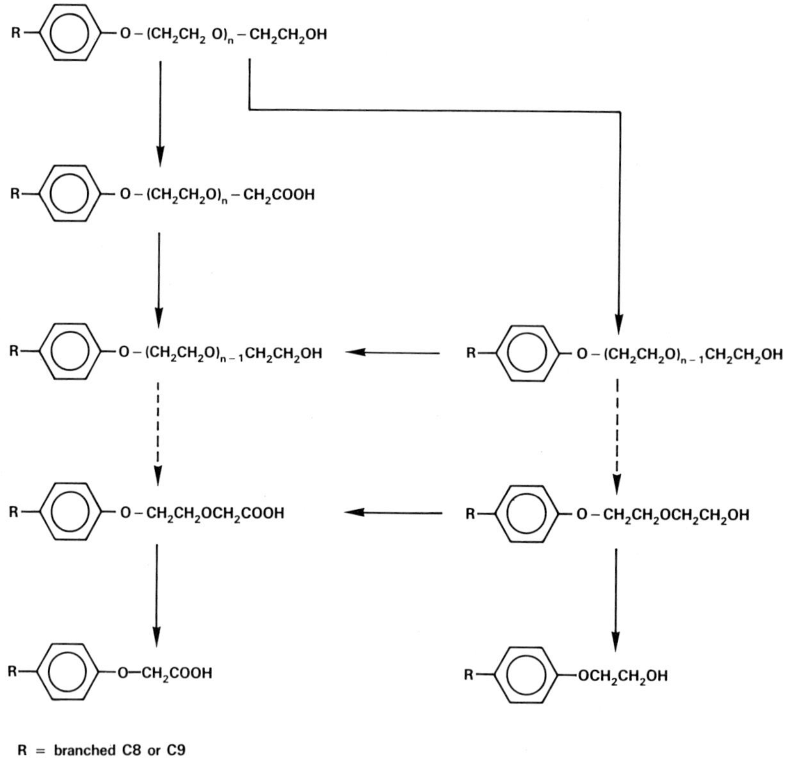

R = branched C8 or C9

Fig. 11. Proposed pathways for biodegradation of APE

nonylphenol nucleus, but in other cases the hydrophile is oxidised to the carboxylic acid. Using a pure culture capable of growth on NP9EO they have identified a hydrolase system which they claim was responsible for the stepwise cleavage of 1 ethylene oxide molecule. Brueschweiler and Gaemperle [163, 164] studying the degradation of APEs demonstrated the formation of NP2EO and also carboxylated NP2EO as a second relatively stable intermediate in laboratory simulation tests. Small amounts of NP were also identified as a metabolite.

The biodegradability of ethoxylates based on iso-octylphenol (8–9 EO) has been evaluated by field trials at a trickling filter plant serving a small community in the U.K. (Mann and Reid [165]). Removal increased from 26% in March to a level of >80% in late August and September. During November and December removal fell until by the end of January it was only 20–25%. The reduction in removal corresponds to a fall in temperature of the sewage filter bed from 15 to 5 °C. The effect of temperature on biodegradability was studied in 'porous pots' at 15, 11 and 8 °C over a period of 122 days by Stiff et al. [166]. At 20 mg/l the percentage removal (analysed by TLC) was high during the period at 15 °C but at 11–12 °C and 8 °C varied between 40–95% and 20–80% respectively suggesting that at 12 °C and below a stable population of APE degrading organisms could not be maintained. Schöberl and Mann [167] studied the biodegradation of NPE in freshwater and sea water at two temperatures over a period of 45 days. Primary biodegradation of 33–36% at 20–23 °C in fresh water and >95% in sea water were reported. At 3–4 °C it was degraded by 37% in fresh water and only 15% in sea water. In a more recent experiment (Kravetz et al. [168]) the impact of simulated cold weather on both primary and ultimate biodegradability of NP9.8E at 5 mg/l in domestic sewage has been studied. Continuous bench scale activated sludge studies have shown primary biodegradation at 12 °C was 98% but at 8 °C was 84% and that the effluent produced at 8 °C had high foaming properties. Ultimate biodegradation studies showed that the conversion of the ^{3}H labelled aromatic ring hydrophobe of NPE into ^3H_2O decreased from 20 to 2% in changing the temperature from 25 to 8 °C. Conversion of the ^{14}C labelled polyethoxylate hydrophile to $^{14}CO_2$ decreased from 60% at 25 °C to 50% at 12 °C and to only 10% at 8 °C. In contrast the biodegradation of the hydrophobe of LAE was essentially unaffected by the lower temperature. In an earlier series of reports Kravetz et al. [138, 169] studied the ultimate biodegradation of NPE in a model activated sludge plant using similarly labelled material as above. The rate of release of tritium as tritiated water indicated that the ultimate degradation of the hydrophobe was slow, however evolution of $^{14}CO_2$ from the POE chain was rapid. Between 35–50% of the hydrophobe (as soluble ^{3}H labelled intermediates) was discharged in the effluent. A further 8–14% of soluble ^{14}C labelled intermediate was also detected in the effluent. The EO:hydrophobe ratio for NP9EO showed a fall from an initial value of 5.4:1 to 2.2–2.4:1 after only 8 hours after which a steady state ratio of 2.4:1 was reached. They postulated a mechanism of biodegradation that involved the stepwise shortening of the POE chain to leave a relatively resistant NP2EO. Yoshimura [48] studied the biodegradation of $C_{16}EO_7$ and NPE_9 in river water and sediments. In stirred sediments >98% AE and NPE disappeared within 5 days whereas it took 10 days to achieve the same extent of degradation in unstirred systems. Nonylphenol acetic acid and nonylphenol ethoxy-carbonic acids with 1–3 mol EO were formed in the biodegradation of NPE in river water (Hellman [170]).

Evidence for the presence of significant amounts of NP, NP1EO, NP2EO and their carboxylated equivalents in sewage treatment and the receiving environment are given in the section on environmental concentrations. Although categorical evidence that NP is an intermediate in NPE biodegradation has yet to be produced, Giger [56] has suggested that the high concentrations of NP in anaerobic sludge are proof that anoxic conditions lead to the conversion of NP2EO and NP1EO into NP. Neufahrt et al. [171] investigated the significance of the incomplete mineralisation of NPE on the receiving environment. He concluded that primary degradation of >95% occurs in well operated activated sludge plants and that the concentrations of NPE, NP, NP1EO, and NP2EO in the outflow are on average 460, 0.5, 5, and 20 $\mu g/l$ respectively. From their studies it was not possible to detect any radical impairment of the aquatic flora and fauna caused by these discharges and they conclude that there is no justification in preventing the use of APE in domestic detergent formulations.

Studies by Khalil et al. [172] showed that NPE at concentrations up to 100 ppm had no effect on the rate of methanogenesis. However Ball et al. [173] indicates that halogenated OPE which can be formed as a result of the chlorination of sewage inhibited a methanogenic consortium at a concentration of 60 ppm. They also report on the aerobic and anaerobic degradation of OPE and halogenated OPEs. Under anaerobic conditions there was partial transformation of OPE to OP and recalcitrant OPEC compounds over a 190 day period. Halogenated residues were transformed into recalcitrant metabolites such as halogenated OP2EC and halogenated OP. No intermediates from alkyl chain or phenol ring degradation were found but under aerobic conditions degradation of OPE and OPEC was said to be complete after 127 days.

Other Nonionics

Biodegradation of a number of nonionic dispersants, which have sugars or sugars and poly EO as the hydrophilic moiety, in sea water were described by Una and Garcia [174]. Pure cultures were isolated from sea water following an oil spill. Degradation was said to be directly related to the polyethoxy content and ranged from about 70% for the Tween series (polyoxyethylene sorbitan fatty acids) to 50% for the Span (no polyoxyethylene) series.

Ecotoxicology of Nonionic Surfactants

There are no well supported general trends in the relationship between measured toxicity and the molecular structure of nonionic surfactants and we are left with a rather general picture of molecules that interfere with membrane processes. This interference is thought to arise by disruption of the complex membrane structure by the surfactancy of the materials rather than disruption of some specific molecular recognition or receptor site.

The nature of the toxic effects of nonionic surfactants depends in part upon the type of organism that is exposed to them and may be behavioural or physiological. For example, the gill lamellae of fish may be affected by oedema and mucosis with a possible narcotic effect on motor control while in bivalves the ion balance, osmoregulation and respiration may be disrupted. In the crayfish *Astacus fluviatilis* it has been shown that the receptor potential of the stretch receptor neuron is inhibited due to selective but reversible action by AE's on the sodium channels in the membrane (Ottoson and Rydqvist [211]). Modification of the habitat itself may also be important; mosquito pupae may drown as a consequence of reduced surface tension by surfactants in water bodies. There does not appear to be any one mechanism other than the general "surfactancy" process to explain these effects.

AE's and APE's, the two major groups of nonionic surfactants, provide an interesting discussion on environmental acceptability. The parent molecules of APE's are generally less toxic than equivalent AE's but the reverse is true of their biodegradation intermediates, where APE intermediates are more toxic than AE intermediates. The former possess a shortened ethoxylate chain and retain the surfactancy inherent in these structures. Furthermore, APE intermediates, which possess a reduced number of ethoxylate groups, are also more toxic than their parent compounds. The cleavage of AE's into alcohol and glycol sub-units, on the other hand, removes their surface activity. Decreasing toxicity appears to be related, however, both to an increase in the ethoxylation of a hydrophobe and also in changing the hydrophobe from an alcohol to an alkylphenol.

Scott Hall et al. [219], for example, carried out exposure studies with the estuarine crustacean *Mysidopsis bahia* and 17 ethoxylated surfactants and concluded that materials with high ethoxylation (30–50 EO units) were consistently of low toxicity. The hydrophobe was not thought to be an important contributor to the toxicity of the pure product. Such information will be of use in qualitative predictions of biological effects of these materials.

Table 8 contains data on the toxicity of AE's and APE's to algae. Lewis [214] compared laboratory and field data on the exposure of natural assemblages of phytoplankton to an OPE and found, by measuring parameters such as population density and diversity index, that observed effect concentrations in the field were thirty two times higher than in the laboratory. Yamane [215] studied the growth inhibition of freshwater planktonic algae (a green alga, *Selenastrum capricornutum*, a blue-green alga, *Nitszchia fonticola* and a diatom, *Microcystis aeruginosa*) and found that APE's were less toxic than AE's. It was also concluded that effects on algae were species-specific; the most sensitive species for one surfactant was not always the most sensitive for another. Discrepancies in some of the comparative data for algae are considerable. For example, the effect concentration of an OPE on the growth of *Selenastrum* ranged from 0.2 mg/l

Table 8. Acute and chronic toxicity of nonionic surfactants to algae

Surfactant	Test organism/system	Acute (mg/L)	Chronic (mg/L)	Reference/Notes
APE				
Rokaphenol N-8	*Scenedesmus quadricauda*	ND	8 d (mg dm^{-3}) N-8; 17 NK-8; 9.5	Solski and Erndt [224]
Rokaphenol NK-8	periphyton	ND	4 weeks N-8; 3.3 NK-8; 2.6	development inhibited
OPE	*Selenastrum*	96 hr EC50 0.21	ND	Lewis [214]
	Mycrocystis	7.40	ND	first acute effects at 0.1 mg/L
	lake phytoplankton community. 82 species identified. 13 'major' species.	ND	10 d exposure in enclosures	no effect below 3.2 mg/L (range tested =.26 to 40 mg/L)

		EC50	
APE			Yamane et al. [215]
Emulgen 910	Selenastrum capricornutum	20	ND
Emulgen 909		50	
AE			
AE(EO:4)	S. capricornutum	2 to 4	ND
AE(EO:9)*		4 to 8	
AE(EO::3)		10	
			*three algae tested together (EC50s);
			Selenastrum = 4 to 8 mg/L
			Nitzschia fonticola = 5 to 10 mg/L
			Microcysistis aeruginosa = 10 to 50 mg/L
Sugar ester			
OWA 1570	S. capricornutum	8	ND
LWA 1540		12	ND
			Guhl and Gode [217]
AEO$_{10}$	algae	0.3	ND
AEO$_{14}$	,,	0.3	ND
AEO$_{33}$	,,	30	ND
NP-10	,,	1	ND
EOPO$_{24}$	,,	0.6	ND
EOPO$_{36}$	,,	0.7	ND
EOPO$_{45}$	,,	0.5	ND
AEO-Bu	,,	0.3	ND
AGlc	,,	6.5	ND

ND = no data

Table 9. Acute and chronic toxicity of nonionic surfactants to *Daphnia* spp.

Surfactant	Test organism	Acute (mg/L)	Chronic (mg/L)	Reference
C_{12-14} mono-methyl dihydroxy ethyl ammonium chloride	*D. magna*	48 hr LD_{50} 0.29 to 0.70	ND	Lewis and Perry [216]
AEO	*D. magna*	24 hr LC0	life cycle	Guhl and Gode [217]
AEO_{10}		1.4	3	
AEO_{14}		2.0	1	
AEO_{30}		700	10	
APEO				
NP10		10	1	
AEO-PO				
$EOPO_{24}$		2.0	ND	
$EOPO_{36}$		5.0	ND	
$EOPO_{45}$		2.8	ND	
AEO-Bu		0.7	0.3	
AGlc		59	1.9	
APE	*D. magna*	48 hr LC_{50}		Solski and Erndt [224]
Rokaphenol N-8		18.3	> 1.65	
Rokaphenol NK-8		19.5	> 1.3	

ND = no data

(Lewis, *op. cit.*) to 20 mg/l (Yamane *op. cit.*). The toxic effect of nonionics on algae is less than with charged surfactants.

The copepod *Daphnia magna* has been found to be particularly sensitive to nonionic surfactants in both acute and chronic studies (Table 9). For example Lewis and Perry [216], in work to determine the 48 h LD_{50} for an AE, found that *D. magna* was *the* most sensitive organism studied, providing a range for effects from 0.29 to 0.7 mg/l. These workers also demonstrated that the relationship between acute and chronic toxicity data for *D. magna* to a range of nonionic surfactants was not uniform. Toxic effects were reduced by end-capping of the AE. Lewis and Suprenant [218] subsequently provided 48 h LC_{50} data for six species of aquatic invertebrates (Table 10) using a C14–15 linear AE under static conditions in accordance with US EPA methods. Comparing their results with earlier work with *D. magna*, the copepod remained the most sensitive species. The authors concluded that the use of *D. magna* in screening toxicity tests would "provide a conservative estimate of surfactant toxicity for aquatic invertebrates and also for bluegill and fathead minnows ... If daphnids are absent then species either taxonomically or functionally similar to the sensitive *Gammarus*, *Dero* and *Dugesia* should first be examined if surfactants are of concern".

Toxicity data for fish are presented in Table 11. Guhl and Gode [217] tested several nonionic surfactants with varying structural characteristics against a non-identified fish species. It was observed that juvenile stages were only slightly more sensitive than adults, even during long term exposure. Alteration of the hydrophilic moiety modified toxicity; AE's with increasing ethoxylation of 10 EO, 14 EO and 30 EO yielded acute toxicity values of 1–3, 2.5 and 100 mg/l, respectively. Furthermore, toxicity was reduced by capping the terminal alcohol group. These authors concluded that fish were more sensitive than daphnids to nonionic surfactants.

Yoshimura [220] determined the LC_{50} of nonylphenol ethoxylate and its biodegradation intermediates to the Japanese kilifish (*Oryzias latipse*). The range of values was 1.4 to 110 mg/l but little chemical information was reported on the relative composition of the materials under study.

Turner et al. [221] working with rainbow trout (*Oncorhynchus mykiss*) determined the LC_{50} of two linear AE's and their degradation products in both clean and effluent water. It was shown that the C14–15 7 EO material was more toxic than its 11 EO oligomer. Toxicity was reduced by more than 95% by passing the materials through biological filters.

The use of nonionic surfactants as adjuvants in agrochemical formulations has prompted studies of their phytotoxicity to higher plants and data are presented in Table 12. AE's are extensively used in this field. Caux [222] studied a range of Tritons against *Lemna minor*, monitoring frond number, biomass and chlorophyll content. No effects were found at concentrations below 3.2 mg/l. Less pronounced effects occurred with higher ethoxylated products, one of the consistent structure-activity relationships that has been found across a wide range of genera. Solski [223] demonstrated that growth inhibition occurred after 30 days exposure to 6.5 and 8.3 mg/l of an AE.

AE's are more readily biodegradable than APE's and background levels in the environment are generally well below concentrations that could result in biological effects. Reiff [212] exposed rainbow trout to two AE's and one octylphenol ethoxylate before and after a river die-away experiment and demonstrated that biodegradation had effectively removed toxicity. Abram [213] used a model trickling filter to demonstrate the biodegradation and subsequent reduction in toxic effects of two AE's.

In conclusion, the majority of nonionic surfactants in normal use in domestic detergents yield LC_{50} values within the range 1–10 mg/l, well above the concentration of these materials present in sludge leaving sewage treatment plants. With further dilution in receiving waters and metabolism by microbes these relatively readily biodegradable materials should pose no environmental threat.

Addendum

Two recent reviews have dealt comprehensively with chronic and sub lethal toxicities of the major surfactant groups, including nonionics, to aquatic algae (Lewis, 1990) and animals (Lewis, 1991).

Lewis MA (1990) Ecotoxicol Environ Safety 20:123

Lewis MA (1991) Wat Res 25:101

Table 10. Acute and chronic toxicity of nonionic surfactants to macroinvertebrates

Surfactant	Test organism	Acute (mg/L)	Chronic (mg/L)	Reference
C14–15 AE		48 hr LC_{50}		Lewis and Suprenant [218]
	Paratanytarsus parthenogenica (midge)	5.0	ND	
	Gammarus sp. (amphipod)	1.4	ND	
	Asellus sp. (isopod)	6.2	ND	
	Dugesia sp. (flatworm)	1.0	ND	
	Dero sp. (oligochaete)	2.6	ND	
	Rhabditis sp. (nematode)	6.8	ND	
fatty acid ethoxylates	*Hydra attenuata* (hydrozoan)	total lethality after 24 hours at 2×10^{-1} mM	ND	Bode et al. [225]
APE Ortho X-77	*Chironomus riparius* (midge)	48 hr LC_{50} 8.6	ND	Buhl and Faerber [226]
APE Rokaphenol N-8 Rokaphenol NK-8		48 hr LC_{50} (mg/dm^{-3})		Solski and Erndt [224]
	Poecilia reticulata (copepod)	N-8; 16.4 NK-8; 12.7	>1.6 >1.3	
	Gammarus pulex (amphipod)	N-8; 20.9 NK-8; 24.3	ND ND	
	Tubifex tubifex (oligochaete)	ND ND	N-8; >3.3 NK-8; >6.5	
Hyoxid 1011	*Coleoptera Cybister tripunctatus*	0.5 hr LD 100		Shirgur [227]

adult	50.0	ND
larva	5.0	ND
C. cognatus	40.0	ND
C. sugillatus	35.0	ND
Heteroptera		
Sphaerodema rusticum	10.0	ND
Anisops barbata	1.0	ND
Ranatra elongata	0.75	ND
Laccotrephes ruber	0.75	ND
Odonata, naiads		
Bradynopyga geminata	$>10^4$ (no effect after 96 h)	ND
Idictinogomphus rapax	8×10^4	ND
Sympetrum sp.	7×10^4	ND

	48 hr LC_{50}		
Mysidopsis bahia (estuarine crustacean)			Scott Hall et al. [219]
linear $APEO_{1.5}$	2 to 7	ND	
linear $APEO_9$	2 or less	ND	
linear $APEO_{50}$	least toxic of series	ND	
tp-$NPEO_{1.5}$	2 or less	ND	
tp-$NPEO_9$	2 or less	ND	
tp-$NPEO_{15}$	2 to 7	ND	
tp-$NPEO_{40}$	>100	ND	
tp-$NPEO_{50}$	least toxic of series	ND	
$OPEO_{1.5}$	2 to 7	ND	
$OPEO_5$	2 or less	ND	
$TDAEO_{9.75}$	2 to 7	ND	
$TDAEO_{10}$	2 or less	ND	

ND = no data

Table 11: Acute and chronic toxicity of nonionic surfactants to fish

Surfactant	Test organism	Acute (mg/L)	Chronic (mg/L)	Reference/notes
AEO	unspecified	48 hr-LC0		Guhl and Gode [217]
AEO_{10}		1 to 3	ND	
AEO_{14}		2.5	ND	
AEO_{30}		100	ND	
APEO				
NP-10		5.5	ND	
AEO-PO		–		
$EOPO_{24}$		1.5	ND	
$EOPO_{36}$		1	ND	
$EOPO_{45}$		1	ND	
AEO-Bu		0.3	ND	
AGlc		3.7	ND	
C_9APE_9	Japanese killifish (*Oryzias latipse*)	48 hr LC_{50} 1.4 to 110.0	ND	Yoshimura [220]
APE X-77	bluegill (*Lepomis macrochirus*)	24 hr LD_{50} 5.5 (range 5.1–6.0) 96 hr LD_{50} 5.5 (range 4.9–6.1)	ND	Watkins et al. [228]
a. Dobanol 45-7	rainbow trout (*Oncorhynchus mykiss*)	96 hr LC_{50} a i. 0.78 (range 0.72–0.84) a ii. 0.75 (range 0.46–1.21) b i. 1.08 (range 0.83–1.41) b ii. 1.10 (range 0.97–1.24)	ND ND ND ND	Turner et al. [221] diluent = i. unchlorinated ground water ii. effluent form a sewage trickling filter
b. Dobanol 45-11				

	fathead minnows (*Pimephales promelas*) 96 hr LC$_{50}$	bluegill (*Lepomis macrochirus*) LC$_{50}$; 24 hr	96 hr	
Neodol 45-7	* i. 1.2 ii. 1.38 iii. 2.48			Maki and Rubin [229] acute toxicity completely eliminated by secondary waste-water treatment *diluent = i. carbon-filtered tap-water ii. creek water iii. secondary effluent
Surfonic N-40	ND	1.5	1.3	Macek and Krzeminski [230]
Igepal CO-520	ND	2.8	>2.4 <2.8	ND
Triton X-45		3.5	>2.8 <3.2	ND
Surfonic N-95		7.8	7.6	ND
Igepal CO-630		8.9	7.9	ND
Triton X-100		16.2	12.0	ND
Igepal CO-880		>1000	>1000	ND
Triton X-305		1080	531	ND
Neodol 25-3		1.8	1.5	ND
Neodol 25-9		2.1	2.1	ND
Alfonic 1012-60		6.4	6.4	ND
Surfonic TD-90		7.8	7.5	ND
Tergitol 15-S-9		4.7	4.6	ND

ND = no data

Table 12. Phytotoxicity of nonionic detergents

Surfactant	Test organism	Acute (mg/L)	Chronic (mg/L)	Reference/notes
	Lemna minor (duckweed)			
Triton X-15		ND	14 d	Caux et al. [222]
Triton X-35		ND	exposure	
Triton X-100		ND	no response	parameters =
Triton X-114		ND	at 1.0 µg/L	
				a. growth(biomass)
			at 10.0 µg/L	b. frondfluorescence and
			frond	chlorophyll content
			development	c. conductivity of the testmedia
			depressed by	d. specificion leakage
			25–50%(except	
			with X-114)	
				Solski and Erndt [223]
		ND	4 weeks	
			exposure	
a. Rokaphenol N-8	*L. minor*		a. 8.3	increase in biomass inhibited
b. Rokaphenol NK-8			b. 6.5	
	Ceratophyllum demersum (rigid hornwort)		a. 3.3	,, ,,
			b. 2.6	

ND = no data

References

1. Maag HH (1981) In: Stache H (ed) Tensid taschenbuch. Hanser, Munchen
2. Bocker T, (1989) Tenside 26:318
3. Donbrow M (1987) In: Schick MJ (ed) Nonionic surfactants—physical chemistry. Surfactant Series No 23. Dekker, New York
4. Rosen MJ (1989) Surfactants and interfacial phenomena. Wiley, New York
5. Schick MJ (1987) (ed) Nonionic surfactants—physical chemistry. Surfactant Series No 23. Dekker, New York
6. Swisher RD (1987) Surfactant biodegradation. Surfactant Science Series No 18. Dekker, New York
7. Llenado RA, Jamieson RA (1971) Anal Chem 53:174R
8. Llenado RA, Neubecker TA (1983) Anal Chem 55:93R
9. Thomas JDR (1982) Anal Tech Enviro Chem 133 Pergamon Ser Environ Sci
10. Garti N, Kaufman VR, Aserin A (1983) Sep Purif Methods 12:49
11. Sedlak RI, Booman KA (1986) Soap/Cosmetics/Chemical Specialities April, 44 and 107
12. Kubeck E, Naylor CG (1990) J Am Oil Chem Soc 67:400
13. Sheldon LS, Hites RA (1978) Environ Sci Technol 12:1188
14. HMSO (1986) The sampling and initial preparation of sewage and waterworks sludges, soils, sediments, plant material and contaminated wildlife prior to analysis. In Methods for the Examination of Waters and associated materials, 2nd edn, HMSO, London.
15. Marcomini A, Stelluto S, Pavoni B (1989) J Environ Anal Chem 35:207
16. Jones P, Nickless G (1978) J Chromatog 156:87
17. Schmitt TM, Allen MC, Brain DK, Guin KF, Llenado RA, Osburn QW (1990) J Am Oil Chem Soc 67:103
18. Neufahrt A, Hofman K, Tauber G (1984) J Comun Jorn Esp Deterg 15:123
19. Saito T, Hagiwara K (1982) Fresenius Z Anal Chem 312:533
20. Thurman EM, Willoughby T, Barber LB, Thorn KA (1987) Anal Chem 59:1798

21. Valls M, Fernandez P, Bayona JM, Albaiges J (1989) In: Quaghebeur D, Temmerman I, Angeletti G (eds) Organic contaminants in waste water, sludge and sediment. Elsevier, London
22. Marcomini A, Capri S, Giger W (1987) J Chromatogr 403:243
23. Marcomini A, Stelluto S, Pavoni B (1989) J Environ Anal Chem 35:207
24. Gorenc B, Gorenc D, Keber I, Pihlar B (1986) Vestn Slov Kem Drus 33:11
25. Kozarac Z, Zutic V, Cosovic B (1976) Tenside Surf Det 13:260
26. Stephanou E, Giger W (1982) Environ Sci Technol 16:800
27. Shiraishi H, Otsuki A, Fuwa K (1982) Bull Chem Soc Jpn 55:1410
28. Veith GD Kiwus LM (1977) Bull Environ Contam Toxicol 17:631
29. Ahel M, Giger W (1987) Environ Sci Technol 21:697
30. Wickbold R (1971) Tenside Surf Det 8:61
31. Wickbold R (1972) Tenside Surf Det 9:173
32. HMSO (1982) Methods for the examination of waters and associated materials, 'Analysis of Surfactants in Waters, Wastewaters and sludges
33. Waters J, Longman GF (1977) Analytica Chim Acta 93:341
34. Waters J, Garrigan JT, Paulson AM (1986) Wat Res 20:247
35. Gledhill WE, Huddleston RL, Kravetz L, Nielsen AM, Sedlak RI, Vashon RD (1989) Tenside Surf Det 26:276
36. Boyer SL, Guin KF, Kelley RM, Mausner LM, Robinson HF, Schmitt TM, Stahi CR, Setzkorn EA (1976) Environ Sci Tech 11:1167
37. Nozawa A, Oknuma T, Sekine T (1976) Analyst 101:543
38. Favretto L, Stancher B, Tunis F (1978a) Analyst 103:955
39. Favretto L, Stancher B, Tunis F (1978b) Proc Intern Congr Natural Resources (Del Bianco, Udine), 717
40. Favretto L, Stancher B, Tunis F (1981) Intern J Environ Anal Chem 10:23
41. Favretto L, Stancher B, Tunis F (1983) Intern J Environ Anal Chem 14:201
42. Assanelli P, Burzio F, Calcinai D, Favretto L, Gemiti F, Gori M, La Noce T, Luciani L, Perin G, Princi M, Ruffo C, Stancher B, Tonti AM, Tunis F, Valentiniis G (1984) Riv Ital Spostanze Grasse 61:633
43. Wee VT (1981) Adv Identif Anal Org Pollut Water 1:467
44. Tobin RS, Onuska FI, Brownlee BG, Anthony DHJ, Comba ME (1976) Wat Res 529
45. Allen MC, Linder DE (1981) J Am Oil Chem Soc 58:950
46. Nozawa A, Ohnuma T (1980) J Chromatogr 187:261
47. Kudoh M, Ozawa H, Fudano S, Tsuji K (1984a) J Chromatogr 287:337
48. Yoshimura H (1986) J Am Oil Chem Soc 63:1590
49. Giger W, Stephanou E, Schaffner C (1981) Chemosphere 10:1253
50. Stephanou E (1984) Chemosphere 13:43
51. Rivera J, Ventura F, Caixach J, De Torres M, Figueras A (1987) Intern J Environ Anal Chem 29:15
52. Ventura F, Figueras A, Caixach J, Espalder I, Romero J, Guardiola J, Rivera J (1988) Wat Res 22:1211
53. Ahel M, Giger W (1985a) Anal Chem 57:1577
54. Ahel M, Giger W (1985b) Anal Chem 57:2584
55. Holt MS, McKerrell EH, Perry J, Watkinson RJ (1986) J Chromatogr 362:419
56. Giger W, Brunner PH, Schaffner C (1984) Science 225:623
57. Wahlberg C, Renberg L, Wideqvist U (1990) Chemosphere 20:179
58. Stephanou E (1986) Comm Eur Communities (Eur). Org Micropollut Aquat Environ 155
59. Chlebicki J, Garncarz W (1980) Tenside Surf Det 17:13
60. Crisp PT, Eckert JM, Gibson NA (1979) Anal Chim Acta 104:93
61. Grasso G, Bufalo G (1986) Atomic Spectroscopy 7:93
62. Le Bihan A, Courtot-Coupez J (1977) Anal Lett 10:759
63. Van Hoof FM, Van Craenenbroeck WJ, Dewaele JK (1985) Intern J Anal Chem 19:155
64. Julia-Danes E, Casanovas AM (1979) Tenside Surf Det 16:317
65. Schneider E, Levsen K, Dahling P, Rollgen FW (1983) Fresenius Z Anal Chem 316:277
66. Szymanowski J, Szewczyk H, Hetper J (1981) Tenside Surf Det 18:333
67. Tsuji K, Konishi K (1974) J Am Oil Chem Soc 51:55
68. Weber R, Levsen K, Louter GJ, Boerboom AJH, Haverkamp J (1982) Anal Chem 54:1458
69. Reinhard M, Goodman NL, Barker JF (1984) Environ Sci Technol 18:953
70. Rivera J, Fraisse D, Ventura F, Caixach J, Figueras A (1987) Fres Z Anal Chem 328:577
71. Sheldon LS, Hites RA (1979) Sci Total Environment 11:279
72. Shiraishi H, Otsuki A, Fuwa K (1985) Biomed Mass Spectrometry 12:86

73. Stephanou E (1985) Intern J Environ Anal Chem 19:155
74. Stephanou E, Reinhard M, Ball HA (1988) Biomed Environ Mass Spectrom 15:275
75. Ventura F, Caixach J, Figueras A, Espalder I, Fraisse D, Rivera J (1989) Wat Res 23:1191
76. Yashuhara A, Shiraishi H, Tsuji M, Okuno T (1981) Environ Sci Technol 15:570
77. Kudoh M, Kotsuji M, Fudano S, Tsuji K (1984b) J Chromatogr 295:187
78. Yoshimura H, Sugiyama T, Nagai T (1987) J Am Oil Chem Soc 64:550
79. Bear GR (1988) J Chromatogr 459:91
80. Aserin A, Frenkel M, Garti N (1984) J Am Oil Chem Soc 61:805
81. Benning M, Locke H, Ianniello R (1989) J Liquid Chromatog 12:757
82. Escott REA, Brinkworth SJ, Steedman TA (1983) J Chromatogr 282:655
83. Nakamura K, Morikawa Y, Matsumoto I (1981) J Am Oil Chem Soc 58:72
84. Pellzetti E, Minero C, Maurino V, Sclafani A, Hidaka H, Serpone N (1989) Environ Sci Technol 23:1380
85. Rothman AM (1982) J Chromatogr 253:283
86. Marcomini A, Giger W (1987) Anal Chem 59:1709
87. Kalinoski HT, Jensen A (1989) J Am Oil Chem Soc 66:1171
88. Stolzenberg GE, Zaylskie G, Olson PA (1971) Anal Chem 43:908
89. Hei RD, Janisch NM (1989) Tenside Surf Det 26:288
90. Holmqvist P (1977) Anal Chim Acta 90:35
91. Jones DL, Moody GJ, Thomas JDR, Birch BJ (1981) Analyst 106:974
92. Tsubouchi M, Yamasaki N, Yanagisawa K (1985) 57:783
93. Crisp PT, Eckert JM, Gibson NA, Webster IJ (1981) Anal Chim Acta 123:355
94. Chester TL (1989) Chromatogr Sci 45:369
95. Geissler PR (1989) J Am Oil Chem Soc 66:685
96. Matsumoto K, Tsuge S, Hirata Y (1987) Mass Spectroscopy 35:15
97. Onuska FI, Terry KA (1988) J High Resolut Chromatogr Commun 11:874
98. Brown D, De Henau H, Garrigan J, Gerike P, Holt M, Keck E, Kunkel E, Matthijs E, Waters J, Watkinson RJ (1986) Tenside Surf Det 23:190
99. Brown D, De Henau H, Garrigan J, Gerike P, Holt M, Keck E, Kunkel E, Matthijs E, Waters J, Watkinson RJ (1987) Tenside Surf Det 24:14
100. Giger W, Brunner PH, Ahel M, McEvoy J, Marcomini A, Schaffner C (1987) Gas, Wasser, Abwasser 66
101. Bock KJ (1973) Tenside Surf Det 10:178
102. Zoller U, Romano R (1983) Environment International 9:55
103. Klopp R (1987) gwf-Wasser/Abwasser 128:117
104. Narkis N, Ben-David B, Schneider-Rotel M (1987) Tenside Surf Det 24:200
105. Narkis N, Weinberg H (1989) Tenside Surf Det 26:400
106. Brunner PH, Capri S, Marcomini A, Giger W (1988) Water Res 22:1465
107. Kunkel E (1987) Tenside Surf Det 24:280
108. Ball HA, Reinhard M (1985) In Water chlorination; chemistry, environmental impact, and health effects; Jolley RL, et al., (ed), Lewis Publishers: Chelsea MI, 5:1505
109. Reinhard M, Goodman NL, Mortelmans KE (1982) Environ Sci Technol 16:351
110. Giger W, Ahel M, Koch M (1986) Vom Wasser 67:69
111. Patterson SJ, Scott CC, Tucker KBE (1967) J Inst Wat Poll Contr 3:3
112. Zoller U (1989) Tenside Surf Det 26:394
113. Gerike P, Winkler K, Jakob W (1989) Tenside Surf Det 26:270
114. Office of Health Studies (1978) Environmental Health Department, Japanese Environmental Agency, Chemicals Assessment Annual Report, 96
115. Office of Health Studies (1982) Environmental Health Department, Japanese Environmental Agency, Chemicals Assessment Annual Report, 66 and 78
116. Ahel M, Giger W, Molnar-Kubica E, Schaffner C (1984) Analysis of organic micropollutants in water pp 280–288. G Angeletti and A Bjorseth (eds), Reidel Publ Comp, Dordecht, Holland
117. Ahel M, Conrad T, Giger W (1987) Environ Sci Technol 21:697
118. Schaffner C, Stephanou E, Giger W (1982) Comm Eur Communities (Rep) Eur Anal Org Micropollut Water 330
119. Favretto L, Stancher B, Tunis F (1980) In: Technological, environmental and economic trends in detergency, Proc Intern Symp, Rome 2, 308
120. Marcomini A, Capel RD, Lichtensteiger T, Brunner PH, Giger W (1989a) J Environ Qual 18:523
121. Arpino A, Ruffo C, Jacini G (1973) Riv Ital Sost Grasse 50:345
122. Arpino A, Ruffo C, Jacini G (1974) Riv Ital Sost Grasse 51:140

123. Pitter P, Mangerova B (1974) Acta Hydrochim Hydrobiol 2:465
124. Fischer WK, Winkler K (1976) Vom Wasser 47:81
125. Fischer WK (1980) Tenside Surf Det 17:250
126. Kupfer W (1979) X Jornadas Com Espanol Deterg, Barcelona 141
127. Otsuki A, Shiraishi H (1979) Anal Chem 51:2329
128. Cabridenc R (1980) Tenside Surf Det 17:242
129. Saito T, Hagiwara K (1983) Fresenius Z Anal Chem 315:201
130. Berth P, Gerike P, Gode P, Steber J (1988) Tenside Surf Det 25:108
131. Gerike P, Winkler K, Schneider W, Jacob W (1989) Tenside Surf Det 26:21
132. Sheldon LS, Hites RA (1979) Env Sci Technol 13:574
133. Gilbert PA, Kleiser HA (1986) Paper presented at 32nd Referate-Tagung Waschereiforschung, Krefeld, April 1986
134. Kravetz L (1990) ACS Symposium Series 433:96
135. Gledhill WE (1975) Appl Microbiol 30:922
136. Naylor CG, Castaldi FJ, Hayes BJ (1988) J Am Oil Chem Soc 65:1669
137. Henkel (1981) Fatty alcohols: Raw materials, methods, uses. Henkel KGaA, Dusseldorf, West Germany 1981 pp 193
138. Kravetz L, Chung H, Rapean JC, Guin KF, Shebs WT (1978) Presented at American Oil Chemists' Society 69th Annual meeting, St Louis, May 1978
139. Nooi JR, Testa MC, Willemse S (1970) Tenside Surf Det 7:61
140. Larson RJ, Games LM (1981) Environ Sci Technol 15:1488
141. Larson RJ (1984) Household Pers Prod Ind 21(3):55–58, 84; (4):73
142. Larson RJ, Wentler GE (1982) Soap Cosm Chem Specialities 58:33
143. Vashon RD, Schwab BS (1982) Environ Sci Technol 16:433
144. Neufahrt A, Lotzsch K, Gantz D (1982) Tenside Surf Det 19:264
145. Steber J, Wierich P (1983) Tenside Surf Det 20:183
146. Steber J, Wierich P (1985) Appl Environ Microbiol 49:530
147. Griffiths ET, Hales SG, Russell NJ, White GF (1987) Biotech and Appl Biochem 9:217
148. Federle TW, Schwab BS (1989) Appl Environ Micro 55:2092
149. Federle TW, Ventullo RM (1990) Appl Environ Micro 56:333
150. Grant MA, Payne WJ (1983) Biotechnol Bioeng 25:627
151. Wagener S, Schink B (1987) Wat Res 21:615
152. Wagener S, Schink B (1988) Appl Env Microbiol 54:561
153. Steber J, Wierich P (1987) Wat Res 21:661
154. Birch RR (1984) J Am Oil Chem Soc 61:340
155. Cox DP (1978) Advances in applied microbiology 23:173
156. Kawai F (1987) Crit Rev Biotech 6:273
157. Watson GK, Jones N (1977) Wat Res 11:95
158. Jenkins LDL, Cook KA, Cain RB (1979) J Appl Bacteriol 47:75
159. Thelu J, Medina L, Pelmont J (1980) FEMS Microbiol Lett 8:187
160. Dwyer DF, Tiedje JM (1983) Appl Environ Microbiol 46:185
161. Schink B, Stieb M (1983) Appl Environ Microbiol 45:1905
162. Schoberl P, Kunkel E, Espeter K (1981) Tenside Surf Det 18:64
163. Brueschweiler H, Gaemperle H (1982) XIII Jornadas Com Espanol Deterg, Barcelona, 1982, 55
164. Brueschweiler H, Gaemperle H, Scwager F (1983) Tenside Surf Det 20:317
165. Mann AH, Reid VW (1971) J Am Oil Chem Soc 48:794
166. Stiff MJ, Rootham RC, Culley GE (1973) Wat Res 7:1003, 1407
167. Schöberl P, Mann H (1976) Arch Fischereiwiss 27:149
168. Kravetz L, Chung H, Guin KF, Shebs WT, Smith LS (1984) Tenside Surf Det 21:1
169. Kravetz L, Chung H, Guin KF, Shebs WT, Smith LS (1982) Household Pers Prod Ind 19 (3; 4):46, 72; 62
170. Hellman H (1985) Freenius Z Anal Chem 322:42
171. Neufahrt A, Hofmann K, Tauber G (1987) Comun Journ Com Esp Deterg 18:183
172. Khalil EF, Whitmore TN, Gamal-El-Din H, El-Bassel A, Lloyd D (1988) Appl Microbiol Biotechnol 29:517
173. Ball HA, Reinhard M, McCarty PL (1989) Environ Sci Technol 23:951
174. Una GV, Garcia MJN (1983) Eur J Appl Microbiol Biotechnol 18:315
175. Gerike P, Jakob W (1988) Tenside Surf Det 25:166
176. Gerike P, Jasiak W (1986) Tenside Surf Det 23:300
177. Schöberl P, Bock KJ, Huber M, Huber L (1988) Tenside Surf Det 25:86
178. Patterson SJ, Scott CC, Tucker KBE (1970) J Am Oil Chem Soc 47:37

179. Ichikawa Y, Kitamoto Y, Hosoi N (1978) J Ferment Technol 56:403
180. Baggi G, Beretta L, Galli E, Scolatico C, Treccani V (1977) In: Chater KWA (ed) The oil industry and microbial ecosystems. Proc Inst Pet Warwick 1977, Heyden, London
181. Cook KA (1979) Wat Res 13:259
182. Larson RJ (1984) In: Klug MJ, Reddy CA (eds) Current perspectives in microbial ecology p 677, American Soc for Microbiol, Washington D.C.
183. Plucinski J, Pawlaczyk-Szpilowa M, Sebastion M, Staroojciec O, Karpinska-Smulikowska J (1983) Environ Protection Engineer 9:49
184. Fincher EL, Payne WJ (1962) Appl Microbiol 10:542
185. Borstlap C, Kortland C Fette, Seifen (1967) Anstrichm 69:736
186. Sturm RN (1973) J Am Oil Chem Soc 50:159
187. Pitter P (1973) Cool Czech Chem Comm 38:2665
188. Ohmata S, Kojima N, Sakai T (1974) Hakko Kyokaishi 32:196
189. Harada T, Nagashima Y (1975) J Ferment Technol 53:218
190. Ogata K, Kawai F, Fukaya M, Tani Y (1975) J Ferment Technol 53:757
191. Haines JR, Alexander M (1975) Appl Microbiol 29:621
192. Kawai F, Hanada K, Tani Y, Ogata K (1977) J Ferment Technol 55:89
193. Kawai F, Fukaya M, Tani Y, Ogata K (1977) J Ferment Technol 55:429
194. Kawai F, Kimura T, Fukaya M, Tani Y, Ogata K, Ueno T, Fukami H (1978) Appl Environ Microbiol 35:679
195. Kawai F, Yamanaka H (1986) Arch Microbiol 146:125
196. Jones N, Watson GK (1976) Biochem Soc Trans 4:891
197. Hosoya H, Miyazaki N, Sugisaki Y, Takanashi E, Tsurufuji S (1978) Agric Biol Chem 42:1545
198. Child J, Willett A (1978) Biochim Biophys Acta 538:316
199. Willetts A (1981) Biochim Biophys Acta 677:194
200. Wiegant WM, de Bónt JAM (1980) J Gen Microbiol 120:325
201. Pearce BA, Heydéman MT (1980) 118:21
202. Caskey WH, Taber WA (1981) Appl Environ Microbiol 42:180
203. Schöberl P (1983) Tenside Surf Det 20:57
204. Strass A, Schink B (1986) Appl Microbiol Biotechnol 25:37
205. Dwyer DF, Tiedje JM (1986) Appl Environ Microbiol 52:852
206. Hosoi N, Morimoto K, Ozaki C, Kitamoto Y, Ichikawa Y (1978) J Ferment Technol 56:566
207. Toraya T, Honda S, Fukui S (1979) J Bacteriol 139:39
208. Willetts A (1979) Biochim Biophys Acta 588:302
209. De Bont JAM, Van Dijken P, Van Ginkel KG (1982) Biochim Biophys Acta 714:465
210. Tanaka K, Pfennig N (1988) Arch Microbiol 149:181
211. Ottoson D, Rydqvist B (1978) Acta Physiol Scand 103:9
212. Reiff (1976) In: Koeman JH, Strik JJTWA (eds) Sublethal effects of toxic chemicals on aquatic animals. Elsevier Amsterdam
213. Abram FSH, Brown VM, Painter HA, Turner AH (1977) In: IVth Yogoslav symposium on surface active substances. Dubrovnik
214. Lewis (1986) Environ Toxicol Chem 5:319
215. Yamane AN, Okada M, Sudo R (1984) Wat Res 18:1101
216. Lewis MA, Perry RL (1979) In: Branson DR, Dickson KL (eds) Aquatic toxicology and hazard assessment. 4th Conference ASTM STP 737
217. Guhl W, Gode P (1989) Tenside Surf Det 26:282
218. Lewis, Suprenant (1983) Ecotox Env Safety 7:313
219. Scott Hall W, Patoczka JB, Mirenda RJ, Porter BA, Miller E (1989) Arch Environ Contam Toxicol 18:765
220. Yoshimura K (1986) JOACS 63:1590
221. Turner AH, Abram FS, Brown VM, Painter HA (1985) Wat Res 19:45
222. Caux PY, Weinberger P, Carlisle DB (1988) Environ Toxicol Chem 7:671
223. Solski A, Erndt E (1987) Pol Arch Hydrobiol 34:551
224. Solski A, Erndt E (1987) Acta Hydrobiol 29:387
225. Bode H, Ernst R, Arditti J (1978) Environ Pollut 17:175
226. Buhl KJ, Faerber NL (1989) Arch Environ Contam Toxicol 18:530
227. Shirgur GA (1979) J Anim Morphol Physiol 26:1
228. Watkins CE, Thayer DD, Haller WT (1985) Bull Env Contam Toxicol 34:138
229. Maki AW, Rubin AJ, Sykes M, Shank RL (1979) Journ WPCF 51:2301
230. Macek KJ, Krzeminski SF (1975) Bull Env Contam Toxicol 13:377
231. Leenheer JA, Wershaw RL, Brown PA, Noyes TI (1991) Environ Sci Technol 25:161

Quaternary Ammonium Surfactants

Robert S. Boethling and David G. Lynch

U.S. Environmental Protection Agency, Office of Toxic Substances, TS-798, 401 M St., SW., Washington, DC 20460 USA

Structural Features 146
Production Methods 146
Uses. 148
Consumption. 149
Discharge to the Environment 149
Physical/Chemical Properties 150
Analytical Methods. 152
Environmental Fate of QACs 153
   Fate in Wastewater Treatment 154
   Fate in Receiving Waters 157
   Fate in Soil and Ground Water 161
Toxicity of QACs 161
   Microbial Toxicity in Wastewater Treatment
   and Receiving Waters 161
   Ecotoxicity. 164
   Clinical Toxicology. 169
QAC Residues in the Environment. 169
Summary and Conclusions. 174
References 175

Summary

This article discusses surface-active quaternary ammonium compounds (QACs), an important class of industrial chemicals with a broad spectrum of commercial and consumer uses. Major emphasis is placed on the environmental chemistry and ecotoxicology of the most important classes of QACs. The discussion of environmental chemistry encompasses physical/chemical properties, analytical methods and monitoring data, wastewater treatment and environmental fate. Also addressed are production methods, volumes and uses. A summary of data on human health effects is presented. Existing data on the environmental chemistry and toxicity of QACs are critically reviewed, with the aim of determining where data are adequate for risk assessment, and where significant data gaps exist. This analysis shows that many QACs are readily biodegraded in sewage treatment and in receiving waters, but that their fate in benthic sediments is poorly understood. The bioavailability of sorbed or complexed QACs is another critical issue that deserves more emphasis in future research.

Structural Features

Quaternary ammonium compounds (QACs) are tetrasubstituted ammonium salts of the following generalized structure:

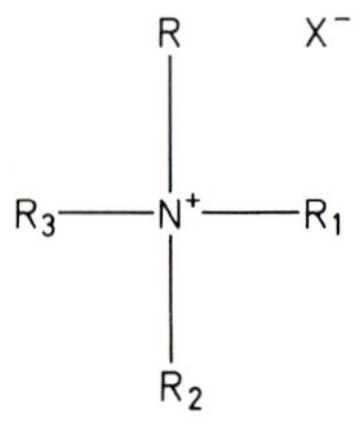

The main distinguishing features of surface active QACs are the permanently charged pentavalent nitrogen and the presence of at least one alkyl chain, approximately C_{10} or longer, which imparts hydrophobicity to the molecule. The R groups may contain N–C or O–C bonds, and they may be branched or linear, substituted or unsubstituted, saturated or unsaturated, and aromatic or aliphatic. Table 1 contains a list of structures and abbreviations for QACs considered in this review.

Production Methods

Although production methods may vary considerably, QACs are usually prepared by reacting the tertiary amine of choice with an alkylating agent such as an alkyl ester, as shown below [1].

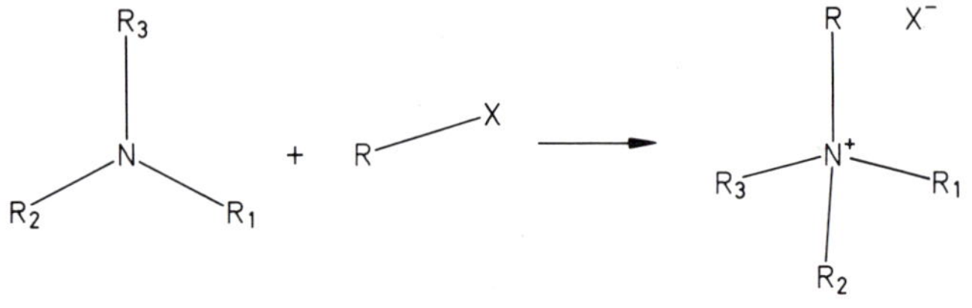

Common alkylating agents include methyl chloride, dimethyl sulfate, diethyl sulfate and benzyl chloride.

On an industrial scale QACs are synthesized in glass or stainless steel reactors. Tertiary amine and solvent (e.g. isopropanol and/or water) are charged to the reactor and heated to 80–100 °C, and the alkylating agent is added. Excess alkylating agent is then removed by evaporation, and the product is purified in

Table 1. Structures and abbreviations for selected quaternary ammonium compounds

General structure	Abbreviation	Quaternary ammonium compound
[R-N⁺(pyridinium)] X⁻	A_nPB A_nPC	alkyl (C = n) pyridinium bromide alkyl (C = n) pyridinium chloride
[R-N⁺(CH₃)₃] X⁻	TTMAC A_nTMAB A_nTMAC	tallowtrimethylammonium chloride alkyl (C = n) trimethylammonium bromide alkyl (C = n) trimethylammonium chloride
[Ph-CH₂-N⁺(CH₃)₂-R] Cl⁻	A_nDMBAC	alkyl (C = n) dimethylbenzylammonium chloride
[R-N⁺(CH₃)₂-R] X⁻	DTDMAC DTDMAMS DA_nDMAC	ditallowdimethylammonium chloride ditallowdimethylammonium methylsulfate dialkyl (C = n) dimethylammonium chloride
imidazolinium structure with $^{\ominus}OSO_3CH_3$	IQAMS	imidazoliumquaternaryammonium methylsulfate (R = tallow alkyl)
R-C-NHCH₂CH₂-N⁺(CH₂CH₂OH)-CH₂CH₂NH-C-R, $^{\ominus}OSO_3CH_3$	EEQAMS	ethoxylatedethanaminiumquaternary ammoniummethylsulfate (R = tallow alkyl)

one or more steps to remove starting materials or byproducts and to improve color. QACs are difficult to separate and purify, however, and this makes their production in solid form commercially nonviable. They are generally available as solutions or dispersions in aqueous or lower aliphatic alcohol–water mixtures (e.g. isopropanol–water).

A recent patent permits the formulation of QACs as finely divided dispersions or in solution in an organic medium serving as part of the final product, offering an economical and desirable means of production [2]. This process would result in, for example, a formulation containing a QAC consisting of C_{12-14} monoalkyl and C_{1-4} trialkyl or mixed alkyl and hydroxyalkyl ammonium halide, carboxylate, or methyl sulfate, and a nonionic surfactant, such as a primary C_{14-15} aliphatic alcohol condensed with 7 to 15 moles of ethylene oxide per mole of alcohol.

Uses

The affinity of QACs for negatively charged surfaces makes them suitable for a wide variety of uses (Table 2). The primary market is as fabric softeners [1]. The softeners are commonly available in three forms: rinse cycle-added (4 to 8% dispersion), dryer-added formulation impregnated in nonwoven sheet or polyurethane foam, and wash cycle-added, consisting of combined detergent, softener and antistatic agent. Wash and rinse cycle softeners are typically formulated using liquid QACs consisting of C_{16-18} alkylammonium chlorides, including distearyldimethylammonium and dialkylimidazolium salts. Dryer-added softeners are formulated using a non-corrosive anion such as methyl sulfate.

QACs are finding increased application in multifunctional liquid laundry detergents (wash cycle-added), where growth from 25 to 30% of the U.S. market occurred from 1985 to 1986. The convenience they add to the use of liquid laundry detergents, eliminating the need to add antistatic and fabric softening products separately to the wash, is being found desirable by a growing number of consumers [3].

Antistatic sprays contain QACs such as DTDMAC in the range of 1 to 2%. QACs are also highly potent germicides and deodorizers, and see widespread use, in part because they are relatively nontoxic to humans, odor free, and stable in storage. Alkyldimethylbenzylammonium chlorides, usually having alkyl groups in the C_{12-22} range, are the most common active ingredients in these products. A deodorizing cleaner, for example, typically contains about 3% QAC [4]. Other major uses include emulsifiers in water-based asphalt, manufacture of organomodified clays (used in drilling muds), hair-care preparations such as shampoos and cream rinses, textile dye retarders, and industrial lubricants and corrosion inhibitors [5].

New markets include wood preservatives and phase transfer catalysts, and are developing areas of technology. Substantial growth is also predicted for

Table 2. Major uses and consumption of cationic surfactants in 1987 [7]

Use	Consumption $\times 10^3$ metric tonnes	
	United States	Western Europe
Softeners and Detergents	79.8	72
Mineral processing and oil: flotation aids, drilling muds, etc.	30.4	16.5
Textiles/fibers: dye assistants, dye levelers, etc.	17.1	16.5
Road chemicals: asphalt emulsifiers	15.2	12
Biocides	11.4	10.5
Others	36.1	22.5
Total	190	150

cationic starch and guar. The former is used as a retention aid in paper manufacture [6].

Consumption

Recent data on consumption of cationic surfactants in the United States and Western Europe are given in Table 2, adapted from Roes and de Groot [7]. Consumption is expected to grow at an annual rate of 4 to 5% in the U.S., Western Europe and Japan. These rates are higher than expected for anionic and nonionic surfactants, which substantially exceed cationics in volume.

Discharge to the Environment

Most uses of QACs can be expected to lead to their release to wastewater treatment systems. If all uses, with the exception of drilling muds, result in the release of QACs to publicly-owned treatment works (POTWs), it can be estimated that approximately 80 million pounds of QACs were sewered in the United States in 1979 [8]. Since the total volume of water flowing through POTWs discharging to surface waters was 3.6×10^{13} L that year [9], the average concentration of QACs in domestic sewage should have been approximately 1 mg/L. Using the same approach, Gerike et al. [10] arrived at a figure of 1.4 mg/L for the approximate level of QAC fabric softening agents (mainly DTDMAC) in German sewage. This difference in predicted QAC concentrations is expected, since Germany is the world leader in per capita consumption of fabric softeners [11].

IQAMS and EEQAMS (as well as their more highly ethoxylated derivatives, the polyethoxylated ethanaminium quaternary ammonium compounds) are also used mainly as fabric softeners. Procter and Gamble [12] estimated that consumer uses of IQAMS and EEQAMS should yield concentrations of these compounds in U.S. sewage of about 0.17 and 0.28 mg/L, respectively. In a similar fashion, Woltering and Bishop [13] estimated approximate levels of C_{12-18} monoalkyl quaternaries in U.S., French, British and German sewage as 0.11, 0.30, 0.42 and 0.50 mg/L, respectively.

Calculations for linear alkylbenzene sulfonates (LAS) indicate that the level of LAS in domestic sewage in the United States should have been about 9 mg/L in 1979; however, LAS levels are typically on the order of 3 or 4 mg/L. It is likely, therefore, that actual levels of QACs in sewage are lower than predicted. QAC monitoring data, although somewhat limited, generally confirm this suspicion. These data are discussed in detail in the section QAC Residues in the Environment.

Physical/Chemical Properties

Measured values of environmentally important physical/chemical properties are summarized in Table 3. As the table shows, available data are sparse. Despite having a permanent positive charge, QACs as a class have rather low water solubility. This is especially true for fabric-softening dialkyl quaternaries such as DTDMAC, which have solubilities in the low milligram per liter range. Vapor pressures should be extremely low, however, so that QACs are not expected to volatilize significantly from soil or water.

The most important property of QACs from an environmental perspective is that they are strongly sorbed by a wide variety of materials. Sorption is rapid in well-mixed test systems. Games et al. [14] determined equilibrium constants (K_d) for A_{18}TMAC sorption to activated sludge and influent wastewater solids (Table 3), and found that more than 98% of the A_{18}TMAC initially present at 20 mg/L was removed from solution within 90 min of its addition to a

Table 3. Measured physical/chemical properties of surface-active quaternary ammonium compounds (QACs)

QAC	MW	Water sol, mg/L[a]	log K_{ow}[b]	BCF	K_d
A_{10}TMAB	235.5		−0.74		
A_{12}TMAC	263.5	5100			423[c]
A_{14}TMAB	291.5		−0.45		
A_{16}TMAC	319.5	440			
A_{16}TMAB	363.9				17 000; 71 000[d]
A_{18}TMAC	347.5	130	1.50		66 000; 226 000[d] 18 000–49 000; 2 800–9 000[e]
DA$_8$DMAC	305.5	8 100			
DA$_{10}$DMAC	361.5	700			
DA$_{12}$DMAC	417.5	77			
DA$_{14}$DMAC	473.5	12			
DA$_{18}$DMAC	585.5	2.7			3 800; 11 000; 12 000[f]
DTDMAC	529.5–585.5[g]		2.69	256; 94; 32; 13[h]	
IQAMS	577.0–633.1[g]	19[i]	2.15[i]		
EEQAMS	639.1–695.5[g]	35[i]	2.48[i]		
A_{16}PB	383.9			21, 22, 13[j]	

[a] 30 °C; from Kunieda and Shinoda [107], except as noted
[b] Woltering et al. [98], except as noted
[c] Sediment from Little Miami River and/or Rapid Creek [108]
[d] Ohio River sediment and EPA$_{18}$ sediment, respectively [15]
[e] Ranges of five values for deactivated (mercuric chloride treated) sludge, activated sludge and seven values for raw domestic wastewater, respectively [14]
[f] Ohio River, Rapid Creek and EPA$_{18}$ sediment, respectively [15]
[g] C_{16-18} alkyl (tallow range)
[h] Inedible fish (bluegills) tissue, well water; inedible tissue, river water; whole body, well water; whole body, river water [39]
[i] Temperature not stated [12].
[j] Whole-body BCFs for clams, minnows and tadpoles, respectively [103]

semicontinuous activated sludge (SCAS) system. Sorption was also rapid in flask experiments, equilibrium being reached within 30 min in both wastewater and activated sludge.

QACs sorb as strongly to natural sediments as they do to sewage and activated sludge solids. Larson and Vashon [15] found that sorption was rapid, equilibrium being reached within a few hours. Interestingly, K_d values were higher for the monoalkyl quaternaries A_{16}TMAB and A_{18}TMAC than for DA_{18}DMAC, despite the much lower water solubility of the latter (Table 3). This suggests that sorption of QACs to sediments involves more than a simple surface area-dependent or solute-partitioning phenomenon, in which the chief variables are the hydrophobicity and organic carbon content of the solute and adsorbent, respectively.

Indeed, organic cations such as QACs appear to be adsorbed to clay minerals mainly by electrostatic attraction. This explains observations such as those of Weber and Coble [16] and Barbaro and Hunter [17], who reported little or no effect of kaolinite on biodegradation of compounds containing quaternary ammonium nitrogen, in contrast to the inhibitory influence of montmorillonite. The latter is an expanding lattice clay, unlike kaolinite, with a much higher cation exchange capacity. Moraru et al. [18] have shown that A_{16}TMAB intercalates into montmorillonite, with the QAC forming two layers having parallel orientation of the alkyl chains. Thus, organic cations apparently can be protected from microbial attack by adsorption on the inner surfaces of the expanding lattice. It may be that this mechanism of sorption is not available to the much bulkier dialkyl quaternaries (such as DA_{18}DMAC), for steric reasons.

Numerous studies indicate that QACs are also rapidly and strongly sorbed by a wide variety of other materials. In addition to activated sludge, sediments, and clay, QACs sorb to minerals, including halides, sulfides, oxides, and sulfates [19], proteins [20–25], and cell walls of microorganisms [26–30]. Sorption to mineral substrates and protein have been reviewed by Ginn [31] and Swisher [32], respectively. The substantial capacity of microorganisms to sorb QACs is well documented in the older literature. Specific rates of A_{16}TMAB uptake at saturation in excess of 200, 300, and 400 mg/g dry weight have been reported for yeast [28], *Staphylococcus aureus* [26, 27] and *Escherichia coli* [27, 29], respectively. These figures are consistent with the data of Neufahrt et al. [33], who found levels of DA_{18}DMAC on activated sludge in the range of 250 to 470 mg/g dry matter, in an activated sludge plant dosed continuously with 10 mg/L DA_{18}DMAC plus 40 mg/L LAS.

A related property of QACs is that they form 1:1 complexes with anionic materials, especially anionic surfactants. The complexes are relatively hydrophobic, but largely ionic in character [34]. QACs should ordinarily exist in this form in surface waters as well as domestic sewage, since anionic surfactants are produced and released to the environment in much higher quantities than are QACs [8], and similar removal efficiencies are expected for both. One result is that QACs may not be detectable by some analytical methods, possibly leading to a failure to demonstrate their presence in environmental samples, or to confusion of complexation with degradation in fate studies. Analytical methods are discussed in the next section.

It has often been stated that these "electroneutral salts" precipitate near the equivalence point, but insolubility seems unlikely at QAC levels in the microgram per liter range, which have been reported in several river monitoring studies [35–40]. In fact, the solubilities of 1:1 complexes of nonylxanthate (nonyldithiocarbonate) with A_{16}TMAB and A_{12}TMAB have been measured, and found to be approximately 1 and 10 mg/L, respectively [34]. Uncertainty over so fundamental a property as solubility clearly demonstrates the need for a careful study of the properties of QAC-anionic surfactant complexes under environmental conditions.

Analytical Methods

Numerous analytical methods have been developed for QACs. The most commonly used methods are colorimetric, in which QACs are reacted with anionic dyes, the QAC-dye complexes are extracted into an organic solvent, and the absorbance of the solution is measured spectrophotometrically [41]. These methods are unsuitable for monitoring QAC levels in sewage or environmental samples or for laboratory studies in which anionic surfactants are also present, because the affinity of QACs for anionic surfactants is often greater than their affinity for the dyes. For this reason, the method of Waters and Kupfer [42] or modifications of it have largely replaced the older methods in environmental studies. In this procedure, samples are passed through an anion-exchange column to remove the anionic component from the QAC-anionic surfactant complex before colorimetric analysis for QAC as the disulfine blue-QAC complex. The results are reported as disulfine blue-active substances (DBAS).

One drawback to this method is that long-chain primary, secondary or tertiary alkylamines, which are possible metabolites in QAC biodegradation [43], as well as other nitrogen-containing compounds that may be present in environmental samples, also react with disulfine blue [35]. To avoid this problem, QACs may be removed from QAC-disulfine blue complexes by cation-exchange chromatography and analyzed by thin-layer chromatography (TLC), permitting the separation of QACs and other cationic compounds. The combined disulfine blue-TLC procedure allows the determination of both DBAS and specific QACs [37, 44].

Michelsen [45] also developed a method suitable for QACs in wastewater, based on TLC with quantitation by densitometry. QACs are isolated by N_2-bubble stripping (sublation) and separated from anionic surfactants by ion exchange. After elution from the cation exchanger, the cationic material is concentrated and separated by TLC, and the plates are sprayed with Dragendorff reagent to detect the individual compounds.

All of these procedures are long and tedious. However, Wee and Kennedy [38] recently described a new method that appears to be superior to existing methods in sensitivity, specificity and ease of performance. This method uses high-performance liquid chromatography (HPLC) with conductometric detection in a nonaqueous medium, made possible because QACs are both soluble and ionized in organic solvents. Before HPLC analysis, QACs are extracted into a nonpolar medium after addition of LAS to the water sample, which enhances the extraction. Using this method, QACs can be determined at submicrogram quantities without derivatization. The detection limit for environmental samples is about 0.02 μg, or 2 μg/L in terms of the QAC concentration in the sample.

Wee [40] used the HPLC method to determine the concentration of DTDMAC in river water and sewage treatment plant influent and effluent, and compared the results to those obtained using a modified Waters-Kupfer DBAS procedure, and a DBAS-TLC procedure. These data are discussed in more detail in the Sect. "QAC Residues in the Environment". The comparison showed that the DBAS procedure was less specific and sensitive than either DBAS-TLC or HPLC, and tended to overestimate DTDMAC levels in environmental samples. The DBAS-TLC and HPLC methods were comparable in specificity, but HPLC was more sensitive and convenient to perform than either DBAS or DBAS-TLC.

Environmental Fate of QACs

Biodegradation is the ultimate fate of QACs released to wastewater treatment and the environment. The biodegradability of QACs has been studied extensively in the laboratory, especially in the last 10 years. Degradability ranging from 0 to 100% has been reported, often for the same or structurally related compounds. To some extent, this is a reflection of widely differing test systems and methods used to measure biodegradation. But in large measure it is also a result of failure to appreciate two key aspects of the environmental behavior of QACs. First, as mentioned previously, QACs are strongly sorbed by a wide variety of materials, and form complexes with anionic substances. This tendency has sometimes made it difficult to distinguish among sorption, complexation, and biodegradation as mechanisms to account for the disappearance of the parent compound, particularly in wastewater treatment systems. Second, acclimation may profoundly influence biodegradability, especially for QACs other than monoalkyl quaternaries, since evidence shows that the other QACs are likely to be somewhat more resistant to degradation.

Fate in Wastewater Treatment

In acclimated activated sludge systems, the efficiency of QAC removal at nontoxic QAC levels should generally exceed 90%. Removal will normally correspond to biodegradation, but not necessarily to ultimate degradation. Degradation will occur mainly on sludge solids, since sorption is much faster than biodegradation. QAC biodegradation in anaerobic sludge digestion has not been investigated. This is a potentially serious data gap in view of the expected partitioning of QACs in wastewater treatment, and the likely subsequent treatment of sludge by anaerobic digestion and/or landfilling. Removal data from key aerobic treatability studies are summarized in Table 4.

In several early studies [46–48], neither the analytical method nor the experimental design was adequate to distinguish removal due to biodegradation from removal by sorption, or analytical masking by complexation with anionic sufactants.

Barden and Isaac [49] were probably the first to provide good evidence that QACs could be readily biodegraded with acclimation. Concentrations of $A_{16}PB$ up to 25 mg/L were easily removed by model trickling filters acclimated to 30 mg/L of $A_{16}PB$. The detection of only traces of $A_{16}PB$ in the effluent at 25 and 30 mg/L was not the result of complexation with anionic substances in the sewage feed, since measurements of HPB in the feed always showed only slightly less than the added amount. Moreover, extraction of the slime on the filters failed to reveal significant quantities of $A_{16}PB$, ruling out sorption per se as the primary mechanism of removal. More recently, Gerike [50] examined the removal of $DA_{18}DMAC$ by model trickling filters. Filters were inoculated with effluent from a POTW, and QAC levels in the synthetic sewage feed were gradually raised from 5 to approximately 20 mg/L. In filters exposed to $DA_{10}DMAC$, $DA_{18}DMAC$ or $DA_{18}DMAC$ plus LAS, QAC disappearance reached 100%, and mean DOC losses were 94, 99, and 84%, respectively. Subsequent extraction of the filters revealed only small amounts of QACs, suggesting again that removal was due mainly to biodegradation.

Most investigators have used model activated sludge systems to study the removal of QACs in treatment. Gerike et al. [10] studied the fate of $A_{16}TMAB$, $DA_{18}DMAC$, $A_{12}DMBAC$ and $DA_{10}DMAC$ in the OECD Confirmatory Test. Without the simultaneous presence of LAS in the feed, removal of $A_{16}TMAB$ in excess of 90% was observed at 15 mg/L, but only if the system was acclimated by gradually raising the $A_{16}TMAB$ concentration to that level. In the presence of 20 mg/L of LAS, removal of all four QACs at 10 mg/L was high (>90%) and steady. With all four QACs, removal apparently corresponded to biodegradation, since minimal levels of QACs were found in sludge from the aeration vessel. However, a tendency to sorb to sludge is suggested by the presence in the return sludge of $DA_{18}DMAC$ at up to 12% of the total throughput. Gerike [50] later calculated the maximum amounts of $A_{16}TMAB$, $A_{12}DMBAC$ and $DA_{10}DMAC$ that could have been removed on a continuous basis by sorption alone, using the

Table 4. Summary of selected studies on fate of QACs in wastewater treatment

QAC	[QAC], mg/L	Test system	QAC removal, percent	QAC ultimate degradation, percent[a]	Ref.
$A_{16}PB$	25	TF[b], A	100		[49]
$A_{16}TMAB$	5–15	CAS, A	91–98		[10]
	10	CAS, A, C	98		[10]
	15	CAS, A		107	[10]
$A_{18}TMAC$	20	SCAS	>98 from soluble phase in 90 min; primary degradation $t_{\frac{1}{2}}=2.5$ h		[14]
	0.1–1.0	SCAS[c], C		60–90[d] depending upon position of ^{14}C	[14]
ADMBAC	10–16	CAS, A, C	97; 94 primary degradation		[52]
$A_{12}DMBAC$	5	CAS, A, C	96		[10]
		CAS, A, C		83	[53]
$A_{14}DMBAC$	20	CAS, A	>70	probably substantial	[83]
DTDMAC	10–16	CAS, A, C	96–97; 62–82 primary degradation		[52]
	2.1	SBAS, C	>90 from soluble phase in 4 d; 70 primary degradation in 39 d	probably substantial	[55]
$DA_{10}DMAC$	5	CAS, A, C	95		[10]
	15	CAS, A, C		0.3	[10]
	20	TF, A	100	94	[50]
$DA_{18}DMAC$	10	CAS, A, C	95		[10]
	20	CAS, A, C		108	[10]
	20	TF, A	100	99	[50]
	20	TF, A, C	100	84	[50]

[a] Defined as percent DOC removal with respect to QAC, except as noted
[b] TF = trickling filter, A = acclimated, CAS = continuous-flow activated sludge (OECD Confirmatory Test), C = cationic/anionic surfactant complex, SCAS = semicontinuous activated sludge test, SBAS = semibatch activated sludge
[c] Incubation period extended from 24 to 172 h; suspended solids level reduced to approximately 1 g/L.
[d] Percent of theoretical $^{14}CO_2$ evolution

mathematical model of Wierich and Gerike [51]. These calculations indicated that removal by sorption could account for only 8 to 29% of the total elimination of the three QACs listed above, suggesting that biodegradation was responsible for most of the observed QAC loss.

May and Neufahrt [25] and Janicke and Hilge [52] obtained similar results, also using the OECD Confirmatory Test. May and Neufahrt [25] examined $DA_{18}DMAC$ degradation in mixtures consisting of $DA_{18}DMAC$ (2.8 mg/L) and secondary alkane sulfonate (8.0 mg/L), or $DA_{18}DMAC$ (1.8 mg/L) and dodecylbenzenesulfonate (5.8 mg/L), secondary alkane sulfonate (2.6 mg/L) and a nonionic surfactant (0.5 mg/L). Removal of $DA_{18}DMAC$ exceeded 90% in both systems. Janicke and Hilge [52] investigated the treatability of DTDMAC and A_{8-18} DMBAC. After acclimation to 10 to 16 mg/L of DTDMAC or $A_{8-18}DMBAC$ in the presence of an equivalent amount of alkylbenzenesulfonate (ABS), filtered effluents contained 3.5 to 4.5% of the influent DTDMAC and 3% of the influent $A_{8-18}DMBAC$. Net elimination—sorption plus biodegradation—was 96% for DTDMAC and 97% for $A_{8-18}DMBAC$. As in the preceding studies, the authors found relatively low levels of QACs in the sludge, and concluded that the QACs were biodegraded.

Gerike et al. [10] also observed complete disappearance of $A_{16}TMAB$ and $DA_{18}DMAC$ as measured by loss of DOC. DOC removal was somewhat lower in the $A_{12}DMBAC$-amended test system (54%), but a subsequent study reported a value of 83% [53]. DOC removals for $A_{16}TMAB$ and $A_{12}DMBAC$ were also high in the Zahn-Wellens test. Together with the failure to detect substantial amounts of QACs in sludge [10], these findings suggest that $A_{16}TMAB$, $DA_{18}DMAC$, and $A_{12}DMBAC$ underwent extensive ultimate degradation. For $DA_{10}DMAC$, primary degradation did not coincide with ultimate degradation, however. In contrast to its high removability as measured by loss of parent compound, DOC loss amounted to only 0.3%.

Use of ^{14}C-labeled QACs has resolved many of the uncertainties concerning their fate in biological treatment systems, particularly with respect to ultimate degradation and the role of sorption. Krzeminski et al. [54] conducted a detailed study of the environmental impact of Hyamine 3500, an alkyldimethylbenzylammonium chloride. That compound was extensively degraded at 10 mg/L in a semicontinuous activated sludge (SCAS) test, but only if the test system was exposed to increasing levels of the QAC over a period of about 2 weeks. Proof of ultimate degradation was provided by the detection of 80% of the influent ^{14}C as $^{14}CO_2$, and 16% of the ^{14}C in the effluent.

Games et al. [14] combined SCAS studies with measurements of sorption in sewage and activated sludge to determine the fate of $A_{18}TMAC$ in wastewater treatment. Primary degradation was slower than sorption but still rapid, occurring with a half-life of 2.5 h in an acclimated system. Thus, biodegradation was primarily associated with sludge solids. Ultimate degradation followed apparent first-order kinetics at initial $A_{18}TMAC$ concentrations of 0.1 and 1.0 mg/L, levels close to those normally expected in sewage. Similar results were obtained with a 1:1 complex of $A_{18}TMAC$ and LAS. Conversion of $A_{18}TMAC$ to $^{14}CO_2$ was extensive and established the ultimate biodegradability of both the methyl groups and the alkyl chain.

Sullivan [55] studied the biodegradation and sorption of ^{14}C-labeled DTDMAC in semi-batch activated sludge reactors. In these experiments DTDMAC was added only once at 2.1 mg/L as a complex with LAS, but synthetic sewage was fed daily to maintain organic loading rates characteristic of conventional and extended aeration activated sludge treatment. Results were similar to those of Games et al. [14]. Degradation of DTDMAC was considerably slower in both systems than was A_{18}TMAC degradation in the latter study, however, suggesting that dialkyl quaternaries may be more recalcitrant than the monoalkyl derivatives in wastewater treatment.

The recent work of Ruiz Cruz [56] adds substantially to the list of QACs for which treatability data are available. Removal was high in both the Spanish Official test (a SCAS method) and the OECD Confirmatory Test for nearly all test compounds, including an isoquinolinium and two imidazolium QACs, classes not previously tested. Of course, the high removal of these compounds is not a guarantee of ultimate biodegradability. Data for QACs not previously tested as either the chloride or bromide salt are summarized in Table 5.

Fate in Receiving Waters

Several generalizations can be made regarding the fate of QACs in receiving waters. First, monoalkyl quaternaries as a class should be most rapidly and alkylpyridinium QACs least rapidly biodegraded, with alkyldimethylbenzyl and dialkyl quaternaries intermediate in degradability. Second, aquatic environments that receive effluent from sewage treatment plants should be acclimated to some QACs, such as DTDMAC, and biodegradation rates for these may be quite rapid. Finally, sorption of QACs to suspended solids could play an important role in their biodegradation, but despite recent progress, a complete understanding has not been achieved.

Screening tests have been used in many studies of QAC biodegradability [15, 53, 56–62], but these reports must be interpreted with caution since river dieaway studies have shown that screening tests may underestimate QAC bio-

Table 5. Removal of selected QACs in the Spanish Official and OECD Confirmatory tests [56][a]

QAC	QAC removal, percent	
	Spanish official	OECD confirmatory
A_{12}TMAC	97	96
DA_{12}DMAC	96	96
A_{18}DMBAC	94	95
p-tert-octylphenoxyethoxyethyldimethylbenzyl-ammonium chloride	30	45
A_{12}PC	93	91
A_{14}PC	94	92
dodecylisoquinolinium chloride	90	92
didodecylimidazolium chloride	89	92
dodecylhydroxyethylimidazolium chlorhydrate	94	97

[a] All QACs were present in the synthetic sewage feed at 5 mg/L, without anionic surfactant. Both the Spanish Official Test (a SCAS method) and the OECD Confirmatory Test (a continuous-flow method) require acclimation before determination of removal

degradability [63]. The high toxicity of QACs accounts for this at least in part, since test chemical concentrations are typically in the milligram per liter range. Anionic surfactants seem to mitigate toxicity in screening tests, just as they do in bench-scale treatability tests. This effect is illustrated in Fig. 1, which shows that A_{18}TMAC alone inhibited endogenous metabolism at 10 mg/L (note the negative slope), but that toxicity was eliminated in the presence of an equal concentration of LAS. Nevertheless, taken together, the screening data do suggest the order of relative biodegradability given in the preceding paragraph.

The most useful information on the fate of QACs in receiving waters has come from die-away studies. Kaplin et al. [64] reported that A_{18}TMAC, A_{17-20}TMAC, A_{18}DMBAC and an unidentified alkyldimethylbenzylammonium chloride disappeared from river water without added bottom sediment within 30 to 40 days. Several QACs disappeared from the water column more rapidly in microcosms containing sediment, but sorption probably accounted for some removal. Baleux and Caumette [65] studied the biodegradation of several QACs in river water and sewage. Among the QACs were A_{12}PC, A_{16}PB, A_{16}TMAC, DA_{18}DMAC, dioctadecyl(ethoxy)$_{15}$methylammonium chloride, diisobutylphenoxyethoxyethyldimethylbenzylammonium chloride (Hyamine 1622), and an unspecified alkylimidazoline. A_{16}TMAC and the alkylimidazoline disappeared completely from river water in 7 to 14 days, depending upon the source of the water. Hyamine 1622 apparently was not degraded. The slow degradation of many of these QACs may have been the result of toxicity at the

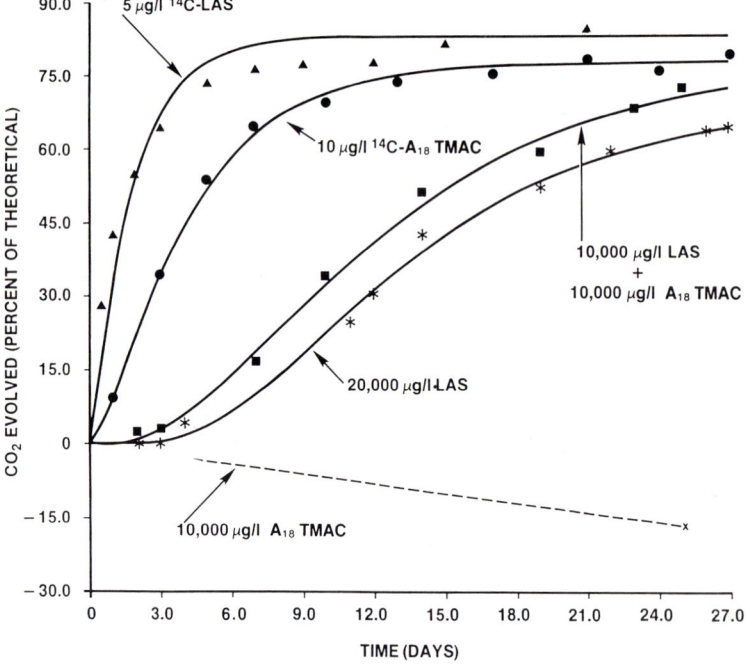

Fig. 1. Ultimate degradation of A_{18}TMAC and LAS in screening tests and natural waters

high concentrations employed (15 to 20 mg/L), but other tests failed to demonstrate a significant correlation between recalcitrance and toxicity.

Ruiz Cruz and Dobarganes Garcia [66] carried out the most extensive study to date of the influence of chemical structure on biodegradation of QACs in river water. Although removal by sorption cannot be ruled out, the kinetics of QAC loss observed in this study and low level of suspended solids in the river water (approx. 1.4 mg/L) suggest that biodegradation was the primary mechanism of removal. Of a series of QACs having C_{12} alkyl chains, A_{12}TMAC and dodecylimidazolium chloride were the most rapidly degraded. Following these QACs, in order of decreasing degradability, were A_{12}DMBAC, DA_{12}DMAC, A_{12}PC and dodecylisoquinolinium chloride. Lag periods and half-disappearance times with unacclimated river water ranged from less than one to several days. These results suggest that monoalkyl quaternaries will be most rapidly degraded and alkylpyridinium quaternaries least rapidly degraded in receiving waters.

River die-away studies by Krzeminski et al. [54], Larson and Perry [67], and Larson and Vashon [15] also suggest that monoalkyl quaternaries should be more rapidly degraded in receiving waters than other QACs. Larson and Perry [67] used an electrolytic respirometer to follow ultimate degradation of A_{16}TMAC in Ohio River water. A_{16}TMAC was degraded with a half-life of 2.7 days after a 2-day lag, at A_{16}TMAC levels below the apparent toxicity threshold of 40 mg/L. Larson and Vashon [15] measured production of $^{14}CO_2$ from ^{14}C-labeled QACs, with initial concentrations ranging from 1 to 100 μg/L. Ultimate degradation of A_{16}TMAB and A_{18} TMAC in river water occurred with no detectable lag, and half-lives of 2 to 3 days (Fig. 1). In contrast, DA_{18}DMAC was not appreciably degraded in 9 weeks in sediment-free water. Moreover, Krzeminski et al. [54] found that more than 3 weeks were required for complete ultimate degradation of ^{14}C-labeled Hyamine 3500 (an alkyldimethylbenzylammonium chloride), at initial QAC concentrations from 0.01 to 1.0 mg/L.

Vives-Rego et al. [68] studied the biodegradation of anionic, nonionic and cationic surfactants in sea water. Half-lives for loss of DBAS were in the range of 4 to 9 days, even though the QAC tested, A_{16}TMAB, showed signs of toxicity at the high concentration employed (20 mg/L).

Biodegradation rates in the environment are also influenced by numerous factors other than chemical structure. Ruiz Cruz [69, 70] investigated the influence of several experimental variables on disappearance of QACs from river water, at an initial QAC concentration of 5 mg/L. The variables included biomass, acclimation, aeration, temperature, and organic and inorganic nutrients. Acclimation may be the most important of these variables. Degradation of A_{12}PC, A_{12}TMAC, A_{16}TMAC, A_{18}TMAC, A_{12}DMBAC, A_{14}DMBAC, A_{16}DMBAC, A_{14}PC and A_{18}DMBAC proceeded more rapidly and without a lag in water samples previously exposed to these QACs. However, acclimation was lost if sufficient time passed before water was respiked with test compound.

Studies of the effects of acclimation on QAC biodegradation have recently been extended to a laboratory model stream and field sites. Ventullo and Larson [71] observed effects of preexposure to A_{12}TMAC similar to those reported by

Ruiz Cruz [69, 70]. Upon initial exposure, ultimate degradation of A_{12}TMAC proceeded with a half-life of 90 h after a lag phase of 24 h; but degradation proceeded with no lag and a half-life of 20 h after the respike. Heterotrophic activity studies also showed that chronic exposure to A_{12}TMAC resulted in significant increases in maximum degradation rate (V_{max}) and number of A_{12}TMAC degraders, relative to unexposed controls. Larson and Bishop [72] found that A_{12}TMAC was degraded with half-lives ranging from <2 to 15 h in the water column of a model stream dosed continuously with A_{12}TMAC for 15 months, whereas half-lives often exceeded 5000 h in the control. Responses to A_{12}TMAC exposure were similar in the epilithic and sediment compartments. Acclimation of the epilithic microbial community was examined more closely at a field site, the Little Miami River in Ohio. In situ exposure to A_{12}TMAC produced an adapted community that degraded this QAC with half-lives that averaged about 2 h, which was 14 to 50 times lower than in the unexposed control.

The tendency of QACs to sorb strongly to sludge, sediments and other materials has been discussed. What has been less clear is the effect sorption has on biodegradability. Investigators in several older studies [16, 17, 54] found that sorption inhibited QAC biodegradation. But Larson and Vashon [15] reported the opposite effect of sediment on degradation of DA_{18}DMAC, and no significant inhibition of A_{16}TMAB and A_{18}TMAC degradation.

As discussed previously, differences in the sorbent (montmorillonite, an expanding-lattice clay, in the first two studies mentioned above) may be partly responsible for this apparent conflict. In any case, it seems clear that QACs are not necessarily unavailable for biodegradation in the presence of sediment. This question was addressed directly by Shimp and Young [73]. They found that biodegradation rates for A_{12}TMAC were a function of the total amount of chemical present, sorbed and unsorbed, in sediment slurries, whereas in cores containing settled sediment with overlying water, sorbed material was not directly available for biodegradation. Shimp [74] extended this work to acclimation of sediment and water-column microbial communities to A_{12}TMAC degradation. Higher levels of A_{12}TMAC in the water column were required to elicit an adaptive response in settled sediments as compared to overlying water, but once acclimation occurred, the response was more prolonged than in the water column. Thus it appears that although sorption may reduce bioavailability, it can also increase the robustness of the adaptive response by buffering QAC levels in interstitial water.

A potentially important observation made by Ruiz Cruz [70] relative to the influence of experimental variables concerns the effect of exposure of test vessels to sunlight. Disappearance of A_{12}TMAC and A_{12}DMBAC—normally complete within several days—appeared to stop after loss of 50 to 60% of the QAC initially present. This suggests that photodegradation may have resulted in the accumulation of toxic degradation products, or products not further degraded, but still detectable by the analytical method used. That QACs may be photooxidized was also suggested by Neufahrt (cited in [11]), who observed a reduction in the carbon and hydrogen contents of $DA_{18}DMAC$ from 75 to 55.4% and 13.8 to 8.9%, respectively, with 16 to 72 h of exposure to UV light. In this case the resulting products were readily biodegradable. The potential for photodegradation of QACs should be studied in detail.

Fate in Soil and Ground Water

Data on the biodegradability of QACs in surface soils and subsurface environments are very limited. Half-lives for aerobic ultimate degradation of $A_{18}TMAC$ were reported to be 1.4, 5.8, 5.3, and 3.2 to 8.7 days for compost, organic loam, sandy loam and silt loam, respectively [72]. On the other hand, Larson and Vashon [15] found much slower degradation of $A_{12}TMAC$ in aerobic sludge amended soil ($t_{1/2} = 1$ mo). Degradation is therefore slower in surface soil than in water, but may still be rapid enough to prevent their accumulation. Only monoalkyl quaternaries have been tested, however.

Larson [75] and Federle and Pastwa [76] studied the biodegradation of QACs in subsurface samples. $A_{12}TMAC$ was degraded in aerobic ground water/subsurface soil slurries with a half-life of 1 week [75]. Federle and Pastwa [76] studied the aerobic biodegradation of $A_{18}TMAC$ and $DA_{18}DMAC$ as a function of depth in samples of subsurface sediment from below a laundromat discharge pond and a nearby control pond. Neither QAC was mineralized to a significant extent in control sediment in 3 months, but extensive mineralization occurred in the laundromat profile. $A_{18}TMAC$ was more rapidly and extensively degraded than $DA_{18}DMAC$. These results provide another illustration of the effect of acclimation on biodegradation rates. But the lack of degradation in the control profile is disturbing, and suggests that QAC discharges could adversely impact pristine sites.

It should be noted that photochemical instability, just discussed, could also affect the fate of QACs in surface soil. This would be particularly relevant for soil amended with wastewater sludge containing QAC residues. However, the potential for photodegradation of QACs must first be confirmed experimentally.

Toxicity of QACs

Microbial Toxicity in Wastewater Treatment and Receiving Waters

An extensive literature, not reviewed here, amply documents the potent germicidal activity of many QACs [77]. This property probably accounts for the strong emphasis placed on the issue of toxicity in wastewater treatment in the older literature on environmental fate and effects of QACs. It can now be stated that, under normal circumstances, QACs are unlikely to pose a significant risk of toxicity to microorganisms in wastewater treatment systems. The reasons are twofold. First, microbial populations in POTWs can be expected to be acclimated to low levels of commonly used QACs, as shown previously. This is significant because the high toxicity of some QACs reported in the older literature is not observed in properly acclimated, high-biomass test systems. Second, the presence of anionic surfactants in sewage will greatly reduce QAC toxicity.

Nevertheless, sudden discharges of QACs resulting in temporarily high levels in treatment plants could upset plant function. This may be especially true for the nitrification process, which appears to be somewhat more sensitive to inhibition by QACs than sewage purification in general. Slug doses are not likely to be a problem for QACs in consumer products, but other uses could lead to such discharges. Recent evidence also suggests that ecologically significant responses are possible in aquatic environments unacclimated to QACs, at QAC levels below 1 mg/L.

An early study by Pitter [46] illustrates all of the major points outlined above in regard to wastewater treatment. Pitter studied treatability of $A_{16}PB$ and $A_{16}TMAB$ in a bench-scale activated sludge system. The QAC concentration was raised in steps from 1 to 20 mg/L. Without ABS, no effect of $A_{16}PB$ or $A_{16}TMAB$ was observed at 1 mg/L, but at 3 mg/L both QACs strongly inhibited nitrification. Even at 6 mg/L the ability of sludge microorganisms to reduce the biochemical oxygen demand (BOD) of synthetic sewage was reduced by only 6%, but nitrification was blocked completely. But when the QAC concentration was suddenly increased from 6 to 20 mg/L, sludge flowed from the system.

In contrast, Pitter [46] observed no inhibition of treatment processes by 6 mg/L of $A_{16}PB$ and $A_{16}TMAB$ in the presence of an equivalent amount of ABS. He therefore concluded that QACs would not be toxic in wastewater treatment systems at concentrations up to 6 mg/L in the presence of equivalent or higher levels of anionic surfactants, which is the situation likely to prevail in domestic sewage treatment under normal circumstances. But sudden loading of the system with 20 mg/L of $A_{16}PB$ or $A_{16}TMAB$ and an equivalent amount of ABS again resulted in large amounts of sludge in the effluent, a drastic reduction in the oxidizing ability of the microbial population (BOD removal), and complete inhibition of nitrification.

Similar results have been obtained for these and other QACs—including $DA_{18}DMAC$, the major component DTDMAC—by other investigators [10, 49, 78]. That nitrification may be especially sensitive to inhibition by QACs is also suggested by earlier reports [79, 80], but interpretation of these studies is difficult because the authors provided little information on test methods. In contrast, Reynolds et al. [81] reported that a pure culture of nitrifying bacteria did not appear to be more sensitive to $A_{16}TMAB$ and $A_{12}DMBAC$ than many commonly used indicators of toxicity, such as respiration.

The EC_{50} values reported by Reynolds et al. [81] for inhibition of respiration by A_{16}TMAB and A_{12}DMBAC were generally in the range of 10 to 40 mg/L. Other investigators have obtained similar values [49, 82, 83]. Larson and Schaeffer [84] used a novel method, inhibition of ^{14}C-glucose uptake, to determine the toxicity of chemicals to activated sludge. The EC_{50} for unacclimated sludge was 28 mg/L for A_{14-18}TMAC, in line with the respirometric data. Pitter and Horska [85] used still another method, inhibition of dehydrogenase activity in unacclimated sludge fed glucose or peptone, with similar results. The EC_{50} values were 9 and 19 mg/L for A_{16}PB and A_{16}TMAB, respectively. Overall, the data suggest that QACs are unlikely to manifest significant toxicity in wastewater treatment at the levels normally expected (≤ 1 mg/L).

It also seems unlikely that QACs will inhibit anaerobic digestion under normal circumstances, but here the evidence is weaker. In studies by May and Neufahrt [25] and Janicke and Hilge [52], sludge from laboratory activated sludge systems in which treatability of QACs had been studied was incubated anaerobically, and gas production measured over time. May and Neufahrt [25] observed no difference between digestion of control sludge and sludge from systems acclimated to mixtures of DA_{18}DMAC and anionic surfactants, and Janicke and Hilge [52] found no inhibitory effects of DTDMAC or A_{8-18}DMBAC. But since biodegradation resulted in the removal of approximately 90% of the influent QAC in the activated sludge process in both studies, relatively little of the starting material was sorbed to the sludge and subsequently transferred to the anaerobic digestors. Thus the possibility exists that more recalcitrant QACs—or even normally degradable QACs, in the event of discharges of slug doses to the system—could reach inhibitory levels.

Studies of QAC toxicity in low-biomass test systems, most commonly involving measurement of the reduction in the biochemical oxygen demand (BOD) of natural or synthetic sewage, have generally indicated much higher toxicity under these conditions [35, 46–48, 57, 82]. In most cases, however, the microbial populations were not acclimated by previous exposure to QACs. Moreover, the significance of such studies for wastewater treatment operations, where biomass levels are much higher, is doubtful. Similar statements can be made in reference to studies in which effects on bacterial growth were determined [65, 66].

Data from test systems such as these may be more relevant to aquatic environments expected to receive QACs. In fact, in both of the studies just cited, river water was used as a source of microorganisms. For example, Ruiz Cruz and Dobarganes Garcia [66] incorporated QACs into agar plates and inoculated with river water. A_{12}PC inhibited colony development at 1 mg/L, but all four QACs tested (including A_{12}PC, A_{16}PB and A_{16}TMAC) reduced colony counts at levels below 10 mg/L. That QACs may elicit ecologically significant responses at concentrations below 1 mg/L was recently confirmed by Ventullo and Larson [71], who determined first-effect concentrations (FECs) for inhibition of heterotrophic activity in freshwater bacterial communities. FECs for TTMAC,

A_{12}TMAC and DTDMAC were approximately 0.1, 0.1 and 1.0 mg/L, respectively. A_{16}TMAB was also toxic to bacteria in sea water [68], but here this result is not so surprising, in view of the high (20 mg/L) surfactant concentration employed.

Toxicity probably accounts for the failure of some investigators to detect biodegradation of QACs, and for the frequent observation that biodegradability in low-biomass test systems is a function of initial concentration [30, 58, 59, 64, 67, 69]. In fact, toxicity may account for the widespread failure to observe significant QAC biodegradation in low-biomass biodegradability screening tests, where concentrations of test chemical are typically on the order of 20 mg/L [14, 15, 53, 61]. Figure 1, taken from Larson and Wentler [62], is typical, and shows that A_{18}TMAC was not biodegraded and apparently was toxic at 10 mg/L in a biodegradability screening test.

Ecotoxicity

As a class, QACs are not only potent germicides, but also acutely toxic in the milligram per liter range and lower to aquatic organisms, including algae, fish, mollusks, barnacles, rotifers, starfish, shrimp and others [35, 39, 54, 86–98]. Cooper [99] has reviewed the environmental toxicity of QACs, with emphasis on the older literature. Data from Taft [86] on the toxicity of alkyldimethylbenzylammonium chlorides to invertebrates are typical (Table 6).

Vallejo-Freire et al. [88] showed that a wide range of QACs were acutely toxic to snails (*Australorbis* sp.) at concentrations of 1 to 10 mg/L. The dose-response curves were very steep, with 0 and 100% mortality occurring within a 10-fold change in concentration. However, Cooper [99] estimated that LC_{50} values for 2 of 15 compounds were approximately 0.3 mg/L, and that no-effect levels could therefore be even lower. Moreover, if the dose-response curves for

Table 6. Toxicity of alkyldimethylbenzylammonium chlorides to invertebrates [86]

Species (group)	Toxicity
Balanus (barnacle)	2.5 mg/L, no effect; 5 mg/L, 100% mortality "within a short time"
Euglena (protozoan)	20 mg/L, 100% mortality in 19 h
rotifers	1 mg/L, 100% mortality in 19 h; 10 mg/L, 100% mortality in 2 h
Amphipoda (amphipods)	2.5 to 10 mg/L, 100% mortality "in a few hours"
Cypridae (ostracods)	50 mg/L, active after 5 h
Planaria (flatworm)	10 mg/L, 100% mortality in 1 h
Ascaris (roundworm)	20 mg/L, alive after 18 h
Enchytrae albidus (annelid)	1 mg/L, 100% mortality in ⩽4 h
Asteroidea (starfish)	2.5 mg/L, 100% mortality in 24 h

species listed in Table 6 are similar, no-effect levels for some species could be as low as 0.01 mg/L.

More recent data on fish and invertebrates are generally consistent with the older studies. Some of these results are given in Table 7. In addition, Woltering et al. [98] have published 48-h LC_{50} values for toxicity of $A_{12}TMAC$ to a wide variety of freshwater invertebrates. These data suggest that daphnids (*Daphnia magna*) may be more sensitive than many other invertebrates.

An aquatic safety assessment by Lewis and Wee [39] adds substantially to the database on aquatic toxicity of QACs, especially in regard to the effects of suspended solids on observed toxicity. Tests were conducted in which freshwater and marine species representing three trophic levels were exposed to dialkyl quaternaries, in surface waters containing suspended solids as well as high-quality waters free of solids. Acute toxicity data for fish and invertebrates are summarized in Table 8. The 96-h LC_{50} values for QAC toxicity to bluegills (*Lepomis macrochirus*) in high-quality water varied from 0.64 to 3.0 mg/L, but toxicity was much reduced in Town River water. A similar effect was observed for *Daphnia magna*. Woltering et al. [98] later reported that data for monoalkyl quaternaries showed the same trends. Marine species appeared to be less sensitive to DTDMAC ([39] Table 8), but mysid shrimp (*Mysidopsis bahia*) were as sensitive as daphnids.

Table 7. Acute toxicity of QACs to fish and invertebrates

Test species	QAC	LC_{50} (mg/L)	Ref.
Daphnia magna (water flea)	ATMAC[a]	1.2–5.8	[89]
Idus melanotus (golden orfe)	ATMAC[a]	0.36–8.6	[89]
Planorbis corneus (water snail)	ATMAC[a]	0.73–23	[89]
Lepomis macrochirus (bluegill sunfish)	Hyamine 3500[b]	0.5	[54]
Carrasius auratus (goldfish)	Hyamine 3500[b]	2	[54]
Idus melanotus	$A_{12}DMBAC$	3.5[c], 8[d]	[90]
Rasbora heteromorpha (harlequin fish); *Salmo truta* (brown trout); *Idus melanotus*; *Carrasius auratus*	(unidentified)	mean = 2.04–3.51	[91]
Lepomis macrochirus	various	mean = 1.65 (range = 0.33–5.90)	[99]
Salmo gairdneri (rainbow trout)	various	mean = 3.80 (range = 0.34–12.3)	[99]

[a] $A_{10}TMAC$, $A_{12}TMAC$, $A_{14}TMAC$, $A_{16}TMAC$, $A_{18}TMAC$, and $A_{20/22}TMAC$, were tested. The ranges of LC_{50} given exclude the values for $A_{10}TMAC$
[b] An alkyldimethylbenzylammonium chloride
[c] LC_0
[d] LC_{100}

Table 8. Acute toxicity of DTDMAC to freshwater and marine fish and invertebrates [39]

Species	Dilution water	Suspended solids (mg/L)	Endpoint	Effect level (mg/L)[a]
Freshwater				
Lepomis macrochirus (bluegill)	well[b]	0	96-h LC_{50}	1.04
	reconstituted	0	same	0.62–3.00
	reconstituted[c]	0	same	1.23
	Town River	2–84	same	10.1–>24
Daphnia magna (water flea)	well	0	48-h LC_{50}	1.06
	reconstituted	0	same	0.19–0.48
	reconstituted[b]	0	same	0.16
	White River[b]	3–5	same	3.1
Marine				
Crassostrea virginica larvae (oyster)	estuarine water, salinity 16–26%		48-h EC_{50}	2.0
Penaeus duorarum (pink shrimp)	same		96-h LC_{50}	36.0
Mysidopsis bahia (mysid shrimp)	same		same	0.22
Callinectes sapidus (blue crab)	same		same	>50
Cyprinodon variegatus (sheepshead minnow)	same		same	24.0

[a] Nominal level of active ingredient
[b] QAC tested was $DA_{18}DMAC$, the major component of DTDMAC
[c] QAC tested was DTDMAMS, not DTDMAC

Although Marchetti [100] stated that the toxicity of cationic surfactants decreases with increasing water hardness, water hardness did not appear to be a factor in the Lewis and Wee [39] study. It is more likely that other factors—especially sorption to suspended solids and the tendency to form complexes with anionic surfactants—were responsible for the reduction in acute toxicity in natural waters relative to high-quality water. Support for this conclusion was provided by experiments in which the effects of added river bottom silt and C_{12} LAS on QAC toxicity to bluegills were determined directly. For example, no mortality was observed at $C_{12}LAS/DTDMAMS$ molar ratios of 0.5/1 to 4/1 (absolute concentrations, in milligrams per liter, of 0.3/1.2 to 2.7/1.2). Knauf [89] obtained similar results with the golden orfe (*Idus melanotus*).

The potential chronic effects of QACs have not been as well studied as acute toxicity. The no observed effect concentration (NOEC) for daphnids exposed to $DA_{18}DMAC$ in river water was 0.38 mg/L [39], about one-tenth the 48-h LC_{50} (Table 8). NOECs for fathead minnows (*Pimephales promelas*) were 0.23 mg/L with river water and 0.053 mg/L with filtered well water. However, Woltering et al. [98] recently reported that acclimation or compensation may occur in *Daphnia* populations chronically exposed to $A_{12}TMAC$. Although there was some initial mortality in the mixing zone in a laboratory model stream, total numbers and biomass were not significantly different from control values after several generations, despite nominal QAC levels that should have been associated with lethality.

There were no significant differences in larval survival and adult emergence of midge (*Paratanytarsus parthenogenica*) exposed to sediments from Rapid Creek relative to the controls [39]. DTDMAC levels in sediments and water were 2 to 67 µg/g and 8 to 92 µg/L, respectively. Similar results were reported for midge in the model stream study [98], where emergence and population density were not affected despite levels of A_{12}TMAC in sediment that were 35 to 75 times above those associated with mortality in water. These data suggest that sorption to sediment reduces bioavailability of QACs to benthic organisms.

Cationic polyelectrolytes have been shown to elicit acute toxic effects in aquatic organisms by disrupting gill membranes, thus interfering with O_2 exchange [101]. QACs presumably act by the same mechanism. However, the only published work on uptake and distribution of QACs is that of Neufahrt et al. [102] and Knezovich et al. [103]. Neufahrt et al. [102] mixed dried activated sludge from a bench-scale wastewater treatment system dosed continuously with ^{14}C-DA_{18}DMAC with fish food, and followed the uptake and tissue distribution of radiolabel in carp (*Cyprinus carpio*). Radiolabel was concentrated in the intestinal tract and gall bladder, with some residual ^{14}C in the gills, probably due to filtration of contaminated dung particles or leftover feed. Knezovich et al. [103] studied the uptake and distribution of A_{16}PB in three species: clams (*Corbicula fluminea*); tadpoles (*Rana catesbeiana*) and fathead minnows. Organisms were exposed to a sublethal concentration (10 µg/L) of ^{14}C-A_{16}PB in the water for 24 h. As expected, radiolabel was concentrated in the gills in all three species, with some residual ^{14}C in the intestine and skin of tadpoles. These results are consistent with the postulated mechanism of acute toxicity.

The data of Knezovich et al. [103] also suggest that A_{16}PB has a low bioaccumulation potential, since fat bodies and tissues of toxicological interest (e.g. liver and kidneys) contained only trace amounts of ^{14}C. In fact, whole-body bioconcentration factors (BCFs) for radiolabel from ^{14}C-A_{16}PB following a 24-h exposure were only 21, 22 and 13 for clams, minnows and tadpoles, respectively. Although the exposure time was very short in this study, the results are consistent with the data of Krzeminski et al. [54] and Lewis and Wee [39]. The latter reported whole-body BCFs for DTDMAC in bluegills of 32 (well water) and 13 (river water) following a 49-day exposure period (Table 3). The data suggest that QACs bind preferentially to the negatively charged surfaces of gill tissues and/or intercalate into gill membranes, but are not transported across membranes. Such behavior is consistent with the known mechanism by which QACs bind to and kill bacteria [77].

Knezovich (personal communication) also examined the effect of sediment and montmorillonite on A_{16}PB bioavailability. The presence of a layer of bottom sediment reduced the amount of ^{14}C in gill tissue of clams and tadpoles relative to sediment-free controls by 36 and 29%, respectively, in 7-day exposures. Montmorillonite (100 mg/L) had a much greater effect, reducing gill uptake by 96% in clams and 99% in tadpoles. However, the clay mineral had a much smaller effect on ^{14}C uptake by tadpole intestine (29% reduction), and settled sediment had no effect at all. The data suggest that sorption of QACs reduces bioavailability to gill tissue, the primary site at which acute toxicity is manifested, but not necessarily to the gastrointestinal tract. The effect of sediment on $A_{16}PB$ availability to gill tissue thus is consistent with earlier observations of reduced toxicity in the presence of solids [39, 98].

QACs are toxic to algae and possibly higher plants as well. Walker and Evans [94] found that $A_{16}TMAB$, $A_{12}DMBAC$, $DA_{10}DMAC$ and $A_{16}PB$ inhibited growth of green algae (*Chlorella* sp.) and great duckweed (*Spirodela oligorhiza*) at 3 to 5 mg/L. The data of Lewis and Wee [39] on algal toxicity of DTDMAC, summarized in Table 9, indicate that algae may be generally more sensitive than fish and invertebrates. An apparent effect of QAC sorption to suspended solids again was observed, but even in water containing solids, algistatic concentrations were below 1 mg/L in some cases.

Recent studies by Lewis and coworkers [95–97] have explored the relationship between laboratory and field data on QAC toxicity to algae. Laboratory-derived EC_{50} values for growth of selected species were compared to effect concentrations for measures of algal community structure [95] and function [96], derived from lake studies in which natural phytoplankton communities were exposed to QACs in situ. In general, laboratory-derived EC_{50} values were many times lower than FECs for measures of community structure and function.

Table 9. Phytotoxicity of DTDMAC to freshwater and marine algae [39]

Algae	Dilution water	Suspended solids (mg/L)	Algistatic concentration (mg/L)[a]
Freshwater			
Selenastrum	distilled	0	0.23
capricornutum	White R. (autoclaved)	3–5	0.71
	Rapid Creek (autoclaved)	131	2.6
	Rapid Creek (filtered)[b]	131	>4.0
	distilled[c]	0	>0.1<0.5
Microcystis	distilled	0	0.32
aeruginosa	White R. (autoclaved)	3–5	0.21
	distilled[c]	0	0.10
Navicula	distilled	0	>1≤10
seminulum	distilled	0	0.5
Marine			
Dunaliella tertiolecta	artificial sea water	—[d]	>0.5≤1.0

[a] Nominal level of active ingredient
[b] Filtered through a 0.45-μm filter, then 131 mg/L of river sediment (dry wt) added to replace suspended solids removed by filtration
[c] QAC tested was DTDMAMS, not DTDMAC
[d] Not reported, but presumably zero

Table 10. Algal toxicity of QACs in laboratory tests and in situ [96]

QAC	Lake photosynthesis studies, 3-h EC_{50} (mg/L)		Laboratory growth studies, 96-h EC_{50} (mg/L)[a]		
	mean	range	Sc	Ma	Np
DTDMAC	6.4	0.4–31.9	0.06	0.05	0.07
A_{12}TMAC	2.2	0.4–6.1	0.19	0.12	0.20
A_{16}TMAB	0.6	0.1–2.6	0.09	0.03	
saturated imidazolium compound	1.9	0.3–3.5	0.60	0.45	
unsaturated imidazolium compound	3.5	11.0–18.4	0.30	0.21	

[a] Laboratory tests were conducted once with each test species
Sc = *Selenastrum capricornutum*; Ma = *Microcystis aeruginosa*; Np = *Navicula pelliculosa*

Selected data are presented in Table 10. Lewis et al. [97] reported similar findings for river periphyton communities.

Clinical Toxicology

QACs appear to be relatively nontoxic to mammals. They are in some instances irritating to skin and mucous membranes of eyes, but in general they have a low order of toxicity. Ethoxylated QACs such as EEQAMS have given negative results in Ames mutagenicity tests and in the Mouse Lymphoma assay. Typical results of acute and chronic toxicity as well as dermal and eye irritation studies are shown in Table 11.

QAC Residues in the Environment

The available data suggest that QACs are ubiquitous contaminants of sewage and surface waters, at least in populated areas. At the same time, levels are considerably lower than for anionic and nonionic surfactants, as would be expected. Monitoring has been conducted at several wastewater treatment facilities treating primarily domestic sewage, and the results generally confirm the expectation of high removal rates ($\geqslant 90\%$). Unfortunately, data on QAC residues in aquatic sediments near sites of industrial or domestic effluent discharge are virtually nonexistent.

Table 11. Clinical toxicity of QACs

QAC	Acute oral LD$_{50}$, mg/kg[a]	Chronic oral	Skin irritation[b]	Eye irritation, mg/kg body wt[b]	Ref.
TTMAC	>500, >600		1%, irritation and tissue breakdown	1%, >30 0.1%, <30	[109]
soyatrimethyl-ammonium Cl	800, 300	1:25 dilution, no toxicity		10%, 14.3	[109]
cocotrimethyl-ammonium Cl	1000, 500			10%, 17.2	[109]
dicocodimethyl-ammonium Cl	1100, 600	1:1000 dilution, no toxicity	2, 5, 10%, mild irritation	2%, 16	[109]
DTDMAC	7000	1 g/kg body wt/day, threshold	2, 5, 10% for 21 d, mild irritation	10%, 11.7	[109]
A$_{12}$TMAC	250–300			1%, 3.6 10%, 59.6	[109]
A$_{16}$TMAC	1000			1%, 3.6 10%, 47.5	[109]
A$_{18}$TMAC	250–300			10%, 11.9	[109]
ADMBAC	0.3–0.35 ML/kg	2.98 ML/kg, acute dermal	0.1%, nonirritating and nonsensitizing	0.1%, nonirritating	[109]
A$_{16}$TMAB	410				[110]
A$_{16}$PC	200				[110]
dodecylimidazoline	3200				[110]
EEQAMS	>15 000 mg active/kg[c]				[12]
PEQAMS[d]	>5000[e]				[12]

[a] Except as noted
[b] Water solution of QAC in all cases. Solution concentrations are in weight percent
[c] 75% active ingredient tested
[d] PEQAMS = polyethoxylated ethanaminium quaternary ammonium methyl sulfate
[e] 90% solids tested

Michelsen [45] analyzed random samples of Swiss sewage having various amounts of input from metallurgical processes and/or the textile industry. Levels of cationic substances, obtained by sublation-TLC, ranged from 0.04 to 0.45 mg/L. Huber [35] and Kupfer [36] described monitoring studies in Germany, and Wee [40] determined levels of DTDMAC in untreated sewage and final effluent from a plant in the United States by three analytical methods: DBAS, DBAS-TLC and HPLC. These data are summarized in Table 12. Kupfer [36] found that levels of cationic surfactants in sewage varied with time of day and day of the week, the highest concentrations being recorded in the evening hours on the "main laundry days" of Monday and Tuesday.

Topping and Waters [37] conducted a detailed monitoring study of QACs in full-scale activated sludge plants in England and Germany. At a plant in Germany, overall removal of DBAS and $DA_{18}DMAC$ was 87.5 and 94.3%, respectively, from initial levels in raw sewage of approximately 1.6 mg/L. Most of the removal—85% for DBAS and 92.5% for $DA_{18}DMAC$—occurred in the biological treatment stage, not in the primary clarifier. At the Alderley Edge plant in England, overall removal was 92.5% for DBAS and 95.7% for $DA_{18}DMAC$. Other data discussed by the authors, from monitoring by the Procter and Gamble Company in Belgium, show that DBAS and $DA_{18}DMAC$ were removed to about the same extent in a trickling filter plant.

Matthijs and de Henau [104] carried out the most comprehensive study to date. Composite samples were collected daily for 8 consecutive days at the Ham plant in Taunton, England and a plant in Heidelsheim, Germany, and analyzed by HPLC for monoalkyl quaternaries. These QACs are used in lower quantities than fabric softening agents such as DTDMAC, so concentrations in sewage and surface water should be lower. The data are summarized in Table 13. Removal was lower in the trickling filter, in contrast to the results cited above: total removal of monoalkyl quaternaries at the Ham plant averaged 97% in activated sludge, but only 80% in the trickling filter. A mass balance study was also performed at the Lüdinghausen plant in Germany, with results similar to those of Topping and Waters [37].

Levels of QACs in receiving waters are typically in the low microgram per liter range. Huber [35] reported concentrations of cationic surfactants of 5 to 20 μg/L (Michelsen-TLC analysis) for the River Main near Frankfurt, West Germany. Kupfer [36] later confirmed these figures. Schneider and Levsen [105] used field desorption in combination with collisionally activated decomposition in a tandem mass spectrometer to determine levels of $DA_{18}DMAC$ in water. The concentrations of $DA_{18}DMAC$ in sewage and surface water samples collected near Bonn, West Germany were 0.35 to 0.48 mg/L and 6 to 12 μg/L, respectively. The most recent European study is that of Matthijs and de Henau [104], who focused on C_{12} to C_{18} monoalkyl quaternaries. Eighty grab samples were collected from rivers in Germany, France and England. German rivers had the highest levels of QACs, with typical concentrations in the range of 3 to 4 μg/L. Over 70% of the samples from France and England had levels below 1 μg/L.

Table 12. QACs in treated and untreated sewage in Germany and the United States

QAC	Analytical method[a]	Mean, concentration μg/L (range) Influent sewage	Mean, concentration μg/L (range) Final effluent	Mean overall removal (percent)	Comment	Ref.
Total cationic surfactants	Michelsen-TLC	80–600 (50–1300)	10–100 (10–200)	83–88	continuous sampling of plant near Frankfurt; functioning poorly	[35]
same	same	200–540 (100–1000)	10–20 (8–25)	94–96	same; functioning well	[35]
same	same	300–600	3–20	94–97	continuous sampling of plant in Taunus	[35]
same	DBAS and Michelsen-TLC	(800–1300)			continuous sampling of Niederrad plant, on River Main near Frankfurt; max. values for laundry days	[36]
same	same	(30–100)			same; min. values for all days	[36]
same	same	540	20	96.3	same; laundry day	[36]
same	same	230	12	94.8	same; non-laundry day	[36]
DTDMAC[b]	DBAS	720 (620–830)	65 (46–80)	91.0	grab sampling at unidentified activated sludge plant	[40]
DTDMAC	DBAS-TLC	290 (250–330)	31 (22–48)	89.2	same	[40]
DTDMAC	HPLC	240 (180–330)	40 (28–56)	83.4	same	[40]

[a] See text for details
[b] Total cationic substances reported as DTDMAC; DBAS alone cannot discriminate between DTDMAC and other cationics

Table 13. Monoalkyl quaternaries in treated and untreated sewage in England (Ham plant) and Germany (Heidelsheim plant) [104][a]

Alkyl group	Ham					Heidelsheim[b]		
		Activated sludge		Trickling filter				
	Influent µg/L	Effluent µg/L	Mean percent removal	Effluent µ/L	Mean percent removal	Influent µg/L	Effluent µg/L	Mean percent removal
C_{12}	49	1.3	97.3	7.2	85.3	<1[c]	<0.5	—
C_{14}	33	1.1	96.7	6.3	80.9	<1	<0.5	—
C_{16}	25	1.0	96.0	5.7	77.2	26	1.1	95.8
C_{18}	26	1.1	95.8	6.6	74.6	50	2.2	95.6

[a] Daily composite samples at both locations (8 days at Ham, 7 days at Heidelsheim)
[b] Activated sludge plant
[c] There were no C_{12} to C_{14} products on the German detergent market at the time of sampling

Wee and Kennedy [38] collected random samples from several rivers in the United States, and reported concentrations (HPLC analysis) of 5 to 30 µg/L for DTDMAC and < µg/L for A_{16}PC, A_{12}TMAC and A_{18}DMBAC. Lewis and Wee [39] subsequently conducted a more extensive study, in which grab samples were collected at various distances downstream from wastewater treatment facilities. Mean DTDMAC levels (DBAS-HPLC method) were <2, 24, 17 and 33 µg/L for Millers River (MA), Otter River (MA), Blackstone River (MA) and Rapid Creek (SD), respectively. Wee [40] compared the results of analyses for DTDMAC in river water by three methods. Levels of cationic substances decreased with distance below a sewage treatment plant regardless of analytical method, but were much higher with the relatively nonspecific DBAS method than with DBAS-TLC or HPLC. Apparent concentrations of DTDMAC in grab samples collected at distances from 7 to 88 km downstream from the plant ranged from 191 to 100, 77 to 50, and 37 to 13 µg/L by DBAS, DBAS-TLC and HPLC, respectively.

Lewis and Wee [39] and Federle and Pastwa [76] provide the only data on QACs in aquatic sediments that we have seen to date. In the first study, sediment samples were collected from Rapid Creek at distances from 0.8 to 88 km downstream from a sewage outfall. DTDMAC levels averaged 23 mg/kg over 18 samples (range = <3 to 67 mg/kg). In the second, DTDMAC was present (HPLC method) at 0.63 mg/L in surface water, and 9.7 (0 to 0.6 m depth) and 7.4 (0.6 to 1.2 m depth) mg/kg dry weight in subsurface sediment, at a pond that had been receiving untreated wastewater from a laundromat since 1962. This sampling location is probably not representative of the majority of sites where effluent discharges containing QACs are expected, however. The relatively high levels reported for Rapid Creek sediment are expected given the sorptive tendencies of QACs.

Summary and Conclusions

Consumption of cationic surfactants was estimated to be 190 000 and 150 000 metric tonnes in 1987 in the U.S. and Western Europe, respectively ([7], Table 2). It is likely that most or at least a large portion of this was sewered. Thus, cationics represent a major class of potential environmental contaminants. This was recognized by the Interagency Testing Committee (ITC) in its 22nd report to the EPA [106], which recommended that IQAMS and EEQAMS undergo testing for chemical fate and health and ecological effects, based on expected widespread release and exposure.

As a class, QACs sorb strongly and rapidly to suspended solids in well-mixed test systems, and form 1:1 complexes with anionic surfactants. Most QACs should also be readily biodegraded in acclimated biological treatment systems, although ultimate degradation may not always be extensive. Overall, QAC removal in wastewater treatment should generally exceed 90%. In receiving waters, mono- and dialkyl quaternaries should be quite rapidly biodegraded in the water column, with half-lives for ultimate degradation in the range of several days or less. However, the aquatic fate of alkyldimethylbenzyl and alkylpyridinium QACs, and other fabric-softening agents like IQAMS and EEQAMS, is not as well characterized. All QACs should enter receiving waters largely as complexes with anionic surfactants, sorbed to sludge solids. QACs that are not rapidly biodegraded are expected to partition to sediments following release to surface waters, but their fate there is not well known.

Whereas the germicidal activity of many QACs is well established, these chemicals will not pose a significant risk to wastewater treatment systems under normal conditions. But slug doses could upset plant function, and nitrification may be especially sensitive. QACs are also toxic in the milligram per liter range and lower to aquatic organisms, including fish, invertebrates and algae. Acute toxicity is reasonably well characterized for mono- and dialkyl quaternaries, but not for other major classes. Data on chronic toxicity are also limited. Results from studies on tissue uptake and distribution of QACs are consistent with the anticipated mechanism of action (interference with gill function). Sorption to sediment apparently reduces acute toxicity and bioavailability to gill tissue, but availability in the gut is still an open question.

In 1982, Huber [11] identified several areas in which definitive studies were needed. Among these were

(i) biodegradability and behavior in anaerobic digestors and benthic sediments;
(ii) chronic ecotoxicity;
(iii) physical/chemical properties and behavior of "electroneutral" surfactant complexes; and
(iv) development of better analytical methods, with the aim of improving analysis of environmental samples.

Through research efforts mainly in the last 10 years, much has been learned about QAC toxicity and behavior in wastewater treatment and the environment. Yet it seems to us that, of the items listed above, only (iv) has received adequate attention.

We therefore call upon interested parties to initiate definitive studies in the other areas. QAC behavior in sediments has received some attention recently [73, 74], but more work is needed. Results from such studies may provide guidance as to the need for further work on chronic toxicity. Studies to define the hazard to deposit-feeding aquatic organisms and detritivores from exposure to sediment-sorbed QACs should be a high priority.

References

1. Reck R (1982) In: Grayson M (ed) Kirk-Othmer encyclopedia of chemical technology, 3rd edn, vol 19. John Wiley, New York, p 521
2. Hardy FE, Talkes BE (1980) U.S. Patent 4 238 373, assigned to the Procter and Gamble Company, 9 December 1980
3. Chem Eng News (1987) 65:21
4. Charles H Kline and Co Inc (1981) Household cleaning products 1981
5. SRI (1984) Specialty chemicals. Strategies for success. Specialty surfactants, vol 9. SRI International, Menlo Park, CA
6. Chem Econ Eng Rev (1986) 18:32
7. Roes IJI, de Groot S (1988) In: Proceedings, 2nd world surfactants congress, "Surfactants in our World-Today and Tomorrow", 24–27 May 1988, Paris
8. SRI (1981) Chemical economics handbook. SRI International, Menlo Park, CA
9. USEPA (1979) Needs survey. Conveyance and treatment of municipal wastewater. Summaries of technical data. U.S. Environmental Protection Agency, Washington, D.C. EPA-430/9-79-002
10. Gerike P, Fischer WK, Jasiak W (1978) Water Res 12:1117
11. Huber L (1982) Tens Deterg 19:178
12. Procter and Gamble (1984) Unpublished information submitted to A Stern (USEPA) by T Mooney (The Procter and Gamble Co), 12 September 1984
13. Woltering DM, Bishop WE (1989) In: Paustenbach DJ (ed) The risk assessment of environmental and human health hazards: a textbook of case studies. Wiley, New York, p 345
14. Games LM, King JE, Larson RJ (1982) Environ Sci Technol 16:483
15. Larson RJ, Vashon RD (1983) Dev Indust Microbiol 24:425
16. Weber JB, Coble HD (1968) J Agric Food Chem 16:475
17. Barbaro RD, Hunter JV (1965) Proc Indust Waste Conf, Purdue Univ 20:189
18. Moraru VN, Markova SA, Ovcharenko (1981) Ukr Khim Zh 47:1058
19. Skrylev LD, Streltsova EA (1979) Zh Prikl Khim 52:1493
20. Putnam FW (1948) Advan Protein Chem 4:79
21. Knox WE, Auerbach VH, Zarudnaya K, Spirtes M (1949) J Bacteriol 58:443
22. Klotz IM (1950) Sci Monthly 70:24
23. Chaplin CE (1951) Can J Bot 29:373
24. Mueller H, Krempl E (1963) Fette Seif Anstrich 65:532
25. May A, Neufahrt A (1976) Tens Deterg 13:65
26. McQuillen K (1950) Biochem Biophys Acta 5:463
27. Salton MRJ (1951) J Gen Microbiol 5:391
28. Fugita T, Koga S (1966) J Gen Appl Microbiol 12:229
29. Salt WG, Wiseman D (1968) J Pharmac (Suppl) 20:14S
30. Mackrell JA, Walker JRL (1978) Int Brodeterior Bull 14:77

31. Ginn ME (1970) In: Jungermann E (ed) Cationic surfactants. Marcel Dekker, New York, p 341
32. Swisher RD (1970) Surfactant biodegradation. Marcel Dekker, New York
33. Neufahrt A, Lötzsch K, Pleschke D, Spaar G (1983) Comun Jorn Com Esp Deterg 14:9
34. Scowen RV, Leja J (1967) Can J Chem 45:2821
35. Huber L (1979) Munchn Beitr J Abwasser Fischerei Flu Biol 31:203
36. Kupfer W (1982) Tens Deterg 19:158
37. Topping BW, Waters J (1982) Tens Deterg 19:164
38. Wee VT, Kennedy JM (1982) Anal Chem 54:1631
39. Lewis MA, Wee VT (1983) Environ Toxicol Chem 2:105
40. Wee VT (1984) Water Res 18:223
41. Llenado RA, Jamieson RA (1981) Anal Chem 53:174R
42. Waters J, Kupfer W (1976) Anal Chim Acta 85:241
43. Cain RB (1977) In: Callely AG, Forster CF, Stafford DA (eds) Treatment of industrial effluents. Hodder and Stoughton, London, p 283
44. Osburn QW (1982) J Am Oil Chem Soc 59:453
45. Michelsen ER (1978) Tens Deterg 15:169
46. Pitter P (1961) Sb Vysoke Skoly Chem Technol Praze, Technol Vody 5:25
47. Uhl A, Sedlmayer H (1965) Brauwelt Jg 105:1529
48. Pauli O, Franke G (1971) Gesundwes Desinfekt 63:150
49. Barden L, Isaac PCG (1957) Proc Inst Civ Engr 6:371
50. Gerike P (1982) Tens Deterg 19:162
51. Wierich P, Gerike P (1981) Ecotoxicol Environ Saf 5:161
52. Janicke W, Hilge G (1979) Tens Deterg 16:117
53. Gerike P, Fischer WK (1979) Ecotoxicol Environ Saf 3:159
54. Krzeminski SF, Martin JJ, Brackett CK (1973) Household and Personal Products Indust 10:22
55. Sulllivan DE (1983) Water Res 17:1145
56. Ruiz Cruz J (1987) Grasas Aceites 38:383
57. Sheets WD, Malaney GW (1956) Sew Indust Wastes 28:10
58. Gawel LJ, Huddleston RD (1972) American Oil Chemists' Society national meeting, 23–26 April 1972. Los Angeles, CA
59. Dean-Raymond D, Alexander M (1977) Appl Environ Microbiol 33:1037
60. Masuda F, Machida S, Kanno M (1978) In: Proc. 7th international congress on surface-active agents. Moscow, U.S.S.R., p 129
60a. Miura K, Yamanaka K, Sangai T, Yoshimura K, Hayashi N (1979) Yukagaku 28:351
61. Larson RJ (1979) Appl Environ Microbiol 38:1153
62. Larson RJ, Wentler GE (1982) Soap/Cosmetics/Chemical Spec 58:33
63. Larson RJ (1983) Residue Rev 85:159
64. Kaplin VT, Zernova LS, Kosogova AS (1968) Ghidrokhim Mater 44:196
65. Baleux B, Caumette P (1977) Water Res 11:833
66. Ruiz Cruz J, Dobarganes Garcia MC (1979) Grasas Aceites 30:67
67. Larson RJ, Perry RL (1981) Water Res 15:697
68. Vives-Rego J, Vaque MD, Sanchez Leal J, Parra J (1987) Tens Deterg 24:20
69. Ruiz Cruz J (1979) Grasas Aceites 30:293
70. Ruiz Cruz J (1981) Grasas Aceites 32:147
71. Ventullo RM, Larson RJ (1986) Appl Environ Microbiol 51:356
72. Larson RJ, Bishop WE (1988) Soap/Cosmetics/Chemical Spec 64:58
73. Shimp RJ, Young RL (1988) Ecotoxicol Environ Saf 15:31
74. Shimp RJ (1989) Environ Toxicol Chem 8:201
75. Larson RJ (1984) Household and Personal Products Indust 21:55
76. Federle TW, Pastwa GM (1988) Ground Water 26:761
77. Lawrence CA (1970) In: Jungermann E (ed) Cationic surfactants. Marcel Dekker, New York, p 491
78. Sayama N (1981) Nippon Eiseigaku Zasshi 35:869
79. Van Beneden G (1952) Bull CEBEDEAU no. 17:159
80. Manganelli R, Crosby ES (1953) Sew Indust Wastes 25:262
81. Reynolds L, Blok J, de Morsier A, Gerike P, Wellens H, Bontinck WJ (1987) Chemosphere 16:2259
82. Manganelli R (1952) Sew Indust Wastes 24:1057
83. Fenger BH, Mandrup M, Rohde G, Kjaer Sorensen JC (1973) Water Res 7:1195

84. Larson RJ, Schaeffer SL (1982) Water Res 16:675
85. Pitter P, Horska M (1968) Sb Vysoke Skoly Chem Technol Praze, Technol Vody F14:19
86. Taft CH (1946) Texas Rpt Biol Med 4:27
87. Pessoa SB (1952) Folia Clin Biol 18:137
88. Vallejo-Freire A, Ribeiro OF, Ribeiro IF (1954) Science 119:470
89. Knauf W (1973) Tens Deterg 10:251
90. Fischer WK, Gode P (1977) Vom Wasser 48:247
91. Reiff B, Lloyd R, How MJ, Brown D, Alabaster JS (1979) Water Res 13:207
92. Kappeler TU (1982) Tens Deterg 19:169
93. Waters J (1982) Tens Deterg 19:177
94. Walker JRL, Evans S (1978) Marine Pollut Bull 9:136
95. Lewis MA (1986) Environ Toxicol Chem 5:319
96. Lewis MA, Hamm BG (1986) Water Res 20:1575
97. Lewis MA, Taylor M, Larson R (1986) In: Cairns J (ed) Multispecies toxicity testing. American Society for Testing and Materials, Philadelphia, PA, p 241. ASTM STP 920
98. Woltering DM, Larson RJ, Hopping WD, Jamieson RA, de Oude NT (1987) Tens Deterg 24:286
99. Cooper JC (1988) Ecotoxicol Environ Saf 16:65
100. Marchetti R (1964) Riv Ital Sostanze Grasse 41:553
101. Biesinger KE, Stokes GN (1968) J Water Pollut Contr Fed 58:207
102. Neufahrt A, Eckert HG, Kellner HM, Lötzsch K (1978) CED Congress, March 1978. Madrid
103. Knezovich JP, Lawton MP, Inouye LS (1989) Bull Environ Contam Toxicol 42:87
104. Matthijs E, de Henau H (1987) Vom Wasser 69:73
105. Schneider E, Levsen K (1986) Comm Eur Commun, EUR 10388 Org Micropollut Aquatic Environ, p 14
106. ITC (1988) Fed Regist 53:18196
107. Kunieda H, Shinoda K (1978) J Phys Chem 82:170
108. Pittinger CA, Hand VC, Masters JA, Davidson LF (1988) In: Adams WJ, Chapman GA, Landis WG (eds) Aquatic toxicology and hazard assessment, 10th vol. American Society for Testing and Materials, Philadelphia, PA, p 138. ASTM STP 971
109. Armak Co (1980) Toxicity data for aliphatic nitrogen compounds. Armak Co, Chicago, IL
110. Deichmann WB, Gerarde HW (1969) Toxicology of drugs and chemicals. Academic, New York, p 690

Phosphate

D. Gleisberg

Hoechst AG, Werk Knapsack, D-5030 Hürth

Introduction 180
Incidence and Distribution 181
Significance for Life. 182
Analysis 184
Toxicity. 184
Production and Use 185
Pathways in the Environment 189
Role in Eutrophication. 192
Elimination from Waters 194
Recycling. 199
Regulations 200
Conclusions 202
References 202

Summary

The element phosphorus exists in nature as compounds with oxygen, the phosphates. Phosphates are omnipresent in nature: in the lithosphere, biosphere, hydrosphere and atmosphere. Because of their unique physico-chemical properties phosphates are essential and indispensable for all organisms. This fact demonstrates that phosphates are non-toxic. The majority of phosphate deposits in four continents of the world consists of marine-sedimentary apatite, a calcium phosphate. About ninety per cent of the steadily increasing phosphate rock production is used in food production. The rest of the raw phosphate rock is used in technical applications in many different fields e.g. water softening, detergents, cleaning agents, food industry, toothpastes, flame retardants, phosphatization, anticorrosion. Also raw phosphates are the basis of many other phosphorus containing products of the chemical industry. Organisms not only take in vital phosphates with their food but they also continuously release them. Thus, phosphates concentrated in deposits are distributed over the world and find their way into surface waters. In former times phosphate contents of surface waters were limiting for plant growth by being too low in proportion to other nutrients. Increasing phosphate concentrations can promote the growth of water organisms, which can finally overburden the limited oxygen content of waters, especially lakes. Existing phosphate elimination processes effectively control phosphate loads of sewage and water inputs of lakes. By using phosphate elimination processes it is possible to recycle phosphates, thus helping the preservation of phosphate resources e.g. by the agricultural use of sewage sludges.

Introduction

To feed a permanently growing world population increasing quantities of raw phosphates are extracted from the earth's mineral deposits, currently running at nearly 170 million tonnes a year. Approximately ninety per cent of these raw phosphates are used by man for the production of foodstuffs. A great deal of it reaches surface waters via the phosphate containing wastes, excrements and erosions and is no longer available for use (Fig. 1). This is the same with technical phosphates. Known phosphate rock deposits which may last a further two centuries will be gradually exhausted in this way [2]. To this general reserve problem a second problem has to be added: Phosphates discharged into surface waters can stimulate plant growth. But an increase in water organisms also increases the oxygen demand needed for their aerobic decomposition after death. This may overtax the limited and differently renewable oxygen balance of the waters to the detriment of the oxygen consuming water fauna and the drinking water supplies quality. This danger first became apparent with the more sensitive lakes, and measures to control phosphates were first taken with these lakes. The so called "eutrophication problem" is mentioned particularly in direct connection with the use of phosphate in detergents. However, restriction of detergent phosphates has never restored surface water quality. The measures should be directed to the expansion and improvement of the sewage treatment the more so as phosphate free detergents also burden waters and may in special cases promote algal growth too [3]. An excellent example demonstrating the sanitation effect of advanced sewage treatment is Lake Constance [4, 5].

Wherever phosphate loads from diffuse sources are larger than from canalized sources phosphates have to be eliminated from the water inputs of the

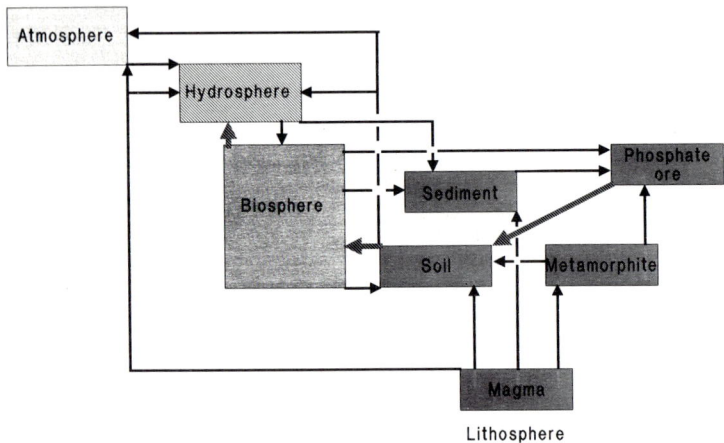

Fig. 1. Pathways of phosphates in the environment [1]

affected lakes, as e.g. the case at the Wahnbach reservoir near Bonn and at Lake Tegel in Berlin [1]. Stagnant waters are mostly threatened by eutrophication and therefore require a particularly large restriction of the phosphate input.

Incidence and Distribution

Phosphates are found everywhere in the environment. This is particularly so in the biosphere, because the living world cannot exist without phosphates. The primary phosphate source is, of course, the soil. Phosphorus amounts to 0.1% of the composition of the Earth's 16 km thick crust. Phosphates are an essential constituent of more than three hundred minerals and, thus, widespread in the lithosphere [6]. The most important phosphate mineral is apatite $(Ca_5(PO_4)_3(F, Cl, OH)$. It is found in crystallized form in the igneous rock and contains about 40% diphosphorus pentoxide P_2O_5. Sediments of the disintegration products of this apatite together with animal teeth and bones formed large deposits of phosphate ores such as those found in North Africa and in North America. The main constituent in bones and teeth is the hydroxyapatite. From the lithosphere and biosphere, phosphates pass into the hydrosphere. There they partially precipitate in the sediments. Seas, springs and groundwaters contain approximately 0.001 to 0.1 mg of phosphours per litre, the so-called mineral waters sometimes considerably more.

The transfer of phosphates from the lithosphere and biosphere to the hydrosphere has increased considerably as a result of man's actions. The main reasons are the growing world population and the consequently increased food production. In order to increase the food production the extraction of phosphate rocks quintupled over the thirty years between 1950 and 1980 (Fig. 2). During the

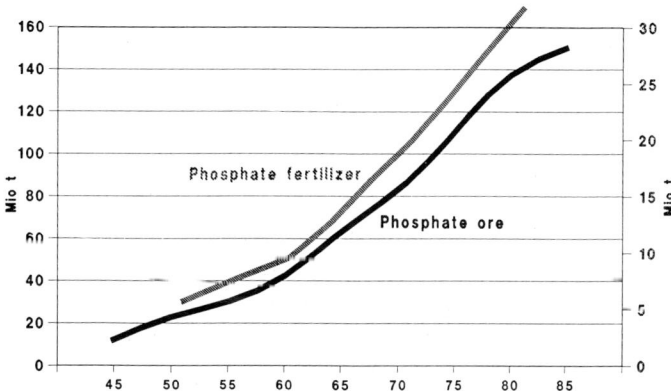

Fig. 2. World production of raw phosphate and consumption of phosphate fertilizer [1]

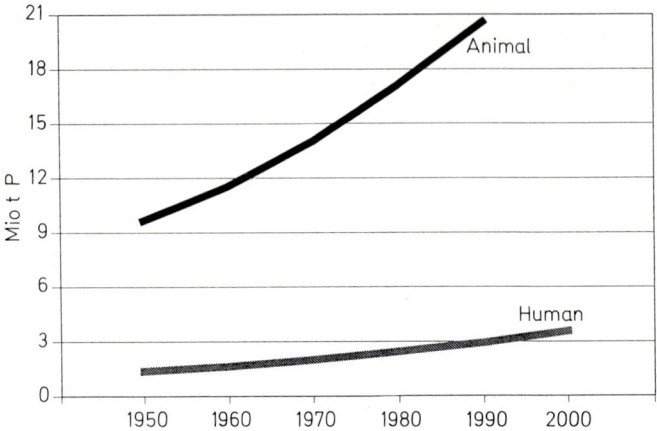

Fig. 3. Development of the phosphorus load from excrements from human beings and animals of the world in the second half of the 20th century [1]

same period world population increased by approximately 75% from 2.5 to 4.45 billion [1]. Parallel to the increase in yields from plant cultivation there is also the addition to livestock breeding. In 1980, phosphates from animal excrement were seven times that from human excrement (Fig. 3). Animal excrement again goes predominantly to plant fertilization. It are mainly bound up in the soil and in plants. But liquid wastes from livestock breeding also reach surface waters directly. Together with fertilizer phosphates, plant wastes and mineral phosphates, animal excrement contributes to the phosphate loads of surface waters due to rain run off and wind erosion.

In many countries the water closet is the pathway by which human excrement reaches surface waters. On the way there is often sewage treatment. For hygienic reasons this step is essential. Sewage treatment also has the task of eliminating phosphates from waste waters to protect waters and preserve phosphate resources.

Significance for Life

In the living world phosphates have important and multiple functions. Nature makes use of the unique properties of the phosphates, essentially based on their high valency and their stability. They are able to form three series of salts and thus act as a buffer and help to maintain a neutral pH value of the body fluids, such as blood. In the form of hydroxyapatite $Ca_5(PO_4)_3OH$ they are essential and stable constituents of bones and teeth, a similar form as they are found in phosphate ores.

Due to their key functions in the photosynthesis of organic compounds from inorganic materials, i.e. as a carrier molecule and energy supplier, phosphates play an important role in the beginning of life. Photosynthesis produces high energy compounds which are decomposed again by respiration:

$$106CO_2 + 16NO_3^- + HPO_4^{2-} + 122H_2O + 18H^+$$
$$P \updownarrow R$$
$$(C_{106}H_{263}O_{110}N_{16}P) + 138O_2$$

The balance between production (P) and respiration (R) is responsible for the regulation of the concentration of free oxygen in the atmosphere.

Phosphates accumulate in the adenosine triphosphate (ATP) of organisms, the carrier of energy necessary for physiological reactions. The total sum of the amounts of ATP built up and split each day in the human body corresponds to the body weight, i.e. about 70 kg. The phosphorus content of the human body is about 1%, and about 80% of this is in the skeleton. Plants contain about 0.1–0.5%, and animals about 0.5%. For the existence of a genetic material such as deoxyribonucleic acid (DNA) at least a divalent bond is needed for the linking of this polyester.

The nucleotide chain is formed by production of a phosphoric acid ester, which is hydrolytically stabilised by the negative electric charge at the phosphate. Only phosphoric acid meets these conditions, and there are no alternatives in nature [7].

The organisms generally have to retain the metabolites of cellular metabolism within the cell walls. This is possible only when the molecules are ionised. This role is assumed by phosphates, and at the same time they play the important function of a catalyst in the reactions, which go in an energetically favourable manner by ester formation and hydrolysis. Biochemistry uses phosphate ester anions that undergo slow hydrolysis in the absence of enzymes, but rapid hydrolysis in their presence. An example of a high energy organic phosphate compound is:

$$O^-NH$$
$$|\|$$
$$^-O-P-N-C-N-CH_2-COO^-$$
$$\|||$$
$$OHCH_3$$

creatine phosphate, the energy reserve of the muscle of vertebrates, which permits quick ATP regeneration (high phosphate group transmission potential).

A key position in several metabolic reactions during which C_2-fragments are converted is held by the acetyl co-enzyme A, a pyrophosphate compound. In the citrate cycle the acetyl group is transferred inward by the acetyl co-enzyme A. These few examples demonstrate convincingly that organic life without phosphates cannot be imagined.

Analysis

Phosphates are found in different forms which are essentially based on orthophosphoric acid. The conventional analytical procedure determines phosphates by photometry as orthophosphates. These, in the presence of antimony and in combination with hexaammonium heptamolybdate in acid solution give, after reduction with ascorbic acid, a blue coloured molybdatophosphate. The intensity of colour is proportional to the orthophosphate content [8]. Phosphate contents are usually reported as P.

In the case of hydrolizable condensed phosphates the sample is boiled and decomposed. Stable organic P-compounds decompose by oxydation (nitric acid/sulphuric acid). By these means the different species can be differentiated.

In sewage treatment plants the phosphate contents, especially of effluents, are determined without differentiation into species, because normally monophosphates are preponderant.

Automatic analysers employ a similar operating procedure. They are used to control precipitant dosage and the efficiency of the phosphate removal process.

In order to differentiate individual phosphate species chromatographic procedure are also used [9, 10]. The recovery of mono-, di- and triphosphate by anion exchange column chromatography was of 97 to 95% with concentrations of 40 ppb P in tap water. Investigating the quantity of monophosphate in the sewage from a municipal sewage treatment plant we find a great increase from untreated waste water to sewage plant effluent after mechanical-biological treatment [9]. Condensed phosphates are biologically hydrolized.

Chromatographic methods are also applied in foodstuff chemistry, in order to differentiate the individual phosphate species [11].

When testing natural water samples with equal orthophosphate contents (by the molybdate method), in order to determine algae contents, a different algal growth was found. The phosphates existed in varying availability for algae, because they were present in different forms. They can be separated by means of molecular sieves. Phosphates are attached to macromolecules and other organic compounds, but when treated with aluminium chloride, precipitant phosphates of all molecular sieve fractions precipitated [12].

Toxicity

According to the EC directive No L 229/19 as well as to the German drinking water treatment regulation, the limiting value for phosphates in drinking water is 5 mg/l P_2O_5 i.e. 2.2 mg/l P. This figures have no relevance to toxicity.

As vital compounds, phosphates are harmless with regard to toxicology. In common with all salts they share the property that excessive amounts affect the

Table 1. Acute toxicity levels of phosphates in animals by oral application [13]

Phosphate	Animal	LD_{50} (mg/kg)
NaH_2PO_4	G. Pig	(>2000)*
$Na_2H_2P_2O_7$	Mouse	2650
$Na_2H_2P_2O_7$	Rat	>4000
$Na_4P_2O_7$	Mouse	2980
$Na_5P_3O_{10}$	Mouse	3210
$Na_6P_4O_{13}$	Rat	3920
Hexametaphosph.	Mouse	7250
$(NaPO_3)_3$ cyclic	Mouse	10300
NaCl (table salt)	Mouse	5890

* Approx. lethal dose

osmotic pressure of the body fluids. The LD_{50} (lethal dose) of different phosphates (Table 1), which are also used in foodstuffs, ranges between 2650 and 10 300 mg/kg body weight (mice and rats) when applied orally. By comparison, for NaCl the LD_{50} is 5890 mg/kg [13]. The acute toxicity levels of phosphates in animals exceed the normal daily P intake of humans by a factor of about fifty. A dietary level of 5–7 g/day of monosodium orthophosphate produced no adverse effects in humans [13].

The assumption that disorders in the behaviour of children (hyperkinetic syndrome) were caused by or sustained by an oral intake of phosphates could not be proved by clinical examination [14].

Production and Use

Phosphate ores are the raw materials for all phosphorus products. Phosphate ore production increased from about 145 million metric tonnes in 1985/86 to about 165 million metric tonnes in 1988 (Table 2). The shares of the biggest producers are given in Table 3.

About 80% of the world phosphate ores is mined from sedimentary deposits, about 18% from igneous deposits and the rest from island deposits. Typical compositions of igneous and sedimentary apatites are shown in Table 4.

By 1995, a phosphate rock demand of more than 200 million tonnes is expected [16]. The economically exploitable resources in known deposits, which contain some 36.6 billion tonnes of ore, are sufficient to last for nearly 50 years, assuming an average annual growth in production of 5% [2].

Additional identified phosphate resources in known deposits, which are insufficiently explored or currently uneconomic, amount to about three times as much as the economically exploitable resources [2].

Table 2. World production of phosphate rock [15]

	Millions of tonnes			
	1985	1986	1987	1988
Production	147.89	144.51	153.97	164.93

Table 3. Share of phosphate rock production of the ten major world producing countries [15]

	Percentage		
	1985	1986	1987
U.S.A.	33.3	27.2	27.8
U.S.S.R.	23.1	24.1	23.9
Marocco	14.0	14.8	14.4
China	4.7	6.6	6.2
Jordan	4.1	4.4	4.7
Tunisia	3.1	4.2	4.4
Brazil	2.8	3.2	3.3
Israel	2.7	2.6	2.6
Togo	1.7	1.6	1.8
South Africa	1.7	2.1	1.8

Table 4. Chemical composition of igneous and sedimentary apatites [2]

	Percentage					
	CaO	MgO	Na_2O	P_2O_5	CO_2	F_2
Igneous deposits						
Kola	55.13	0.42	0.38	40.30	0.42	3.35
Quebec	55.16	0.20	0.25	40.77	0.20	3.42
Sedimentary deposits						
Marocco	54.1	0.81	0.28	38.2	2.32	4.33
Florida	55.1	0.32	0.65	35.8	3.90	4.22

At present, the fertilizer industry uses more than 90% of phosphate ore production. Worldwide, phosphate fertilizer consumption increased by 2 million tonnes P_2O_5 from the years 1982/83 to 1986/87 (Table 5).

Technical phosphates also find their way into foods, e.g. for drinking water treatment, and as additives for food and fodder. These uses amount to about 2% of phosphate ore production. Thus, about ultimately 93% of phosphate ore production is used in human feeding.

According to IFA estimates, the share of raw phosphates used for technical purposes will reduce from 9.5% in 1985 to 9.25% in 1993. Reduction of the use of detergent phosphates contributes decisively to this change (Fig. 4). At present about 3.5% of ore production is used in the production of detergent phosphates.

Table 5. World phosphate fertilizer production and consumption [17]

	(10^3 tonnes P_2O_5)			
	1983/84	1984/85	1985/86	1986/87
Production	35 093.6	36 556.6	34 625.3	36 424.9
Consumption	32 709.1	33 946.7	33 243.7	34 709.3

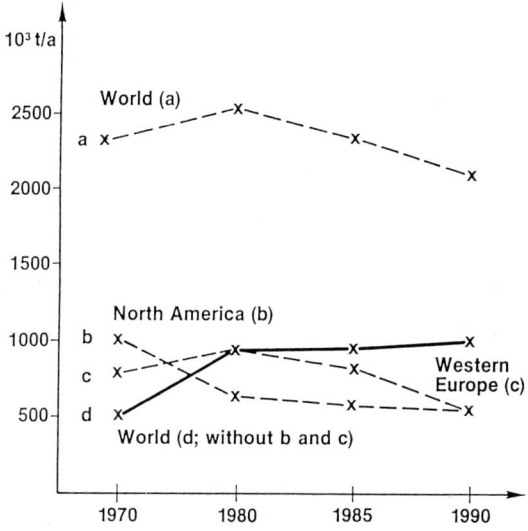

Fig. 4. Consumption of detergent phosphate STPP in P_2O_5 [18]

In order to produce single P-containing fertilizers, the phosphate ore is treated in three different ways:

1. Grinding raw phosphates of a soft consistency, to produce "ground rock" (Hyperphosphate).
2. Calcination, to produce sintered or melted phosphates.
3. Chemical reaction with sulphuric or phosphoric acid to produce super-, or triple super-phosphate.

The byproduct of the steel industry, basic slag or Thomas phosphate, belongs to the single P fertilizer group. Single P fertilizers are usually processed into multiple nutrient fertilizers.

Reactions of raw phosphate with nitric acid produces the complex fertilizer Nitrophosphate.

Usually for production of technical phosphates, the route through phosphorus or pure phosphoric acid is followed. Starting products for the preparation of technical phosphates are phosphorus pentoxide (P_4O_{10}), thermal phosphoric acid, or wet process phosphoric acid which is cleaned by precipitation or solvent extraction.

1. Preparation of thermal phosphoric acid [19]

$$2Ca_5(PO_4)_3F + 8SiO_2 + 15C \longrightarrow 3P_2 + 15CO + 6CaSiO_3 + Ca_4Si_2O_7F_2$$

$$P_4 + 5O_2 \longrightarrow P_4O_{10}$$

$$P_4O_{10} + 6H_2O \longrightarrow 4H_3PO_4$$

2. Wet process phosphoric acid

$$Ca_5(PO_4)_3F + 5H_2SO_4 + 5xH_2O \longrightarrow 3H_3PO_4 + 5CaSO_4 \cdot xH_2O + HF$$

These phosphoric acids and different phosphates find many technical applications in industry and household. The most important fields of application are water softening, cleaning and corrosion prevention.

Pentasodium triphosphate $Na_5P_3O_{10}$ (STPP) has the largest economical importance of the technical phosphates. For water treatment triphosphate (tripolyphosphate) has, in comparison with monophosphate the advantage of not only sequestering the hardness salts, but also of holding them in solution.

Particularly notable is the threshold effect of the polyphosphates which inhibits the calcium carbonate formation in the case of very small understoichiometric amounts. The excellent performance of tripolyphosphate during washing is not only based on these effects but also on a series of other good washing characteristics [18]:

— Removal of incrustations on the fibrous tissue
— Dispersion of pigment soils
— Synergistic intensification of the effect of anionic surfactants
— Contribution to the alkaline reaction
— Carrier for the other detergent powder constituents (builder).

By hydrolysis in the sewage STPP loses its complexing property. It is the only known detergent constituent which can be recycled for reuse.

Complexation of metal ions is one of the properties of phosphates used in food systems. The many applications of phosphates in the food stuff sector reflect their manifold functions in the living world, as well as the properties of phosphates [13]:

— Interaction with organic polyelectrolytes such as protein, pectin, starch (structure, consistency)
— pH alteration
— Buffering/pH stabilization
— Dispersion of food constituents
— Emulsion stabilization
— Hydration/Water binding
— Solubility
— Preservation of food (bacteriocidal action)

Different sodium and potassium salts of mono-, di-, tri- and metaphosphate are mainly used for food products.

Pathways in the Environment

Only a part of the phosphate stream starting by extraction of ores and intended for human use reaches man directly. The largest proportion is consumed in foodstuff production.

According to the Alimentation Report of 1988 adults take in—as an average for men and women—about 1.6 g/day, vegetarians taking more and non-vegetarians taking less than the average amount [20]. The amount of phosphate excreted by man corresponds to the amount taken in. For decades it has remained on the average (± 0.2 g) relatively constant [21]. This demonstrates that phosphates exist in all foodstuffs so that changes in the so called consumption pattern only affect the individual phosphate load to a certain degree (Table 6). Food phosphates are added to some foods during processing and on the average influence the phosphate load by about 0.1 g phosphorus per inhabitant per day, i.e. about 5% of the total phosphate load, found in this range of variation. With food leftovers and kitchen waste some 0.3 g phosphorus per inhabitant per day gets into the household waste water. Thus, 1.9 g per inhabitant per day has to be taken as the daily inhabitant load from foodstuffs, into the domestic sewage [21].

Additionally the following still reach sewage from households (g P per inhabitant and day):

0.2 from dishwashing products
0.1 from laundry dirt [21]
0.1 from household dirt

$0.4 + 1.9$ g $= 2.3$ g P is the average daily inhabitant specific phosphate load in domestic sewage without detergent phosphate.

In 1989 an average of 0.25 g P from detergents per inhabitant per day were fed into domestic sewage in the Federal Republic of Germany. The market share of phosphate containing detergents was reduced to about five per cent during the first six months of 1989 (Fig. 5) as a result of reformulation of most detergents.

Domestic sewage forms part of the municipal sewage, which also receives phosphate from manufacturing and industry as well as from erosion and precipitation (rain and snow) entering the sewers. It is not possible in this case to state inhabitant specific values of phosphate loads, but only estimated total amounts.

Phosphates from trade and industry which entered surface waters in the Federal Republic of Germany in 1987 corresponded to about one third of those

Table 6. Natural phosphate contents of food in milligram phosphorus per 100 g [1]

full cream milk	92
Emmental cheese (45% fat in dry weight)	636
fresh cream cheese (50% fat in dry weight)	170
curd cheese (20% fat in dry weight)	165
a hen's egg	216
veal (fillet)	200
beef (fillet)	164
pork (fillet)	173
salami	167
soft pork sausage	160
rice (unpolished)	325
oat flakes	391
wheat flour (type 550)	95
rye bread	140
rolls	102
white bread	89
noodles	191
herring	250
mackerel	244
trout	242
cauliflower	54
savoy cabbage	56
peas	378
soja beans	591
potatoes	50
edible boletus	115
apples	12
peaches	23
peanuts	372
orange juice	15
white wine	15
beer (light)	28
coffee	192
tea	314

coming from domestic sewage [21]. The treatment of food (for instance dairies, slaughter houses, canning and potato processing factories) play an important part in this connection. Canalized sewage is localized and thus a controllable phosphate source. During sewage treatment phosphates can be eliminated. Therefore, this measure has a high priority in water protection measures.

The control of non localized diffuse phosphate sources, however, causes considerable difficulties. Erosion, precipitation and drainage are the most important sources of diffuse phosphate additions to surface waters. By these paths phosphates coming from soil particles, plant residues, phosphate fertilizers and animal excrement get into surface waters. Of particular importance in this connection is the production of liquid manure during livestock farming.

Fertilization by use of liquid manure has to be carried out in a manner which guarantees that soil and plants are sufficiently able to absorb the phosphate and to minimize erosion. As a first measure the number of cattle units must be limited

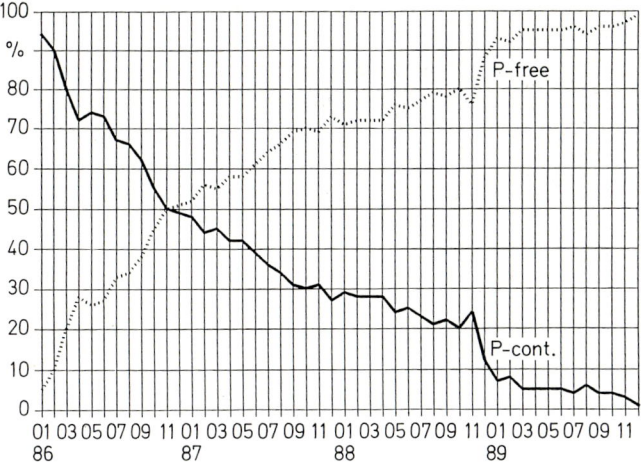

Fig. 5. Development of the market share of phosphate-containing and phosphate-free detergents in the Federal Republic of Germany

Table 7. Yearly excretion of phosphorus of domestic animals [1]

	Phosphorus kg per animal and year
cattle	10
horses	8.8
sheep	1.2
pigs	1.5
hens	0.13

in relation to the hectare of soil to be fertilized. In Switzerland the maximum is four units of heavy livestock per hectare of soil to be fertilized, which still can be reduced depending on the situation [22]. Heavy livestock units are counted as cattle weighing 600 kg, which excrete about 10 kg P per year [1]. Phosphate excretion of other domestic animals are given in Table 7. In the Federal Republic of Germany the number of cattle was around 15.4 million in 1987, and pigs about 24.6 million (statistical yearbook).

Furthermore in 1987/88 approximately 675 000 metric tonnes of phosphate fertilizer calculated as P_2O_5 were still consumed in the Federal Republic of Germany.

However, in both cases only a part of the above mentioned phosphate amounts come into the surface waters through erosion. The time (vegetative phase) and the kind of fertilizer application, the kind of the plants, the soil and the soil treatment influence the erosion results. Plant availability of phosphate fertilizer quickly decreases by ageing, an adsorption process proceeding the stronger the more acid the soil.

An additional barrier for eroded phosphates is a buffer zone with fertilization restriction or ban along the borders of surface waters which exists for instance in Switzerland (Lake Sempach).

In 1987 the share of agriculture and natural sources, including forestry, in the total entry of phosphates into the German surface waters was about 30% of the total [21].

Role in Eutrophication

The weight percentage of the phosphorus in algae is much lower than the percentage of carbon and nitrogen; it is sometimes even less than the percentage of potassium or calcium (Table 8). By calculation, less than one gram of phosphorus is sufficient to produce 100 g algae. This calculated result cannot, however, be transferred into practice because quick recycling in algae cultures occurs, but sometimes phosphates are not be available for the algae or are stored by them in excess. Thus, the calculation only shows that algae can get along with small quantities of phosphorus. Carbon, the main nutrient, is available for algae in large quantities from carbon dioxide. The situation for nitrogen is similar, particularly since special kinds of algae (cyanophytes) can bind atmospheric nitrogen. In former times there was a lack of phosphorus in many surface waters so that phosphorus played the part of the minimum factor controlling plant growth. Lakes naturally but slowly proceed towards eutrophication. Due to the evolution of civilization the contents of all macro- and micro-nutrients in the

Table 8. The percentage of different elements of the dry matter of three algae [23]

	Microcystis (%)	Anabaena (%)	Cladophora (%)
Carbon	46.5	49.7	35.3
Nitrogen	8.1	9.4	2.3
Phosphorus	0.7	0.77	0.56
Potassium	0.8	1.2	6.1
Calcium	0.53	0.36	1.7
Sulfur	0.27	0.53	1.6
Iron	0.27	0.08	0.23
Magnesium	0.17	0.42	0.23
Sodium	0.04	0.18	0.18
Manganese	0.03	0.008	0.10
Zinc	0.005	0.000	0.001
Copper	0.004	0.007	0.019
Boron	0.0004	—	0.0085

surface waters generally increased. As a consequence algae blooms increased in meteorologically favourable periods (temperature, light). Under oxygen depletion the dead and deposited algae are decomposed by bacteria in surface waters and their nutrients are set free. They are partially deposited in the sediment from which the phosphate, bound with trivalent iron, is dissolved again if there is an anaerobic situation which reduces iron. Stagnant waters are most threatened by the oxygen depleting decomposition process due to the limited oxygen supply. Algal growth is limited by turbulence, light shortage, influence of inhibiting substances, lack of trace elements and nutrients. With regard to the nutrient phosphorus, a study of a large number of lakes showed that in most cases the productive level of a lake is determined by its phosphorus load and not by its nitrogen load [24]. This is also true to a certain extent for dams and reservoirs. In studying the response of lakes to eutrophication and restoration measures, it is important to divide the lakes into shallow and deep categories [25]. With rivers, estuaries and seas the conditions are different. There we have insufficient research results. In the open North Sea nitrogen is rather the minimum factor [26]. Also with some inland lakes this can be the case [27]. Mutual effects between N, P, Fe and Mn and other elements (Si) generally influence the growth of algae [28].

The strong reduction in the content of a main nutrient for algae increases the probability of a diminution in the frequency of algal blooms. According to the incidence, the quantity and the elimination, the biggest chance is for phosphorus to become the growth limiting factor. As the surface waters have stored phosphates in their sediments there is, under certain conditions, the possibility to supply the algae from this source. In order to produce any reduction in the massive algae developments the phosphate concentration has to be reduced strongly.

In Lake Constance already between 1924 and 1951 the yearly average of phytoplankton density rose from 80 to 660 cells per milliliter. But in the last thirty years dramatic changes in phytoplankton and oxygen balances in stagnant waters have been happening worldwide. As a consequence of the permanent increase in the sewage load the riparian states around Lake Constance began to instal relief measures in 1960. Sewage treatment plants were built and precipitation purification processes for phosphate removal installed. In the mid 1970s the increase in phosphate content of Lake Constance was halted (Fig. 6) and during the early eighties the treatment measures of sewage installed finally led to a gradual decrease in the phosphate content of Lake Constance [4]. In 1985 the "International Water Protection Commission for the Lake Constance" extended the investment programme and the standards for treatment in order to reach the objective of 30 mg/m^3 ($\mu g/l$) dissolved phosphate-phosphorus during the 1990s [5]. Reductions, with different deadlines, for detergent phosphates in the riparian states are not reflected in changes in the P-concentrations in Lake Constance. Phosphate elimination from waste waters, also comprising detergent phosphates, dominates the non-decisive detergent phosphate contribution. As a "Threshold value" for algal growth research scientists found in the laboratory a

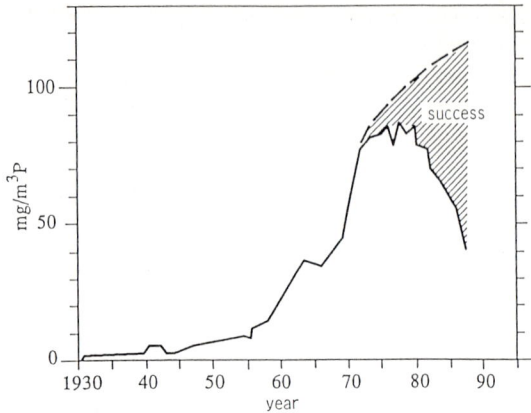

Fig. 6. Development of the P-concentration in the Lake Constance [4]

Table 9. Proposed criteria for the trophic state of lakes [24]

Parameter		Oligotrophic	Mesotrophic	Eutrophic
Total-Phosphorus	$\bar{X}$	8.0	26.7	84.4
mg/m³	$\bar{X} \pm SD$	4.85–13.3	14.5–49	38–189
	Range	3.0–17.7	10.9–95.6	16.2–386

phosphorus concentration of 10 mg/m³ [29]. The actual threshold for algal response was found to be at an inlake concentration of 240 mg/m³ for "Veluwemeer" and 60 mg/m³ for "Schlachtensee" [25]. The content of dissolved phosphate-phosphorus decreased from 1980 until spring 1989 in Lake Walenstadt from 18.5 to 2 mg/m³, in Lake Zurich from 76 to about 59 mg/m³ and in Lake Constance/Lake Uberlingen from 93 to 48 mg/m³. The yearly average of algae biomass was very close together in 1980, although these lakes exhibited three different trophic states. During the following years the algal biomass of the oligotrophic Lake Walenstadt was nearly like that of the mesotrophic Lake Constance [30]. In fact, the P-concentration of a lake does not equate with the general situation of the lake.

The oligotrophic state of a lake (Table 9) is not conclusively necessary for ecological reasons, not always desirable for economical reasons and not always technically attainable.

Elimination from Waters

The traditional mechanical and biological sewage treatment eliminates phosphates bound to solids with settling particles as well as phosphates consumed by the activated sludge (bacteria). About 35% of the phosphates can be eliminated in this way. It is possible to extend both phosphate elimination processes of conventional sewage treatment. Dissolved phosphate can be chemically precipitated. Precipitation purification is the most common and the longest known process in the world for further phosphate elimination. Soluble iron or aluminum salts form insoluble precipitation products with dissolved phosphates which can be separated.

$$AlCl_3 + Na_3PO_4 \longrightarrow AlPO_4 + 3NaCl$$

$$Fe_2(SO_4)_3 + 2Na_3PO_4 \longrightarrow 2FePO_4 + 3Na_2SO_4$$

Mono- and even condensed phosphates are eliminated by the precipitation purification. The anions of the precipitating salts remain in the water. They may replace OH groups of the hydroxides formed by the excess precipitants and, thus, partially remain in the precipitation sludge [31]. In practice the mol ratio Me/P $= \beta$ is about 1.5 relative to the total P-influx [32]. The resulting precipitates show atomic ratios of Me/P > 1 [33]. Basic phosphates are precipitated. Hydrolysis products of the precipitants, which are polynuclear and charged, coagulate finely suspended particles in the waste water. This flocculation process also eliminates considerable quantities of colloidal substances, for instance of the sum parameters BOD (Biochemical oxygen demand) or COD (Chemical oxygen demand), bacteria and heavy metal compounds (coprecipitation). The effect of the precipitation purification therefore goes beyond pure phosphate elimination. It is also used for other purposes besides phosphate elimination, for instance to relieve overburdened municipal sewage plants or to clarify industrial sewage.

The process of precipitation purification consists of the following steps: nucleation, crystal growth, flocculation, phase separation. Nucleation and crystal growth are very fast processes in the precipitation of phosphates with aluminum and ferric salts. This is also the case for the precipitation of hydroxides and oxides from stoichiometric excess aluminum and iron. The adsorption of phosphate to these substances is another fast process so that phosphate binding takes place quickly. The rate determining step in the process is the agglomeration of the so-formed molecular groups to separable particles, this is a slow process. The time and energy needed for the two main parts of the process are different. The fast mixing of the precipitant with the sewage requires high energy input, the slow flocculation process a low energy input. The shearing strength of the built flocs is low.

Besides the precipitants mentioned, ferrous salts are also suitable for precipitation purification but only for application in the biological stage. Aeration oxidizes the ferrous to ferric ions and thereby increases their phosphate elimination effect. Ferrous and even aluminum salts cause inhibition of the nitrification process in the biological stage.

The resulting precipitation flocs are separated by solid/liquid phase separation. In traditional sewage treatment this separation is carried out by sedimentation. Both mechanical and biological sewage treatment stages have a sedimentation tank. Thus, the precipitant can be added during the mechanical stage (pre-precipitation) as well as at the biological stage (simultaneous precipitation). The first alternative removes biodegradable organic substances and relieves the biological stage, the second gives higher utilization of the precipitant by the customary sludge return from the secondary sedimentation tank to the aeration tank (Fig. 7). The simultaneous precipitation is the predominant method used especially in Germany and Switzerland. The efficiency of the separation of the precipitation flocs can be increased by the installation of a separate sedimentation tank (post-precipitation) or of a filtration stage. For the flocculation filtration some precipitant is again added to the sewage from which the bulk of the phosphate has previously been eliminated by simultaneous precipitation. In order to save precipitant the precipitation sludge from the post-precipitation will be transferred into the biological stage (Fig. 7).

The mass of sludge of the traditional mechanical-biological sewage treatment will be increased in the case of simultaneous precipitation by 10–15%, mainly as

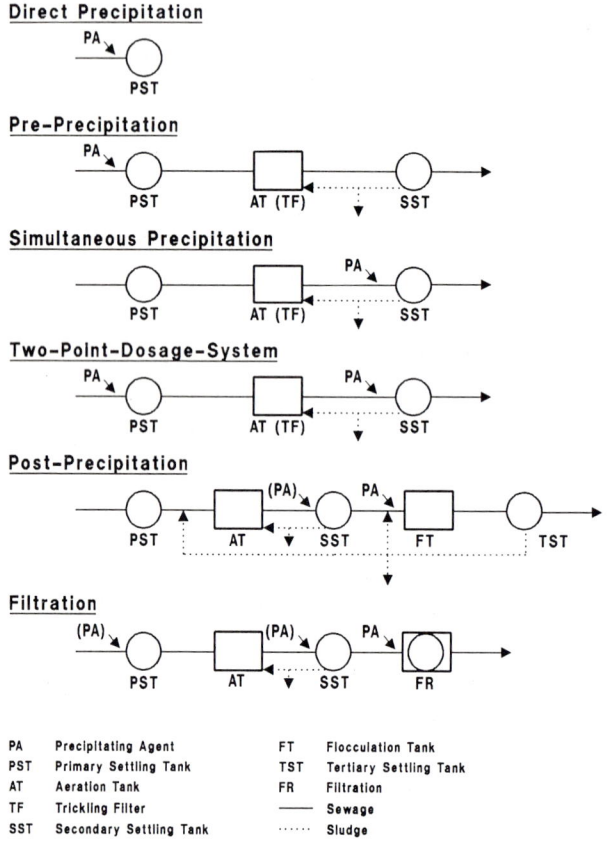

Fig. 7. Methods of precipitation purification in conventional sewage treatment plants

a function of the added precipitant quantity, but the volume is unchanged because the sludge will concentrate better. The pH change due to the precipitant generally is small and depends on the acid capacity of the sewage. On an average basis the phosphate elimination rate over the whole year is 85%. Treatment plants with post-precipitation show a higher efficiency [34]. A still higher elimination rate (approximately 95%) can be achieved by flocculation filtration due to the more effective phase separation compared to sedimentation.

Calcium and phosphate ions in the sewage form a solid phase only very slowly, but at pH values of more than 10.5 suitable rates of formation are achieved. This is essentially accomplished by the addition of lime in order to precipitate the phosphate. The high pH value gives problems at the biological stage. Calcite has an effect as a crystal nucleus for Ca-phosphate and makes phosphate elimination possible even at pH values between 9 and 10.

Ca-phosphate of a conventionally purified sewage precipitates at about pH 10 on the surface of the medium of a fluidized bed. Making use of returning product (medium) a relatively pure granulated calcium phosphate is produced which can easily be dehydrated and used as a raw material for phosphorus production. This phosphate removal by crystallization in a fluidized bed (pellet reactor) was installed in a 10 000 i.e. (inhabitant equivalent) municipal sewage treatment plant in the Netherlands [35]. Another physico-chemical phosphate elimination process uses activated alumina in an additional plant stage after conventional purification. Reactant contact takes place in a stirred vessel or in a filtration bed. The phosphate collected from an acid solution is removed in an alkaline solution and precipitated in a separate tank by means of lime. The lye is recycled into the regeneration stage. The precipitated phosphate is relatively pure and soluble in citric acid [36]. An installation of this type exists in Berlin.

Magnetic sewage treatment processes meet phosphate-emission standards also. Mechanically and biologically pretreated sewage is further treated with Al sulfate or lime, flocculant polymer and magnetite. The flocs which form are attracted by an electromagnetic field to a stainless-steel mesh. The magnetite is recycled. Magnetic separation of the phosphate is speedy and also effective for small particles and takes up only a limited space [37].

At the moment extended biological phosphate elimination is of particular interest. This process is practiced either as a sewage mainstream process in the extended biological purification stage or in a side stream from the conventional sewage treatment plant with a combined biological-chemical process.

In an anaerobic situation aerobic bacteria receive energy for the assimilation of easily degradable substances by liberating phosphates. In the case of the sidestream process a part of the activated sludge from the final clarifier is held in a separate tank in an anaerobic state. The phosphate released by the bacteria to the sewage is precipitated by addition of lime and settled in the precipitation tank. Phosphate-poor activated sludge and separated sewage are recycled into the mechanical-biological sewage treatment plant. The effluent from this plant contains a reduced phosphate level. The reason is that after anaerobic stress bacteria take up phosphate in excess during their stay in the aeration tank. Therefore the phosphate content in the mainstream decreases after passing through a basin, maintained in an anaerobic state, before the aeration tank. By this means the sewage treatment plants in West-Berlin eliminate biologically 3.3–5.5 mg P_{tot} per 100 mg eliminated BOD_5. The maximum P-content of the activated sludge is six percent; therefore precipitation purification for further P-elimination could be required. As nitrates disturb the anaerobic process a denitrification stage (anoxic) has to be added [38].

The principal organism responsible for P-removal is the bacterium *Acinetobacter*, which cannot use glucose but utilises lower fatty acids. In the anaerobic state a facultative anaerobic microflora produces lower fatty acids to serve as a substrate for *Acinetobacter* [39]. Dosage of acetic acid to the anaerobic tank consequently improves phosphate elimination [40]. The presence of sufficient magnesium is necessary for the uptake of excess phosphate [41]. There is an abundance of process variations of the biological phosphate elimination process [42], the operation of which is not completely understood as yet: Bardenpho, Biodenipho, EASC, Phoredox, Rotanox, UCT (Fig. 8).

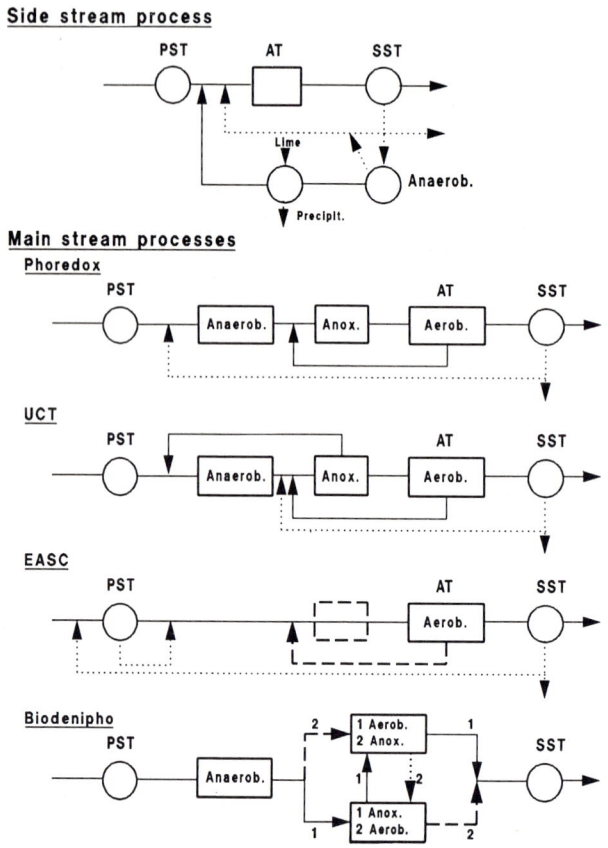

Fig. 8. Methods of extended biological phosphate removal from sewage

Recycling

The elimination of phosphate from waste waters and lake influents also permits its re-use. The fertilizing value of the sewage sludges is considerably increased by the extended phosphate elimination. The precipitated phosphate shows good plant availability [43]. This is an additional incentive for the utilization of sewage sludge in agriculture. However, use has continuously decreased, also in the case of extended phosphate elimination, because of a fear of pollutants. Phosphate precipitation does not cause an increase in the concentration of pollutants. Precipitation sludges, for instance, have considerably lower heavy metal concentrations than conventional sewage sludges [43].

In the Federal Republic of Germany the sewage sludge regulation sets the maximum heavy metal content of sewage sludge and soil as well as the maximum amount of sewage sludge for application. The "Landwirtschaftliche Untersuchungs- und Forchungsanstalt" (Institute for Agricultural Research and Development) in Speyer found that the heavy metal load in the sludges and soils on the average was small. Polychlorinated biphenyls (PCB), if present, are not absorbed by plants for several years even after PCB application [44]. An extensive study by the "Abwassertechnische Vereinigung" (Association for Waste Water Technology) came to the conclusion that the described effects with regard to PCB, dioxines and furanes, in principle, do not jeopardize the use of sewage sludge in agriculture, neither on arable land nor on grassland or on cultivated areas for fodder crops [45].

In 1986, 80% of the sewage sludges were considered suitable for use in agriculture. Provided that this amount of 80% is used agriculturally and applied at a rate of 5 metric tonnes per hectare, in three years according to the German regulation governing the use of sewage sludge about 368 000 hectares could be fertilized. In the Federal Republic of Germany a total area of about 12.2 million hectares is used agriculturally. Calculated on the quantity that is available to plants, sewage sludge has a fertilizing value of about 100 DM/metric tonne. About half of this amount is due to phosphate, ignoring further elimination which is now growing [46]. On average one can calculate 15 kg P_2O_5/tonne sewage sludge and 40 kg P_2O_5/tonne in the case of sewage treatment plants with precipitation purification [47].

Comparative tests in pots with ryegrass (darnel) showed, for post-precipitation products of calcium hydroxide with and without iron chloride, an effect which can be compared to superphosphate and Thomas phosphate. Phosphate that had been precipitated only with iron chloride hardly showed any fertilizing effect, but did so after treatment with calcium oxide. Comparable Al products achieved 80–85% of the yield obtained by the P fertilizers mentioned. A poorer fertilizing effect was shown by the phosphate, precipitated with calcium hydroxide, from the separation with activated alumina. The utilization coefficient of the precipitation products depends on the species of plant. For instance, Al-containing products are readily available for buckwheat [48]. The feasibility of the thermal conversion of sludge phosphate into citrate-soluble calcinated phosphate (Glühphosphat) has been proved. But difficulties can be expected in the electrothermal decomposition of sewage sludge ash in the preparation of elemental phosphorus [49].

In this connection there are better possibilities with the purer products such as the fluidized bed calcium phosphate pellets [35]; but their preparation requires an additional stage in the sewage treatment plant.

Regulations

Up to now in Western Europe agreements and legal regulations requiring a reduction of the use of detergent phosphates have prevailed over rules for phosphate elimination from sewage or for the interception and removal of phosphate influents to surface waters. However, the view is gradually gaining ground that such measures are necessary.

The German Maximum Phosphate Quantity Regulation (MPQR), according to the Detergent Law, (the second phase of which became effective on January 1st, 1984), should achieve a calculated P-reduction in domestic sewage of 23.5% [50].

From 1979 to 1985 the average daily phosphorus load entering five analyzed sewage treatment plants with precipitation was reduced by 20 to 25%; but this reduction was not generally reflected in the effluent P-load [34].

The total phosphorus input from all sources was reduced about 8% by the MPQR instead of the officially predicted 15%. The MPQR remained meaningless for the quality of waters endangered by eutrophication [50]. An effect of the decrease in P-load due to the MPQR on the eutrophic state of lakes and reservoirs could be assumed in only a few individual cases [21], in which other P-reducing measures were also present.

An assessment of the MPQR in comparison to the application of phosphate precipitation shows that the MPQR has negative ecological and economical effects (Table 10). The relative P-input of surface waters in Germany at the

Table 10. Evaluation of the German Maximum Phosphate Quantity Regulation (MPQR) and the increased application of phosphate precipitation [50]

	MPQR	P-precipitation
Effects on water quality	−(−)	+(+)
Economic effects	−(−)	0
Administrative estimation	−	+ +
Efficiency	−(−)	+
Political realizability	+ +	−

Explanation: + + very good; + good; 0 medium; − bad; − − very bad localizable sources did not drop decisively even when the market share of phosphate-free detergents reached nearly two thirds in 1987 (Fig. 9).

Quite a number of the States within the Federal Republic of Germany financially support the phosphate elimination from sewage as a result of programs put immediately into operation after the algal bloom in the North Sea in 1988. They, thus, partially anticipated the Sewage Administrative Regulation issued in September 1989 according to the Water Household Law. This regulation requires a P-effluent value of less than 2 mg/l in sewage treatment plants of between 20 000 and 100 000 i.e. and of 1 mg/l P in the case of sewage treatment plants of more than 100 000 i.e. [51].

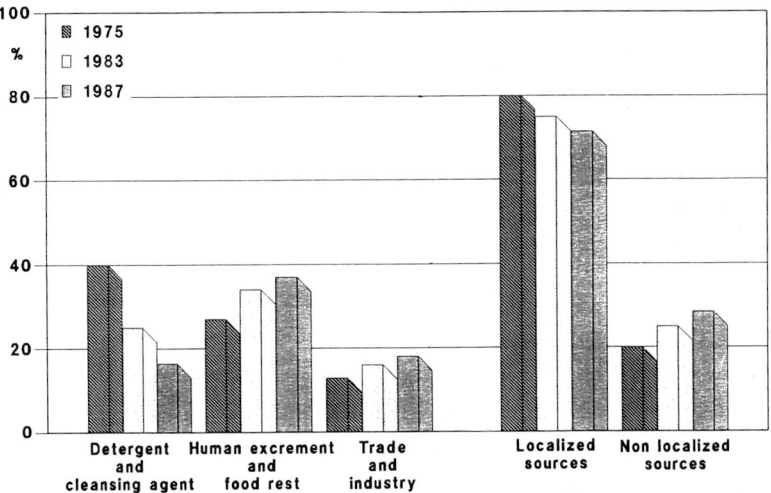

Fig. 9. Relative P-input into surface waters in Germany

Table 11. Phosphate regulations in Europe

Country	Detergent phosphate voluntary	legal	P-elimination programme regional	national
Austria		5.5%	X	
Belgium	Offer P-free			
Switzerland		0.5		X
Denmark		5.5%		X
Italy		1%	X	
Luxemburg	s. Belgium			
Norway		0%		X
Netherlands	>50% P-free			
Sweden	7.5%			X
Finland	7%			X

(X = program established; extent of implementation highly variable)

In the European regions and countries where sewage treatment with additional elimination of phosphate was stimulated obvious progress in the sanitation of surface waters can be observed. But only in 50% of the countries of Western Europe in which legal regulations of detergent phosphates exist are there national programs for phosphate elimination (Table 11).

A rapid development of sewage treatment plants including phosphate elimination, was sponsored by the government in Sweden between 1969 and 1974 by means of subsidies which were connected to the performance of the sewage treatment plants [52]. Today more than three quarters of the whole population are connected to sewage treatment plants with phosphate elimination.

Conclusions

— There is no alternative to living with phosphates, as ubiquitous and vital compounds.
— Phosphates are non-toxic and controllable; hence they present no risk to the environment.
— In the interest of water protection and the preservation of resources phosphate inputs into surface waters have to be reduced significantly.
— There is a wide range of phosphate elimination processes which are technically proved, easy to put into practice, and economic.
— About 5000 sewage treatment plants in the world eliminate phosphates by precipitation purification which may be considered the generally recognized state of technology.
— Phosphates have to be stringently recycled taking into account the growing world population. The raw phosphate resources can then be stretched for some milleniums.
— The possibility of recycling makes phosphates preferable in their technical applications, compared with products which present no possibility of recycling.

Reference

1. Gleisberg D (1988) Chemie in unserer Zeit 22:201
2. U.S. Geological Survey Circular 930-C; Summary Report—Phosphate
3. Booman KA, Sedlack RI (1986) Journal WPCF 58:1092; Beture Setame; Compared impact of household laundry detergents containing or not phosphorus, May (1989), Société d'études d'urbanisme, d'aménagement et d'équipments, St. Quentin
4. Elster H-J (1977) Naturwissenschaften 64:207; Mäckle H (1986) Wasser+Boden 38:603 Internationale Gewässerschutzkommission für den Bodensee; Bericht Nr 40 (1989)
5. Internationale Gewässerschutzkommission für den Bodensee; Bericht Nr 33 (1985)

6. Nriagu JO, Moore PB (eds) (1984) Phosphate Minerals; Springer, Berlin Heidelberg New York
7. Westheimer FH (1987) Science 235:1173
8. DIN 38 405
9. Wernet J, Ebert J (1977) Zf Wasser- und Abwasser-Forschung 1:24
10. Vaeth E, Sladek P, Kenar K (1987) Fresenius Zf Analytische Chemie 329:584
11. Deman JM, Melnychyn P (eds) (1971) Phosphates in food processing; The AVI publishing company, Westport, Connecticut, p 38
12. Steinberg C, Schrimpf A (1980) Vom Wasser 55:295
13. Ellinger RH (1972) Phosphates as food ingredients; Chemical Rubber Company
14. Walther B, Dietrich E, Spranger J (1980) Monatsschr Kinderheilk 128:382
15. IFA, International Fertilizer Industry Association, Paris
16. European Chemical News (1984) February 13:8
17. FAO Yearbook Fertilizer vol 37, 1987 Rome (1988)
18. Kandler J (1987) In: Baldwin AR (ed) Second world conference on detergents. American Oil Chemists Society, p 135
19. Ullmanns Enzyklopädie der technischen Chemie; 4. Auflage (1979) Bd 18
20. Deutsche Gesellschaft für Ernährung e V, Frankfurt, Ernährungsbericht 1988
21. Hamm, A (ed) (1989) Auswirkungen der Phosphathöchstmengenverordnung für Waschmittel auf Kläranlagen und in Gewässern; Academia Verlag Richarz, St. Augustin
22. Baumann P (1984) Gas-Wasser-Abwasser 64:89
23. Centre Européen D'Etudes Des Polyphosphates, Köln; Phosphate und Wasser (1974)
24. Vollenweider RA (1979) Zf Wasser- und Abwasser-Forschung 12:46
25. Sas H (ed) (1989) Lake restoration by reduction of nutrient loading; Expectations, Experiences, Extrapolations; Academia Verlag Richarz, St Augustin
26. Umweltbundesamt; Forschungsvorhaben Wasser 102 04 215
27. Van der Does J, Klapwijk SP (1987) Int Rev gesamten Hyrobiol 72:27
28. Scherfig J, Dixon PS, Appleman R, Justice CA (1973) EPA 660/3-73-015
29. Doemel WN, Brooks AE (1975) Water Research 9:713
30. Stabel H-H, Zimmermann U, Ledergerber HP, Mäckel H (1989) gwf Wasser-Abwasser 130:525
31. Gleisberg D (1979) Berichte aus Wassergütewirtschaft und Gesundheitsingenieurwesen, Institut für Bauingenieurwesen V, Technische Universität München Nr 25
32. Gleisberg D (1988) Korrespondenz Abwasser 35:334
33. Stumm W, Sigg L, (1979) Zf Wasser- und Abwasser-Forschung 12:37
34. Hahn HH, Heiß H-J (1989) gwf Wasser-Abwasser 130:341
35. Eggers E, van Dijk JC (1986/87) Hoechst-Symposium: Entfernung von Phosphaten aus Abwässern und Nutzbarmachung von Klärschlämmen, p 101
36. Donnert D, Eberle SH (1982) Hoechst-Symposium: Weitergehende Reinigung kommunaler Abwässer insbesondere zur Phosphatelimination, p 67
37. Allen DM, Sargent RL, Oberteuffer JA, Nowack HK-H (1978) Kommunalwirtschaft, 9:288
38. Peter A, Sarfert F (1989) Korrespondenz Abwasser 36:242; Kayser R, Ermel G (1982) Hoechst-Symposium: Weitergehende Reinigung kommunaler Abwässer insbesondere zur Phosphatelimination, p 103
39. Wentzel MC, Lötter LH, Loewenthal RE, Marais GR (1986) Water SA 12:209
40. Rensink JH, Donker HJGW, Anink DM (1989) gwf Wasser-Abwasser 130:76
41. Imai H, Endok K, Kozuka T (1988) J Ferment Technol 66:657
42. Yeoman S, Stephenson T, Lester JN, Perry R (1988) Environmental Pollution 49:183
43. Gleisberg D, Taubel N, (1978) Wasser + Boden 29:5
44. Leuning U (1989) Umwelt 19 7/8, 375
45. Keding M, Langenahl T, Witte H (1989) Korrespondenz Abwasser 36:19
46. Malz F (1986/87) Hoechst-Symposium: Entfernung von Phosphaten aus Abwässern und Nutzbarmachung von Klärschlämmen, p 202
47. Malz F, Bortlısz J (1988) abwassertechnik 2·9
48. Baran E (1985) Zusammensetzung und Düngewirkung von Phosphat-Fällungsprodukten aus der dritten Abwasserreinigungsstufe; Dissertation Agrarwissenschaft, Georg-August-Universität Göttingen
49. Schimmel G (1986/87) Hoechst-Symposium: Entfernung von Phosphaten aus Abwässern und Nutzbarmachung von Klärschlämmen, p 179
50. Wicke L, Franke W (1989) Umweltökonomie, Verlag Franz Vahlen, München
51. Bundesminister des Innern (1989) Gemeinsames Ministerialblatt 40:517
52. Kieslinger RM (1982) Korrespondenz Abwasser 29:944

Zeolites

P. Christophliemk, P. Gerike†, and M. Potokar

Henkel KGaA, P.O. Box 1100, D-4000 Düsseldorf 1, FRG

Introduction 206
Nomenclature and Structural Types 206
Physical and Chemical Properties 210
Detergency Properties of Zeolite A 212
Uses and Production Volumes 213
Consumption and Capacities of Detergent Grade Zeolite A 214
Manufacturing Processes 215
Analytical Methods 220
Environmental Fate of Zeolite A 221
Environmental Effects of Zeolite A 222
    Effects in Sewage Treatment Plants 222
    Effects on Aquatic Organisms 223
    Effects on Plant Growth 224
Toxicity of Zeolite A 224
Regulations and Laws 226
References 226

Summary

Zeolites are a well-defined class of crystalline natural or synthetic aluminosilicates with an extensive framework of SiO_4 and AlO_4 tetrahedra, interlinked in different ways by sharing oxygen atoms. The framework cations are exchangeable. Because of their uniform pore sizes from 0.3 to 0.8 nm, zeolites can selectively adsorb or reject molecules based on their size and act as molecular sieves.

Until 1976 zeolites (mainly type Y) were manufactured only on a relatively small industrial scale for use particularly in sorption and catalysis. Although insoluble in water, zeolite A proved to be an excellent detergent builder. By optimization of the detergent formulation and use of cobuilders, the detergency performance of non-phosphate zeolite-containing laundry detergents can reach that of detergents formulated with sodium triphosphate, the traditional detergent builder.

Zeolite NaA is now established as a detergent builder in the U.S., Europe, and Japan with a consumption of more than 700×10^3 t and a production capacity of more than 900×10^3 t in 1991. From an environmental point of view non-detergent applications are much less important.

Among the different zeolites type A has been ecologically and toxicologically scrutinized for a number of years because of its increasing importance as a detergent builder.

A very thorough investigation of the fate and effects of zeolite A as a detergent builder along its disposal pathway has been made—from domestic piping via communal sewage systems into the various stages of sewage works leading to the final receiving water outlets and, via sludge, on arable land. No indications of any adverse effects in any of these stages were found. It has rarely been possible to realize a desired effect—here a lowering of detergent phosphate content and, in the final analysis, a decrease of anthropogenic phosphate input into our surface waters—with so few and so minor, if any, adverse effects.

Health risks or other hazards in connection with preparation, processing, and application have not been found.

Introduction

The chronology of zeolite chemistry is well reported [1–5]. Baron von Cronstedt, a Swedish amateur mineralogist, observed in 1756 that certain minerals, when heated rapidly, released large amounts of water [1]. Therefore he called these minerals "zeolites" from the Greek zeo = I boil and $lithos$ = stone. Already in 1896 such mineral zeolites were used as ion exchangers in the sugar industry. In 1926, McMain recognized their immense practical importance as "sieves" in the molecular range and coined the term "molecular sieve" which nowadays is often used as a synonym for zeolites. The first systematic studies of zeolite syntheses were started in academia in 1938 by Barrer and a few years later also in industry. Since 1954 Union Carbide has distributed samples of the three industrially most important synthetic types of zeolites, A, X, and Y to a large number of companies, particularly within the petrochemical industry, and has commercialized the "Linde Molecular Sieves". This was the beginning of an accelerating world-wide development which has led to about 30 000 scientific publications and more that 10 000 patent applications.

Until the mid-seventies, zeolites (mainly type Y) were only manufactured on a relatively small industrial scale for use particularly in petrochemistry. But with introduction of the first zeolite-based heavy-duty detergent in the Federal Republic of Germany in 1976, production and capacity for detergent grade zeolite (type NaA) has increased rapidly from 50×10^3 t in 1977 to more than 700×10^3 t in 1990. A further increase to more than 1000×10^3 t zeolites for the mid-nineties is expected [6, 7]. Zeolites for non-detergent applications have a production share of about 10% and are from an environmental point of view much less important.

Nomenclature and Structure

According to the common definition, zeolites are natural or synthetic tectosilicates with characteristic structures of their lattices, pores, and cavities containing crystal water, which can be reversibly dehydrated without any essential change in structure [8–14]. The general composition for zeolites is $M_{2/n}O \cdot Al_2O_3 \cdot xSiO_2 \cdot yH_2O$. M is herein an exchangeable cation of valency n. In this simplified oxide formula the chemical composition is referred to $Al_2O_3 = 1$ by definition. The value for x, therefore, directly indicates the ratio of SiO_2/Al_2O_3 which has a particular influence on the ability of the zeolite to exchange cations. In synthetic products the cations M are predominantly ions of alkali metals, alkaline earths, and/or quaternary ammonium compounds. Isomorphous replacement of Si and/or Al (e.g., by P or Ge) in the lattice is also possible and of increasing technical importance.

More than 40 types of naturally occurring zeolites with differing aluminosilicate framework structures are known, some of which are found in large amounts, e.g., in the western U.S.A. However, most types are mineralogical rarities. The number of synthetic types, depending on how exactly they are differentiated, amounts to at least 150 different frameworks, and new ones are added nearly every year. Only very few of these many synthetic products are of technical importance.

Natural and synthetic zeolites are characterized according to certain patterns. The primary structural components are SiO_4- and AlO_4-tetrahedra which are interlinked in different ways by sharing oxygen atoms to form the so-called SBU = Secondary Building Units. Modern zeolite classifications are based on nine secondary building units (Fig. 1): three single rings S4R, S6R, and S8R made up of four, six, and eight tetrahedra, respectively, three double rings D4R (cube),

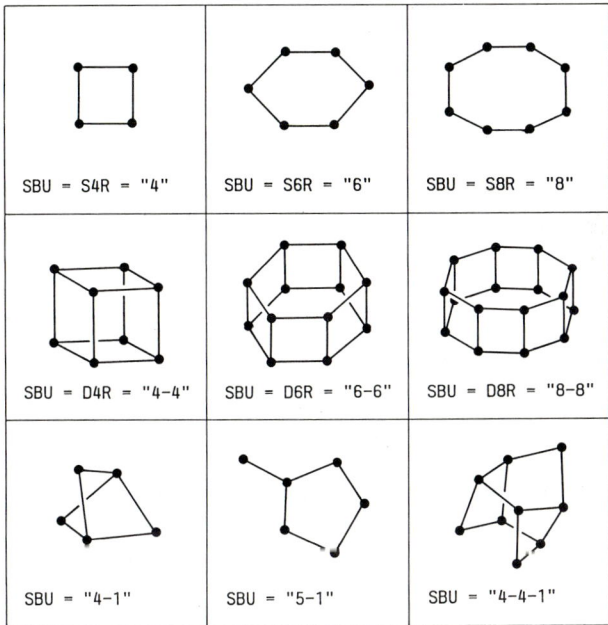

Corner points =
positions of tetrahedral silicons and aluminums

Fig. 1. Secondary Building Units (SBU) of zeolites

D6R (hexagonal prism), and D8R (octagonal prism), as well as three structural units made up of more intricately interlinked rings. This simplified classification concentrates on the number and configuration of tetrahedral aluminums and silicons but without distinguishing between silicon and aluminum atoms. The oxygens are not shown here.

Representative examples of zeolites belonging to the structural groups 1–4 having single rings S4R and S6R and double rings D4R and D6R as SBUs are shown in Table 1. Analcime, phillipsite, and sodalite on a mineral basis are all widespread and synthetically easily accessible. The types X and Y are technically important synthetic analogs of the mineral faujasite. The synthetic zeolite NaA, on the other hand, does not have any mineral counterpart. The unit cells of zeolites are generally very large and less suited to indicate the chemical composition and differences between the zeolites.

Contrary to the ideal composition of the unit cell, the ratio $SiO_2:Al_2O_3$ can in reality vary considerably for different types. Zeolite NaA is one of those types with very low silicate content. The textbook formula is $Na_2O \cdot Al_2O_3 \cdot 2SiO_2 \cdot 4.5H_2O$. The low ratio of $SiO_2:Al_2O_3 = 2.0$ is the main reason for the high cation exchange capacity of zeolite NaA: Stoichiometrically zeolites can be "constructed" by substituting silicon atoms in the three-dimensional SiO_2-lattice by aluminum atoms. Since aluminum is only trivalent, the law of charge neutrality requires that per aluminum atom one additional monovalent alkali atom is added. The higher the aluminum content of the zeolite, the higher its cation exchange capacity. There is a limit to this process. In accordance with the so-called Loewenstein rule, the ratio of $SiO_2:Al_2O_3$ cannot fall below 2 if the tetrahedral structure is to be maintained. Each aluminate tetrahedron then has the full chance of being linked exclusively to silicate tetrahedra. The direct linkage of aluminate tetrahedra by oxygen bridges is unstable. The tetrahedral arrangement and hence the whole zeolite structure collapses when Al–O–Al-bonds are formed.

The zeolite types A, X, and Y have a very similar architecture; as uniform tertiary building units they contain the so-called truncated octahedron, also called "β-cage" or "sodalite-cage", shown in Fig. 2. This polyhedron is bound by 14 faces: 6 squares with octahedral arrangement and two sets of 4 S6R with tetrahedral arrangement. The silicon and aluminum central atoms of the tetrahedra are at the corners of the polyhedron. The linking oxygen atoms are not specifically marked; they lie between the corners, not on but near the connecting lines.

This truncated octahedron therefore displays only the arrangement of the aluminosilicate network. The positions of cations and constitutional water molecules herein are not defined. The truncated octahedra are interlinked by different oxygen bridges. In an idealized truncated octahedron of Zeolite NaA the ratio of $Si:Al = 1.0$ (i.e., ratio $SiO_2:Al_2O_3 = 2.0$) is the lowest possible value. Each silicate tetrahedron is then surrounded only by aluminate tetrahedra and each aluminate tetrahedron only by silicate tetrahedra. Fig. 2 shows furthermore, that in zeolite A eight of these truncated octahedra are connected with

Table 1. Representative examples for zeolites of structure groups 1-4

Structure group	Zeolite type	Typical unit cell contents	Ratio $SiO_2 : Al_2O_3$	Largest pore diameter (nm)
Group 1 SBU = S4R	Analcime Phillipsite	$Na_{16}[(AlO_2)_{16}(SiO_2)_{32}] \cdot 16H_2O$ $(Ca, K_2, Na_2)_5[(AlO_2)_{10}(SiO_2)_{22}] \cdot 20H_2O$	3.6–5.6 2.6–4.8	0.26 0.48
Group 2 SBU = S6R	Sodalite	$Na_6[(AlO_2)_6(SiO_2)_6] \cdot 7.5H_2O$	2.0	0.22
Group 3 SBU = D4R	Zeolite NaA	$Na_{12}[(AlO_2)_{12}(SiO_2)_{12}] \cdot 27H_2O$ $= "Na_2O \cdot Al_2O_3 \cdot 2SiO_2 \cdot 4.5H_2O"$	1.8–2.4	0.42
Group 4 SBU = D5R	Faujasite Zeolite NaY Zeolite NaX	$Na_{13}Ca_{12}Mg_{11}[(AlO_2)_{59}(SiO_2)_{133}] \cdot 23H_2O$ $Na_{56}[(AlO_2)_{56}(SiO_2)_{136}] \cdot 250H_2O$ $Na_{86}[(AlO_2)_{86}(SiO_2)_{100}] \cdot 264H_2O$	4.4–4.6 3–6 2.0–3.0	0.74 0.74 0.74

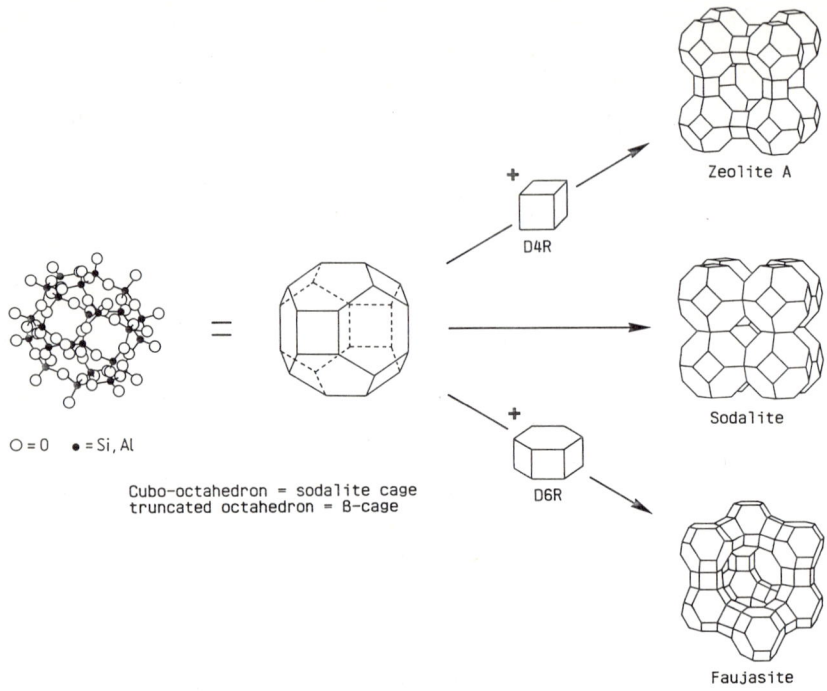

Fig. 2. Different types of aluminosilicates constructed by the same polyhedron

each other by D4R as SBU almost like "yokes", and they surround a large hollow space. This big polyhedral cavity is called the "-cage" and is accessible through 8-ring-pores. In the natural zeolite faujasite and its analogous synthetic types X and Y, however, the truncated octahedra are connected by D6R as "yokes" and hence constitute the bigger faujasite-cage with 12-ring-pores.

Hydroxysodalite is the alkaline synthetic form of sodalite and can result as an undesired by-product of zeolite production, undesired for the very reason that it no longer has any cation exchange capacity. Sodalite is characterized by densely stacked truncated octahedra (similar to a tight sphere packing) without cubic or prismatic "yokes". According to the strict definition, sodalite is not a zeolite but feldspar-like ("feldsparoid"). Eight truncated octahedra are arranged in cubes, enclosing a cavity which again has the form of a truncated octahedron and is, therefore, also called "sodalite cage".

Physical and Chemical Properties

In these and other zeolites, the cavities and pores form a three-dimensional "sieve" with mesh widths in the molecular range. The mesh size of these anionic sieves is considerably influenced by the corresponding cations. Depending on their position in the lattice, their size, and charge, cations have a direct effect upon pore diameter. In Zeolite A, for example, the pore diameter of the 8-ring is 0.42 nm in the sodium modification (= NaA), whereas it narrows down to 0.3 nm when exchanged with potassium ions (= KA), and in the calcium modification (= CaA) it widens to 0.5 nm because of the smaller number of cations in the pore area.

In addition to the molecular structure, the crystal morphology is also important for the resulting properties of the zeolites. Zeolite NaA has distinctive individual crystallites with edges of about 0.1 to 5 µm length, however, typically in the narrower range of 0.3 to 2 µm. Normally the particle size distribution has the shape of a Gaussian curve. These crystallites can have sharp or rounded edges. Figure 3 shows representative examples. For use in detergents small crystallites with rounded edges as in example (**b**) are preferred. The individual crystallites are also called "primary particles". After their agglomeration the resulting units are then accordingly "secondary particles". But these terms are sometimes used differently.

Under normal conditions the microporous system of zeolites is filled with water, which is lost continuously during drying. The molecular-sieve effects are shown mainly by completely dehydrated zeolites. During the adsorption, the micropores fill and empty reversibly. It is possible to tailor the adsorption characteristics in terms of size selectivity or selectivity caused by other interactions, including cation exchange [2, 3, 15–18].

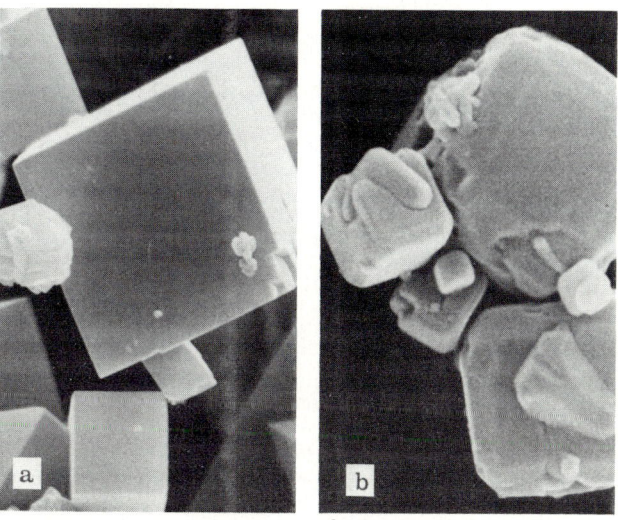

Fig. 3. Scanning electron mircographs of zeolite A. **a)** Particles with sharp edges, magnification 10^4; **b)** particles with rounded edges, magnification 5×10^3

The basic principles of ion exchange within zeolites are well described particularly for aqueous systems [18–20]. The exchange behavior of non-framework cations (selectivity, degree of exchange) depends upon the nature of the cation (the size and charge of the hydrated cation), the temperature, the concentration and, to some degree, the anion species. The selectivity varies with the ratio Si/Al. In zeolite A the selectivity series for unipositive ions is Ag > Na > K > Rb > Li > Cs and Zn > Sr > Ba > Ca > Co > Ni > Cd > Hg > Mg for bivalent cations.

A tremendous amount of literature deals with the catalytic properties of zeolites, depending on type, framework topology, nature of counterions or other sorbed materials, acidity, and reaction conditions.

Detergency Properties of Zeolite A

The effectiveness of zeolite A in the washing process has been examined in a series of basic investigations [21–29]. The known detergency properties are summarized in Table 2.

Due to the small particle size of detergent-grade zeolite A, the water insoluble particles penetrate freely into the mesh of textile fabrics and do not deposit during washing. Under washing conditions the sodium ions can be exchanged for calcium ions from the hard water. The exchange of magnesium ions in the presence of calcium ions occurs only to a small extent. In addition to the high binding capacity for calcium ions, zeolite A has, like sodium triphosphate, a series of other properties important to detergency. They partly correspond to those of sodium triphosphate. Due to the fact that zeolite A is insoluble in water, it also has, however, properties that differ from sodium triphosphate. The adsorptive properties of zeolite A impedes the transfer or readsorption of certain dyestuffs onto the textile surface and decreases the redeposition of some kinds of soil onto the fibers by heterocoagulation with soil particles so that they no more contribute to greying of textiles. In addition, the zeolite particles are a suitable substrate for the crystallization of sparingly soluble compounds, e.g., calcium carbonate or calcium phosphate, the presence of which can be perceived as

Table 2. Detergency properties of zeolite NaA [30]

— High binding capacity for multivalent metal ions, particularly calcium ions
— Enhancement of the action of synthetic surfactants
— Alkaline reaction
— Antiredeposition properties (soil suspension power), particularly by adsorption of molecularly dispersed substances and heterocoagulation with pigments
— Support of the foam depressing action of soaps
— Crystallization surface for sparingly soluble compounds such as, for example, calcium carbonate
— Increase of its effect by the presence of water soluble polycarboxylates or complexing agents (co-builder effect)

encrustation after repeated washing. Trace heavy metal cations, e.g., copper ions, which can adversely affect the bleach performance of perborate by prematurely decomposing it, are also deposited in the form of sparingly soluble salts on the surface of the particles.

Except for the so-called secondary behavior (incrustations, redeposition), the contribution of zeolite A to the detergency performance is somewhat less than that of sodium triphosphate. However, the appropriate optimization of the detergent formulation, particularly by adding co-builders and changing the surfactant system, can make the detergency performance of zeolite-containing laundry detergents comparable to that of phosphate-based detergents [30, 31].

Phosphates have such ideal washing properties that for a long time it was believed that high-quality detergents could not do completely without them. Thus, initially only approximately half of the phosphate in European laundry detergent formulations was replaced by zeolite A. In the meantime the detergent industry has optimized the surfactant composition and combined zeolite A with certain complexing agents thus formulating new, phosphate-free detergents of a quality comparable to that of phosphate-containing products. Such co-builders are, for example, citrate, phosphonates, and polycarboxylates. When these compounds are combined with a zeolite, binary, ternary, or even more complex builder systems with excellent washing properties can be obtained.

By the addition of co-builders, the exchange capacity of zeolite NaA can be better utilized. If the co-builders are added at less than stoichiometric amounts, they take over the function of carrying water hardness; they also have specific effects in dissolving or preventing the formation of sparingly soluble crystalline compounds such as calcite. Already small additions of these carrier substances can accelerate the process of dissolving precipitates by zeolite A to such an extent that the dissolution rates are determined almost exclusively by the water-soluble complexing agent.

Uses and Production Volumes

Initially zeolites were introduced in industry as adsorption materials and, in the early sixties, also as catalysts. Apart from their main use as a detergent builder, zeolites are nowadays being widely used as sorbents ("molecular sieves") for ion exchange and catalysis but without significant environmental impact. Examples are given in Table 3.

By the end of the fifties, worldwide annual sales of zeolites as sorbents totaled no more than 10^6 U.S. dollars, whereas in 1970 they amounted to 40×10^6 and in 1979 already 250×10^6 U.S. dollars. In 1979, about 40×10^3 t of zeolites were manufactured for use in catalysis alone, predominantly for fluidized catalyst cracking (FCC). In 1975 already 95% of all cracking installations

Table 3. Major technical applications of zeolites

Builder in detergents and cleansers	Ion exchange
Adsorption (water, gases)	Water treatment
	Radioactive waste clean-up
Water adsorption/drying	Elimination of ammonium ions (aquaculture)
Insulating of glass panels	
Coolant circuits coating	Catalysis
PUR coating masses	Catalytic cracking
	Hydrocracking
Heat storage	Hydroconversion
Drying/purification of gases	Mobil-process
Gas separation	Fluidized catalyst cracking
Separation of n-/iso-paraffins	"Methanol-to-Gasoline" process
Separation of aromatic compounds	Paraffin isomerization
Separation of paraffins/aromatic compounds	Xylene isomerization
Animal feedstuff	Ethylbenzene synthesis

were operated with zeolite type Y. But it was the use of synthetic zeolites in detergents and cleaners since 1975 that increased production volumes and capacities dramatically within just a few years, particularly in the U.S.A., the Federal Republic of Germany, and Japan. Annual zeolite consumption for partial substitution of phosphates as a detergent builder was about 150×10^3 t in 1981 and about 450×10^3 in 1988 [5–7], whereas the production level for non-detergent applications has been much lower (50–60 $\times 10^3$ t zeolite Y, 2×10^3 t zeolite X, 10^3 t ZSM 5, $< 10^3$ t mordenite annually for use as catalysts).

A detailed list of zeolite manufacturers and their production capacities is given by Rohde [7]. The complex situation within the market for detergent-grade zeolites is described for Europe, Japan, and the U.S.A. in detail by Dietrich and Leonhardt [6]. They expect for the mid-nineties a demand of 900 $\times 10^3$ t $\cdot$ a^{-1} and capacities of more than 1000×10^3 t $\cdot$ a^{-1} detergent-grade zeolites.

Consumption and Capacities of Detergent Grade Zeolite A

Because of different washing conditions and phosphate regulations, the present situation and the medium-term development with respect to the use of zeolite A as a detergent builder are two issues that must be considered separately from each other.

Western and Central Europe. Some 1/10th of the world population presently lives in Europe and consumes 4×10^6 t $\cdot$ a^{-1}, equivalent to 1/4 of all the detergents used worldwide. Of this volume, about 80–95% are heavy-duty detergents. Because of the largely saturated markets, the growing use of liquid laundry detergents is at the expense of powder heavy-duty detergents. Due to frequently high water hardness, the proportion of builders contained in powder products ranges from 20 to 35%. By the end of the eighties, some 40 to 50% of the detergents used in Europe were phosphate-free. This shift towards phosphate-free products will probably be completed by the end of the nineties. In combinations of about 20–25% zeolite A with up to 5% NTA (Switzerland, Netherlands, not in Germany), 3–5% polycarboxylate or 0.5–1.5% phosphonate, zeolite A will constitute the main builder. On the basis of these forecasts we can estimate a need for about 450×10^3 t·a^{-1} of zeolite A in Europe for the early nineties.

Consumption in 1986 was 160×10^3 t while capacities amounted to 240×10^3 t. With a continuing and steady build-up of capacities to more than 650×10^3 t·a^{-1}, long-term supplies of zeolite A are more than secure.

North America. Because of the different washing conditions, heavy-duty detergents here have a less complicated structure than in Europe. They contain an altogether significantly lower share of builders (25–35%).

The market share of liquid detergents is increasing and may reach up to 40% in the nineties. Regional differences are large, and the share of phosphate-free detergents is high and growing. In Canada, phosphate-free detergents are virtually the rule nowadays. The main builders used in North America are phosphate, zeolite A, and NTA as well as combinations of silicate/soda ash plus, optionally, organic co-builders. In 1982, consumption of zeolite A amounted to 140×10^3 t. Because of the above-mentioned trends, this consumption had dropped to 55×10^3 t by 1984. Capacities initially in the range of 200×10^3 t·a^{-1} of zeolite A were therefore downscaled to their present day level of about 150×10^3 t·a^{-1}. The U.S. market for zeolite A is stagnating.

Far East. Although textiles are washed with such frequency as in no other country of the world, the consumption of detergents in Japan amounts to only 835×10^3 t·a^{-1}. Conditions for washing resemble those in the U.S.A. Because of the even lower water hardness of about 50 ppm (as $CaCO_3$), the proportion of builders contained in detergents is only 18–30%, lower even than in the United States. The share of phosphate-free detergents is already about 96% today. Because of the low water hardness, zeolite A without suspending agents and a suitable surfactant system is adequate to match the performance of former phosphate-built laundry detergents. Liquid detergents are as yet of minor significance. Currently, per year consumption of zeolite A is about 150×10^3 t, while capacities come to 175×10^3 t. The market to date is stagnating and relies on its own supplies.

Manufacturing Processes

Because of the large number of different types and modifications of zeolite A and alternative routes for their synthesis, the literature dealing with preparation of zeolites is extensive [2–5, 32–45]. There are three different 2-step processes available for large-scale production, shown in Table 4. Particularly important are the so-called "hydrogel processes", in which during the first step an aqueous gel is prepared. This is crystallized to the desired zeolite in a second step. Usually zeolites are initially produced in their sodium modifications. Any desired cationic modification is prepared afterwards by cation exchange.

The conventional precipitation process according to Scheme I in Table 4 has the highest importance worldwide. The hydrogel is formed by use of only completely dissolved components, namely caustic, water-glass solution, and sodium aluminate solution. A simplified scheme is depicted in Fig. 4.

One of the basic compounds either alumina or silica can also be used in solid form (reactive powder) to produce a heterogeneous gel. The best known examples for that are the acid-clay process [40] according to Scheme II in Table 4 and the zeolitization of metakaolin [3, 38, 39] according to Scheme III. In the course of the acid-clay process colloidal silica is prepared in a first step by acid treatment of clay, which then, as the heterogenous active component, is reacted with highly alkaline aluminate solution. For the zeolitization of metakaolin, clay minerals of the kaolin group with a given molar ratio $SiO_2:Al_2O_3 = 2.0$ are destructured by calcination and afterwards treated with caustic (or sodium silicate). The precipitation process is preferred for industrial purposes, because water glass and alumina trihydrate are available as inexpensive raw materials.

Figure 5 shows, in a generalized view, that the starting materials for zeolite A are sand, common salt, and bauxite.

Table 4. Different technologies for preparation of zeolites in 2-step processes

I. Precipitation process
   a) $NaOH(aq) + Na[Al(OH)_4](aq) + [Na_2O \cdot nSiO_2](aq) \longrightarrow [Na_x(AlO_2)_x(SiO_2)_y \cdot NaOH \cdot aq]$
        Caustic   Aluminate   Waterglass   Hydrogel

   b) Hydrogel $\longrightarrow Na_x[(AlO_2)_x(SiO_2)_y(H_2O)_z] + $ Caustic
                         Zeolite

II. Acid-clay process
   a) Layer silicate (Clay) $+ H_2SO_4 \longrightarrow (SiO_2)$ (amorphous) $+$ Metal salt solution
   b) SiO_2(amorphous) $+ 2Na[Al(OH)_4]aq \longrightarrow [Na_2O \cdot Al_2O_3 \cdot 2SiO_2 \cdot 4.5H_2O]aq$
                      Sodium aluminate   Zeolite NaA

III. Zeolitization of metakaolin (Clay Conversion Process)
   a) $Al_4[Si_4O_{10}](OH)_8 \xrightarrow{550-600\,°C} 2[Al_2Si_2O_7] + 4H_2O$
        Kaolin   Metakaolin

   b) $2[Al_2Si_2O_7] + 4NaOH + aq \longrightarrow 2[Na_2O \cdot Al_2O_3 \cdot 2SiO_2 \cdot 4.5H_2O]aq$
        Metakaolin   Caustic   Zeolite NaA

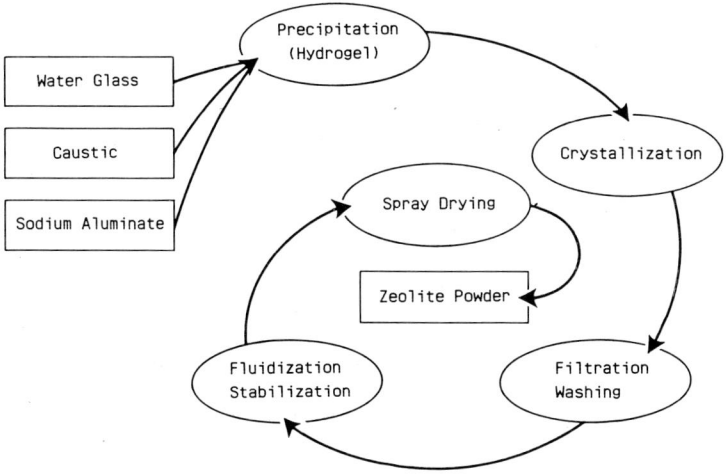

Fig. 4. Production scheme for detergent grade zeolite NaA

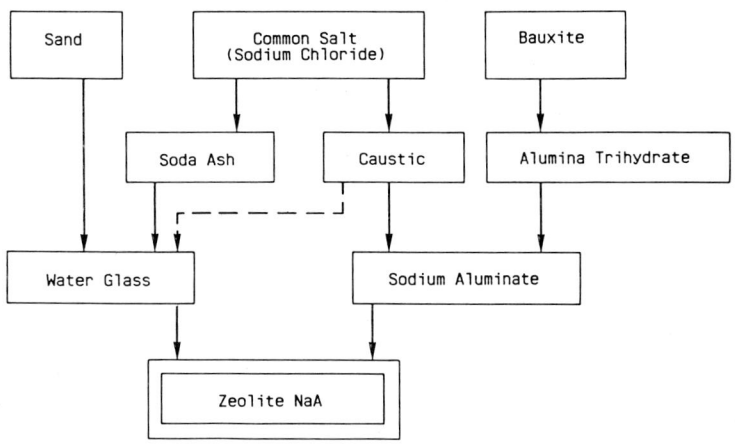

Fig. 5. Raw materials for manufacturing of zeolite NaA

There has been intensive work on processes for the continuous production of zeolite NaA [5, 41–45]. In this case it is almost unavoidable to separate the precipitation and the crystallization reactors because crystallization requires about three to four times as much time as mixing and ageing processes together. In particular, strong agitation forces are required in the mixing reactor to destroy the hydrogel phase, whereas lower flow velocities are already sufficient to avoid sedimentation during crystallization. Among others, the following reactors have been described as acceptable both for precipitation as well as for crystallization in a continuous process: stirring vessels in cascades, multi-chamber reactors, reactors with parts mounted inside to obtain zones of different flow velocity, stirring columns, and stirring reactors. Any quality defects that may occur are due to the formation of oversized agglomerates with particle diameters above 25 μm.

Figure 6 gives a general flow chart for the total process for the continuous production of zeolite NaA on a large industrial scale [5, 44]. In a stirred vessel the silicate component, i.e., the diluted water-glass solution, is heated directly or indirectly to 60–70 °C. The aluminate component is obtained by dissolving aluminum hydroxide (preferably alumina trihydrate because of its good solubility) in caustic soda. During that process only a relatively small amount of fresh caustic must be added; the larger part of alkali needed for dissolving aluminum hydroxide is recirculated. The filtered hot aluminate solution is mixed under vigorous stirring with the silicate component. This agitation is essential to destroy a thick hydrogel formed which would be difficult to handle.

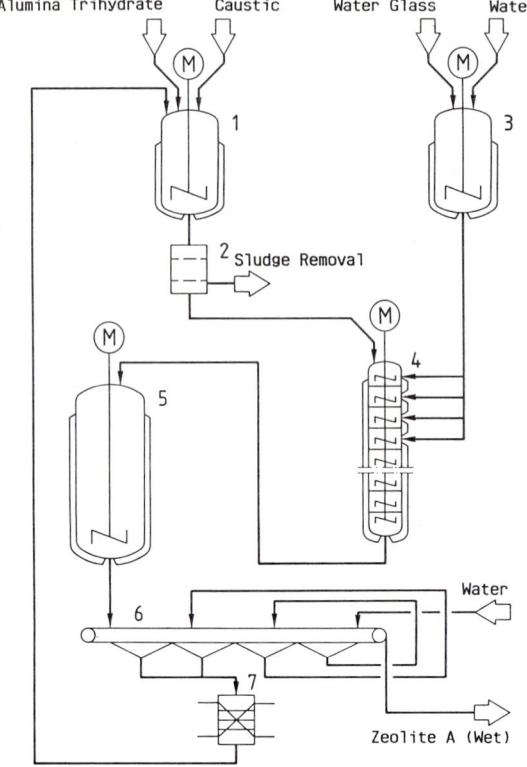

Fig. 6. Flow-chart illustrating the large-scale continuous production of zeolite NaA according to the precipitation/hydrogel process

In the HENKEL-process [44, 45] these difficulties are significantly reduced by a batch-type feeding of one of the components. The flow chart in Fig. 6 illustrates how the silicate component is subdivided into several streams and is fed into the aluminate component down-stream.

It is advisable to keep the hydrogel during the whole subsequent crystallization process in stirrable condition for about 5–10 minutes at precipitation temperatures, thus form as many seed crystals as possible. Thereafter the suspension is heated to 90–95 °C by injecting steam and is stirred lightly for about 40–60 minutes at that temperature. For discontinuous processes, precipitation and crystallization can in principle be achieved in the same reactor. For reasons of optimum use of production equipment, however, it is advisable to employ different parallel reactors for crystallization.

After crystallization the solid is separated from the mother liquor and washed. This separation from the highly alkaline mother liquor does not offer any difficulties in case of a product with a high degree of crystallinity—even if it contains very fine particles—when appropriate separation techniques such as belt filters are used. Thereafter the separated product is washed free from alkali with hot deionized water. Evaporating costs for wash water and solutions make up an essential part of total production costs. In order to keep them as low as possible, counter-current washing facilities are recommended. The total amount of wash water required, however, depends on the desired residual alkalinity and the filter typed used. After separation and counter-current washing, zeolite NaA emerges as a filter cake with 42–48% solids. This filter cake is highly thixotropic and even pumpable after brief agitation.

In order to obtain zeolite NaA of very high purity a number of requirements have to be closely met, above all specific concentration ranges in the 4-component system $Na_2O/Al_2O_3/SiO_2/H_2O$. Adequate ratios of these components can be found in the literature in different graphic forms. Usually one or more components simultaneously with reaction parameters such as the crystallization temperature are kept constant in these diagrams. In spite of all their limitations they are useful aids for determining the right ratios of ingredients and setting favorable crystallization conditions for optimum zeolite synthesis.

Typical ratios and reaction conditions for the production of highly crystalline zeolite NaA with high yields are listed in Table 5. At considerably lower mixing temperatures the precipitation process is slightly delayed and thus, a particularly "homogeneous" hydrogel is produced. However, temperatures of about 60–70 °C are better suited because in that case no expensive heat exchanger is required for concentrating the recirculated mother liquor and the wash water. The time necessary for subsequent crystallization strongly depends on the ratio SiO_2/Al_2O_3 and temperature control. The crystallization time can be shortened by reducing the water content in the composition of the reaction mixture. This is achieved by increasing the temperature and/or the amount and/or concentration of alkali. However, there are limits to these measures because they also promote the formation of hydroxysodalite which is undesired but thermodynamically preferred under these reaction conditions.

Table 5. Typical reaction conditions for preparation of zeolite A, X, and Y, respectively, (Na-modifications) according to the precipitation/hydrogel process

Parameter	Zeolite A	Zeolite X	Zeolite Y
a) Precipitation:			
Molar reactant composition ($Al_2O_3 = 1.0$)	$Na_2O : SiO_2 : H_2O$ 3.4 : 1.8 : 90	$Na_2O : SiO_2 : H_2O$ 4.0 : 3.0 : 140	$Na_2O : SiO_2 : H_2O$ 3.0 : 10 : 150
Temperature	50–70 °C	50–70 °C	20–30 °C
b) Ageing:			
Temperature	50–70 °C	50–70 °C	20–30 °C
Duration	5–10 min	20–30 min	1 d
c) Crystallization:			
Temperature	80–90 °C	85–100 °C	95–100 °C
Duration	40–60 min	8–12 h	1–3 d

The formation of zeolite NaA does not proceed at a uniform speed. After ageing and creation of many seeds the crystallization speed steadily increases, then, after having reached a relatively high rate, drops sharply. The reaction should be stopped at this point because any additional hydrothermal treatment would then lead to the transformation of zeolite NaA to hydroxysodalite, as Fig. 7 indicates.

Analytical Methods

Zeolites are identified best of all by means of their X-ray powder diagram. Crystalline impurities (particularly other types of zeolites) can easily be detected. Data for comparison are available [3]. The chemical analysis usually does not give enough information concerning the type and purity of the framework. Nevertheless it is important for determining the degree of cation exchange and the ratio $SiO_2 : Al_2O_3$. Methods to measure the cation exchange capacity have been described [24].

Zeolite A does not lend itself very well for analysis in environmental samples because of its lack of rare elements or specific characteristic features. Its rather typical form can qualitatively be recognized under the electron microscope and it can be quantified by X-ray diffractometry. However, the detection limit of this method is a content of 2% in sediments, and therefore it is not really useful. Zeolite A can be stained with terbium chloride and then be observed under the fluorescence microscope, but the method has apparently never been developed far enough to render it a truly suitable environmental analysis [46].

For research and field trials, zeolite A was labeled with indium [47] which can be characterized and quantified by neutron activation analysis. The use of

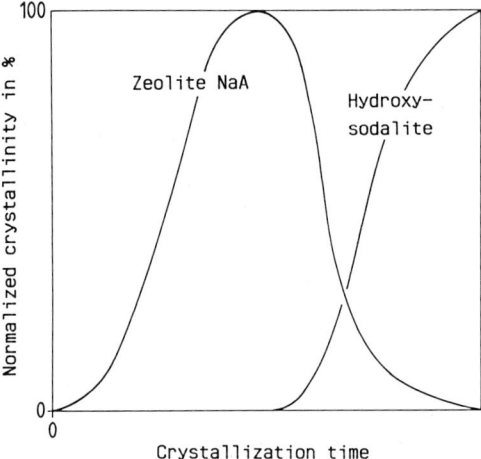

Fig. 7. Typical crystallization sequence for zeolite NaA and hydroxysodalite

indium-doped material made it possible to determine the fate of zeolite A in piping systems, sewer systems, retention in sewage treatment plants, and even bioaccumulation in organisms. Because of the positive outcome of all these investigations, no need existed for the development of an analytical method to monitor the material in environmental samples.

Environmental Fate of Zeolite A

The use of particulate matter such as zeolite A in the washing process raises the suspicion that clogging of domestic wastewater piping systems might occur. This question was investigated thoroughly in two trials involving 7 buildings with 42 households in the first and 25 buildings with 140 families in the second trial. Inspection devices were installed in the piping systems at critical locations. The two trials, lasting 25 and 33 months, respectively, revealed no particular sedimentation and clogging problems due to zeolite A. The material was, of course, identified in sediments, but it contributed no more than 5% to it in horizontal pipes [48].

In the large field trial in the village of Büsnau (FRG) [49], comprising 960 households with altogether 3100 people and lasting 13 months, various aspects of the use of zeolite A were investigated. The first important conclusion was the confirmation of the previous finding that no excessive sedimentation in the house sewer system occurs; the second, that this does not happen in a communal sewage system either. Four inspection devices were installed at different locations with especially low gradients (0.6–1.5%) for this purpose.

The largest fraction of zeolite A, two thirds, was already eliminated in the sand trap and primary settler of the sewage treatment plant, 96% was eliminated after secondary settling, i.e., only 4% left the plant with the final effluent. Thus, most zeolite A used in detergents is found in the primary and secondary sludge of sewage treatment plants, thus increasing its mass but its volume only slightly. These results have been confirmed by other investigators [50–52].

Heavy metals contained in sewage are, to a considerable extent, retained in sewage treatment plants. This is a positive effect because the primary function of sewage works is to keep pollutants from reaching surface waters. The heavy metal retention is probably enhanced by zeolite A [53].

If the receiving water has pH 7 or higher, zeolite A will hydrolyze only very slowly, while at lower pH it will decompose faster [54].

Environmental Effects of Zeolite A

Effects in Sewage Treatment Plants

The influence of zeolite A on aerobic sewage digestion in an activated sludge plant was first investigated in model plants, i.e., the OECD Confirmatory Test [55]. Mobile model units which handled about three times the volume of the former were also placed beside a sewage treatment plant and fed with its settled sewage. Performance of the plants was monitored by their general purification, measured as COD (chemical oxygen demand) removal, MBAS (methylene blue active substance, i.e., anionic surfactant) removal, nitrification and sludge retention.

Initially a purification impairment was noted, followed, after about 10 days, by an improved COD and MBAS removal. This was interpreted as two competitive influences: the first negative, a binding of essential trace heavy metals by zeolite A, the second positive, an improved sludge retention due to this material. This sequence of impairment and improvement was also noted in large-scale experiments [51]. Nitrification was enhanced which is most plausibly explained by the improved sludge retention resulting in an enhanced sludge age. Again, other investigators support these conclusions [52, 56]. In investigations with model trickling filters [55, 57] neither positive nor negative influences could be discerned. This would appear to support the sludge retention and sludge age interpretation since the sludge age in trickling filters is anyhow very high.

The latter finding was also confirmed in large-scale experiments [51]. In the Büsnau field trial no adverse effects of zeolite A on the performance of the sewage treatment plant was observed [49]. Contrary to the previous finding, no increase of the sludge mass was noted in this case.

Since the final and main destination of zeolite A is primary and secondary sludge in sewage treatment plants, its influence on anaerobic sludge digestion also had to be evaluated. Several investigations [50–52, 56–58] yielded no indications of adverse effects, whereby two of these investigations were even conducted with zeolite A loaded with heavy metals [50, 58].

The removal of heavy metals due to ion exchange on zeolite A is thus of no practical importance in sewage treatment. It is, however, relevant if surfactants in formulated detergents are checked for their biodegradability. If the whole detergent is subjected to such tests, the surfactants appear, curiously, poorly degradable. This effect was specifically noted by Huber [59] in the official OECD Screening Test. It is therefore absolutely essential that surfactants are first isolated from formulated detergents before being subjected to biodegradability studies.

Effects on Aquatic Organisms

Up to $500\,\text{mg}\,l^{-1}$ of zeolite A proved, in acute tests, innocuous toward fish (Golden Orfe) and daphniae whereby it was ascertained that zeolite A was kept in suspension, i.e., exposure was assured [60, 61]. Fish were also exposed to heavy metal-(Cu, Cd, Hg)-loaded zeolite A. The heavy metal toxicity was at least an order of magnitude lower than it would have been for the free metal ions. Long-term exposure—four weeks—of fish to $250\,\text{mg}\,l^{-1}$ of zeolite A yielded no indications of subacute effects. The reproductive success of daphniae was not impaired by exposure to $500\,\text{mg}\,l^{-1}$ of zeolite A.

Bacteria, fungi, and yeasts were not inactivated by 32-h exposure to $25\,\text{mg}\,l^{-1}$ zeolite A.

Zeolite A will end up in sediments. Therefore tubifex and mussels (*Unio tumidus*) were grown for four weeks in a sediment containing 250 mg of zeolite A per kg of wet sediment. No lethal effect was observed. No accumulation of the material in these organisms was observed (neutron activation analysis of In-doped zeolite A) [60].

Several authors investigated the influence of zeolite A on algae [60–65]. A growth-inhibiting effect could be observed in nutrient-poor culture media which was interpreted as a depletion of essential trace heavy metals due to the ion exchange by zeolite A rather than a toxic effect. In nutrient-rich media no such effects are observed. This effect is without any practical relevance since zeolite A, as a detergent builder, will never reach oligotrophic water bodies alone but only in conjunction with other nutrients in treated or untreated effluents.

The influence of zeolite A on ecosystems was shown to be minor [66]. Exposure and control were compared with the saprobity index difference S. During zeolite addition an oligotrophying effect was noted. After addition was ceased, the depot of zeolite released nutrients resulting in some continued eutrophication. The biocenotic investigations yielded, altogether, no additional indications of problems with respect to the environmental compatibility of zeolite A.

Effects on Plant Growth

Since a considerable fraction of aerobically and anaerobically digested sewage sludge is disposed on arable land and since zeolite A is, due to its use pattern, destined to end up in sludges, possible effects of zeolite-containing sludges on higher plants needed to be elucidated [67]. Two sludges were applied to four different soils in which grass (*Lolium perenne* and *Lolium italicum*) was grown. Zeolite A per se exerted no specific effects with regard to quality or quantity of the harvest. A small yield increase was noticeable in calcium-poor soils due to an increased calcium availability and perhaps due to an alleviation of manganese excess. Heavy metal contents of the grass were, with the exception of zinc, lowered rather than enhanced.

Toxicity of Zeolite A

Zeolite A was subjected to intensive toxicological tests [68]. Potential exposures in the production plant, during processing, and at the user were decisive for the selection of the test methods. The aim of these toxicological tests was to uncover any potential health hazards that might be connected with the use of zeolite A in detergents. The general safety evaluation of this zeolite was comprised of the following tests: acute oral and dermal toxicity, skin- and mucous membrane compatibility, sensitization potential, absorption after oral administration, subchronic toxicity after oral administration with determination of various elements in the blood and in the inner organs, and chronic toxicity with tumorigenicity.

In order to identify potential exposures in the plant and in the household, the following determinations were made: total dust level in the plant, particle size and particle distribution, and the zeolite A content in the fine dust. Experimentally, the safety of the values so obtained was assessed in studies for silicogenic properties by intraperitoneal and intratracheal application and in long-term inhalation tests with zeolite A detection in the lungs. These tests were complemented by in vitro cytotoxicity tests and tests for zeolite A solubility in human serum. In the tests for the silicogenic potential and in the in vitro tests, a silicosis-producing quartz type served as the control substance.

In the tests for acute toxicity, zeolite A proved to be non-toxic. The tests for skin and mucous membrane compatibility and sensitization did not show any substance-related effects. Of the amount administered orally, approx. 1% silicon is resorbed and excreted via the urinary tract. The aluminum proportion was virtually not detected and was probably not resorbed.

In practice, these findings are of no relevance, because even under the most unfavorable conditions, the daily natural intake of silicon and aluminum is not exceeded. Results of the long-term studies are summarized in Table 6. The organs and the blood, which were subjected to examinations for selected metals within the framework of the 90-day test, did not show any deviations from normal. The total dust measurements carried out in the plant gave average values of 0.89 mg m^{-3} and 2.19 mg m^{-3}. Measurements of the fine dust exhibited mean

Table 6. Long-term studies of zeolite A [68] including of tests for silicogenic properties

Species	Test substance	Dose	Application/ test duration	Findings
Rat	Zeolite A	1000 ppm[a] 5000 ppm 10000 ppm	oral, 90 d	No abnormal effects No abnormal effects Hematuria, ketone bodies, urinary bladder stones, increased Si in the kidneys
Rat	Zeolite A	10 ppm[a] 100 ppm 1000 ppm	oral, 2 years	No indications of any chronictoxic or tumorigenic effects
Rat	Zeolite A	50 mg·kg^{-1}	1x i.p.[b] 2 years	Regional and mediastinal lymph nodes: focal macrophages; mesenterium: macrophages with substance inclusions. Small solitary deposits of the substance with tight fibrous capsule. Organ surfaces: Mild connective tissue deposits
Rat	Quartz DQ 12	50 mg·kg^{-1}	1x i.p. 2 years	Silicosis in the abdominal cavity and inner organs
Rat	Zeolite A, Quartz DQ 12 spray-tower powder	1, 2.5, 5, 10, 50 and 50 mg·kg^{-1}	1x i.p. 11 months	Similar results as above. Deposits on the organ surfaces were of lower degree, especially with spraytower powder Quartz: progressive silicosis in the abdominal cavity
Mouse	Zeolite A	10 mg·kg^{-1}	1x i.p. 18 months	Same results
Mouse	Quartz DQ 12	10 mg·kg^{-1}	1x i.p. 18 months	Silicosis in the abdominal cavity
Rat	Zeolite A	50 mg·kg^{-1}	1x i.t.[c] 2 years	Small foci of alevolitis, accumulation of test substance in the macrophages
Rat	Quartz DQ 12	50 mg·kg^{-1}	1x i.t. 2 years	Severe silicosis
Rat	Zeolite A	20 mg·m^{-3}	inhalation 5 h·d^{-1} 3x weekly 22 months	Substance deposit in phagocytes, (alveolar, peribronchial, perivascular) and in mediastinal lymph nodes. No signs of a further reaction of the connective tissue.

[a] In the diet; [b] intraperitoneal; [c] intratracheal

values of $0.25\ \text{mg m}^{-3}$ and $0.31\ \text{mg m}^{-3}$. The zeolite A values sometimes remained below the detection limit, and on average they reached a maximum of $0.092\ \text{mg m}^{-3}$. Thus there is no accumulation of zeolite A in the fine dusts.

The findings obtained from the tests for cytotoxicity and the studies for toxicity after inhalation classified zeolite A as a neutral dust. In the Federal Republic of Germany, a general dust limit value of $6\ \text{mg m}^{-3}$ has been established as the maximum permissible concentration at the workplace (MAK value) [69]. The solubility of zeolite A in human serum is minimal. On the whole, all the studies listed confirm that zeolite A is a substance which is inert to living organisms. This was also confirmed by other comprehensive studies [70]. Since 1975, cases of mesotheliomas after contact with natural zeolites have become known in Turkey [71–74]. It was detected that ethiologically, these zeolites have a different, fibrous morphology and, unlike zeolite A, are not cubic. Thus zeolite A does not pose a health hazard, either at the workplace or as a component of household products.

Regulations and Laws

There are no regulations dealing directly with the use and consumption of zeolites. But some governments and/or public commissions have formally accepted zeolites as a safe detergent builder and phosphate substitute.

North-America: The International Joint Commission on the Health Implications of Non-NTA Detergent Builders (U.S.A., Canada) published reports on phosphate substitutes. The Task Force on the Health Effects of Non-NTA Detergent Builders stated in their report to the Great Lakes Science Advisory Board of this commission, that use of zeolite A in detergents is not anticipated to lead to adverse human health effects [75].

Italy: A decree of February 1986 specifically states zeolite A as an allowed phosphate substitute [76].

Netherlands: The Dutch government reviewed phosphate substitutes in 1979 and 1983, and came to conclusions that are entirely favorable of zeolites.

Germany: S. Kanowski [77] from the German Environmental Protection Agency (Umweltbundesamt) concluded from [46], that concerning the environmental compatability of potential phosphate substitutes all the relevant questions are answered extensively and with positive result for zeolite A.

References

1. Cronstedt AF (1756) Acad Handl Stockholm 17:120
2. Grubner O, Jiro P, Ralek M (1968) Molekularsiebe, VEB Deutscher Verlag der Wissenschaften, Berlin

3. Breck DW (1974) Zeolite Molecular Sieves, Wiley, New York
4. Barrer RM (1985) Synthesis of Zeolites, In: Drzaj B, Hocevar S, Pejovnik S (eds) Zeolites, Studies in Surface Science and Catalysis, vol 24, Elsevier, Amsterdam
5. Christophliemk P, Fahn R, Ferch R, Kreher A, Worms KH (1983) Siliciumverbindungen, In: Harnisch H, Steiner R, Winnacker K (eds) Winnacker-Küchler, Chemische Technologie, vol 3, Anorganische Technologie II, 4th ed pp 63–75 Carl Hanser Verlag, München
6. Dietrich C, Leonhardt W (1987) Tens Surf Deterg 24:322 (actualized by personal communication with Dietrich C)
7. Rohde D (1988) Chem Ind 111:28
8. Rees LVC (1982) Nature 296:431
9. Smith JV (1984) Zeolites 4:309
10. Puppe L (1986) Chem in unserer Zeit 20:117
11. Dwyer J, Dyer A (1984) Chem Ind 237
12. Liebau F (1983) Zeolites 3:191
13. Liebau F, Gies H, Gunawardane RP, Marler B (1986) Zeolites 6:373
14. Liebau F (1985) Structural Types of Zeolites, Springer-Verlag, Berlin
15. Barrer RM (1978) Zeolites and Clay Minerals as Sorbents and Molecular Sieves, Academic Press, London
16. Ruthven DM (1984) Principles of Adsorption and Adsorption Processes, Wiley, New York
17. Müller U, Tissler A, Unger KK (1988) GIT Fachz Lab 6:635
18. Breck DW, Anderson RA (1978) In: Kirk-Othmer Encyclopedia of Chemical Technology, 3rd edn, vol 15, pp 639–669, Wiley, New York
19. Townsend RP (1984) Chem Ind 1246
20. Puppe L (1979) In: Ullmanns Encyklopädie der technischen Chemie, 4 edn, vol 17, VCH, Weinheim, New York
21. Schwuger MJ, Smolka HG (1976) Colloid Polym Sci 254:1062
22. Schwuger MJ, Smolka HG (1978) Colloid Polym Sci 256:1044
23. Schwuger MJ, Rybinski W von, Krings P (1984) Progr Colloid Polym Sci 69:167
24. Kurzendörfer CP, Liphard M, Rybinski W von, Schwuger MJ (1987) Colloid Polym Sci 265:542
25. Savitsky A (1977) Soap Cosmet Chem Spec 53:29
26. Schwuger MJ (1984) Ber Bunsenges Phys Chem 88:1123
27. Smolka HG, Schwuger MJ (1977) Tenside Detergents 14:222
28. Murakami Y, Ligima A, Ward JW (eds) (1986) New Developments in Zeolite Science and Technology, Elsevier, Tokyo
29. Schwuger MJ, Smulders EJ (1987) In: Cutter WG, Kissa E (eds) Detergency, theory and technology, Surfactant Science Series, vol 20, Marcel Dekker, New York
30. Krings P, Smulders E, Upadek H, Verbeek H (1986) Phosphate Substitute SASIL, Henkel KGaA, Düsseldorf
31. Andree H, Krings P, Upadek H, Verbeek H (1987) In: Baldwin (ed) Proc Second World Conf of Detergents Montreux 1986, Amer Oil Chem Soc 148
32. Barrer RM (1978) Hydrothermal Chemistry of Zeolites, Academic Press, London
33. Barrer RM (1981) Zeolites 1:130
34. Vaughan DEW (1988) Chem Eng Prog 84:25
35. Roland E (1989) In: Weitkamp J, Karge HG (eds) Zeolites as Catalysts, Sorbents and Detergent Builders, Proc Intern Zeolite Sym Würzburg 1988, Studies in Surface Science and Catalysis, Elsevier, Amsterdam
36. Wolf F, Seidig KD (1973) Tonind-Ztg 97:281
37. Giordono N, Recupero V, Pino L, Bart JCJ (1987) Ind Miner 83
38. Bergk KH, Porsch M, Wolf F (1987) Chem Techn 39:251
39. Jung D (1983) In: Ullmanns Encyklopädie der technischen Chemie, 4th ed, vol 24, p 120, VCH, Weinheim, New York
40. Yamane I, Nakazawa T (1986) Pure Appl Chem 58:1397
41. Koch O (1980) Seifen, Öle, Fette, Wachse 106:321
42. Huber Corp (1977) U.S.-Patent 4.041.135
43. Henkel KGaA (1981) U.S.-Patent 4.278.649
44. Henkel KGaA (1981) U.S.-Patent 4.267.158
45. Henkel KGaA (1981) Eur Patent 37.018
46. Müller G (1979) In: Umweltbundesamt, Materialien 4/79, Die Prüfung des Umweltverhaltens von Natrium-Aluminium-Silikat Zeolith A als Phosphatersatzstoff in Wasch- und Reinigungsmitteln, Erich Schmidt Verlag, Berlin

47. Graffmann G, Roland WA, Schmidt RD, Smolka HG, Schneider J, Vogg H, In Ref [46], p 39 and (1979) Chemiker-Ztg 4
48. Roland WA, Graupner W, Holtmann W (1979) In Ref [46] p 39 and (1979) gwf wasser/abwasser 120:55
49. Wagner R (1979) In Ref [46], p 58
50. Malz F, Jendreyko H (1979) In Ref [46], p 49
51. Scherb K (1979) In Ref [46], p 52
52. Dwyer M, Yeoman S, Lester JN, Perry N (1990) Environ. Technol. 11:263
53. Roland WA, Schmid R (1978) Tenside Detergents 15:281
54. Allen CE, Cho SH, Neubecker TA (1983) Water Res 17:1871
55. Fischer WK, Gerike P, Kurzyca G (1979) In Ref [46], p 24 and (1978) Tenside Detergents 15:281
56. Holman WF, Hopping WJ (1980) JWPCF 52:2887
57. Baumann ER, Hopping WD, Warner FD (1981) Water Res 15:889
58. Roland WA, Schmid RD (1978) Vom Wasser 50:177
59. Huber L (1979) In Ref [46], p 45
60. Fischer K, Gode P (1979) In Ref [46], p 18 and (1977) Vom Wasser 49:11
61. Maki AW, Acek KJ (1978) Envir Sci and Techn 12:573
62. Hamm A, Raff J (1979) In Ref [46], p 72
63. Clasen J (1979) In Ref [46], p 78
64. Koppe P, Nusch EA (1979) In Ref [46], p 84
65. Gode P (1983) Z Wasser-Abwasser-Forsch 16:210
66. Guhl W (1987) Z Angew Zool 74:385
67. Schlichting E, Monn L (1979) In Ref [46], p 105
68. Gloxhuber C, Potokar M, Pittermann W, Wallat S, Bartnik F, Reuter H, Braig S (1983) Fd Chem Toxic 21:209
69. Deutsche Forschungsgemeinschaft (1988) Maximale Arbeitsplatzkonzentrationen und Biologische Arbeitsstofftoleranzwerte 1988, VCH, Weinheim
70. Suzuki Y (1983) Int Pneumokoniose-Konferenz, D-Bochum, 20–23 Sept 1983
71. Artvinli M, Baris YI (1979) JNCI 63:17
72. Ataman G (1978) Compt Rend Acad Ser D 287:207
73. Baris YI (1975) Hacettepe Bull Medicine/Surgery 8:167
74. Baris YI, Sahin AA, Ozesmi M. Kerse I, Ozen E, Kolacan B, Altinörs M, Göktepeli A (1978) Thorax 33:181
75. Task Force on the Health Effects of Non-NTA Detergent Builders (1980) Report to the Great Lakes Advisory Board of the International Joint Commission on the Health Implications of Non-NTA Detergent Builders, pp 57–69 Windsor, Ontario
76. Republic of Italy, Ministry of Health, Ministry of industry, commerce and craft, Ministry of Environment (1986) Decree of Feb 15, 1986 Art 1, Gazz Ufficiale della Repubblica Italiana, Serie generale 45:11
77. Kanowski S (1986) Tenside Detergents 23:2

Citrate

Hazen L. Hoyt and Herman L. Gewanter
Pfizer Inc., Chemical Division, Eastern Point Road, Groton, CT 06340

Introduction 229
Physical and Chemical Properties of Citrate 230
Citrate Production 231
Citrate in Detergent Applications. 232
Natural Occurrence of Citrate. 233
Aquatic Chemistry of Citrate 235
Biodegradation of Citrate in Natural Waters and Soils 236
Biodegradation of Citrate in Sewage 237
Safety of Citric Acid and Citrate Salts 238
Eutrophication Effects 240
Conclusion 240
References 241

Summary

Because of its performance, safety and biodegradability, citrate is the prime example of an environmentally acceptable detergent builder. Citrate is widely used in heavy-duty liquid laundry detergents, usually in the form of the neutral sodium salt, where it exhibits good water hardness sequestration performance and excellent solubility and compatibility characteristics. Citrate is ubiquitous in nature; it is a natural constituent and common metabolite of many plants and animals, appearing in small amounts in most organisms and in natural waters and soil. It is commercially produced by microbial fermentation processes. Food-grade forms of citric acid and its sodium, potassium and calcium salts are "generally recognized as safe" for use in foods. Citrate has been shown to be safe for a variety of organisms at levels much greater than would be expected to be added to the environment from many commercial uses including detergents. The rapid biodegradation of citrate in solution under natural conditions and in sewage treatment processes has been demonstrated.

Introduction

The use of citrate in foods and its history of safety as a food additive are well known. Food-grade forms of citric acid and citrate salts are "generally recognized as safe" for use in foods. The environmental acceptability of citrate did not become an issue until the early 1970s when legislative restrictions on phosphates in detergents fueled an industry-wide search for alternative builders. Citrate is now widely used in phosphate-free detergents and cleaners, especially in liquid formulations. Because of its safety and biodegradability, citrate is the prime example of an environmentally acceptable detergent builder.

Citric acid is a natural constituent and common metabolite of many plants and animals, appearing in small amounts in most organisms and in natural waters and soil. It is commercially produced by microbial fermentation processes. The annual world-wide utilization of citrate products is believed to be approaching 500,000 metric tonnes as anhydrous citric acid. The food, beverage and pharmaceutical uses of citric acid account for over 85% of this volume. Detergent producers are the largest industrial users consuming 25–50,000 metric tonnes annually. The major market areas for citric acid in detergents are the United States, Europe and Japan.

Citrate has been studied for many years and the scientific literature concerning its properties, occurrence and safety is voluminous. Citrate has been shown to be safe for a variety of organisms at levels much greater than can be expected for many uses including detergents. The rapid biodegradation of citrate under natural conditions and in sewage treatment processes has been demonstrated. Since the detergent use of citrate is a major contributor to the environment, the following discussion reviews information concerning the environmental safety of citrates in detergents.

Physical and Chemical Properties of Citrate

Citric acid and sodium citrate are used to manufacture detergent products. The physical properties of the commercial forms of these citrate products are summarized in Table 1 [1–4].

Citric acid is a relatively strong organic acid as indicated by its pKa values. A 0.1 N (0.6%) citric acid solution has a pH of 2.2. It displays properties typical of polybasic acids, forming a variety of salts with alkali metal ions and amines. Citric acid is an effective buffer over the pH range 3 to 6. Free citric acid will not volatilize from solution even at high temperature and low pH. Crystalline anhydrous citric acid is stable to light and air, but will cake if exposed to humid conditions.

When heated to 175 °C or above, citric acid will degrade. Compounds such as aconitic acid, acetonedicarboxylic acid, acetone, aconitic anhydride, itaconic anhydride, citraconic anhydride, itaconic acid, citraconic acid, mesaconic acid, tricarballylic acid and methylsuccinic acid may be formed depending on the degradation pathway [1].

Digestion of citric acid with fuming sulfuric acid yields acetonedicarboxylic acid. Oxidation with potassium permanganate produces acetonedicarboxylic acid and, above 35 °C, oxalic acid. When fused with potassium hydroxide or

Table 1. Physical properties of citrate

Anhydrous citric acid		
Formula	$C_6H_8O_7$	CH_2COOH
Molecular Weight	192.13	|
Equivalent Weight	64.04	$HOC-COOH$
Density	1.67	|
Melting Point	153 °C	CH_2COOH
Solubility, 25 °C		Acid Dissociation
g/100 ml		Constants
Water	181	$pKa_1 = 3.08$
Ethanol	59.1	$pKa_2 = 4.74$
Ether	0.75	$pKa_3 = 5.40$
Sodium citrate dihydrate		
Formula	$Na_3C_6H_5O_7 \cdot 2H_2O$	CH_2COONa
Molecular Weight	294.11	|
Dehydration Point	168 °C	$HOC-COONa \cdot 2H_2O$
Melting Point	325 °C	|
Loss on Drying	10–13%	CH_2COONa
Solubility, 25 °C		
g/100 ml		
Water	71	
Ethanol	Insoluble	
Ether	Insoluble	

oxidized with nitric acid, citric acid decomposes to form oxalic acid and acetic acid [1].

Citric acid solutions are corrosive to carbon steels unless used with an appropriate corrosion inhibitor. Citric acid also is corrosive to aluminum, copper and copper alloys. It is not corrosive to stainless steels that are the preferred materials of construction for processes involving citric acid. Solutions of citric acid, especially dilute solutions below 10% under warm conditions, are susceptible to mold growth.

Sodium citrate dihydrate is stable in air and maintains a constant water content under normal storage conditions. Dehydration begins to occur at 150 °C. Decomposition will occur at the melting point. Sodium citrate solutions will exhibit a pH of about 8.5 and are subject to microbial growth unless properly handled and protected. Sodium citrate solutions are mildly corrosive to carbon steels, but not to stainless steels.

Citrate is a chelating agent for di- and trivalent metal ions, especially Fe(III), Ca(II), Mg(II), Al(III), Cu(II) and Zn(II). Its effectiveness as a chelating agent is pH dependent. Typically, citrate will chelate metal ions at a 1:1 mole ratio in the pH range of 3 to 9.

Citrate Production

Citric acid is currently produced by microbial fermentation processes. Reviews of fermentation technology and processes used to manufacture citric acid have been published elsewhere [1, 5]. It also may be produced by extraction from citrus fruit and pineapple waste and by chemical synthesis, but these methods are not used to manufacture citric acid for commerce.

The fermentation production of citric acid is accomplished with specific microorganisms grown on defined media. The microorganisms produce citric acid naturally via the TCA or Krebs cycle. Surface culture or submerged fermentation processes have been employed. Carbohydrates, such as molasses and normal paraffins, can serve as fermentation substrates. When the citric acid concentration in the fermented solution reaches a maximum, various processes are used to break the fermentation cycle and the microorganisms are separated from the broth by filtration or centrifugation. Citric acid is then recovered from the clarified broth by precipitation as the insoluble calcium salt, or by a solvent extraction process. Treatment of the calcium citrate precipitate with sulfuric acid, or reextraction of the citric acid from the solvent solution using water, is followed by purification steps to generate a concentrated citric acid solution effluent. A sequence of evaporator-crystallizer steps is used to process the citric acid effluent and recover crystalline anhydrous citric acid. Sodium, potassium, ammonium and calcium citrate may be produced from the citric acid effluent using conventional neutralization and crystallization processes. While the manufacture of citrate products is straight-forward in principle, the commercial processes employed today are sophisticated combinations of fermentation and separation technologies and engineering techniques.

Crystalline, USP/FCC grades of citric acid and citrate salts are avilable for use in foods, beverages and pharmaceuticals. Citric acid also is available in industrial-grade 50% solutions for use in other applications.

Citrate in Detergent Applications

Citrate is currently the preferred builder for phosphate-free, heavy-duty liquid laundry detergents and liquid household cleaners. Citrate functions as a builder principally by sequestering water-hardness ions, thereby inhibiting the formation of "sparingly soluble calcium and magnesium salts of synthetic surfactants" and preventing "the formation of insoluble soap deposits that act as adhesives for particulate matter" [6]. The citrate ion is an excellent sequestrant for water-hardness ions over the pH range of 2 to 10.

The calcium sequestration performance of trisodium citrate, as determined by various techniques, is summarized in Table 2 [7]. The carbonate precipitation data in this table shows that citrate binds approximately one mole of calcium per mole of citrate at 25 °C. The binding is weaker at higher temperatures. At 50 °C a mole of citrate binds about two-thirds of a mole of calcium. The stronger binding of calcium by citrate at lower temperatures makes citrate a more effective builder in detergents formulated for cold wash conditions. From the test data it seems

Table 2. Calcium sequestration capacity of trisodium citrate

Procedure	Temp (°C)	Value (mg $CaCO_3$/g TSC·$2H_2O$)
Carbonate Precipitation	25	360
	50	215
LAS Turbidity	25	250
	50	230
Calcium Electrode	25	230–240

appropriate to use the value 230 mg calcium carbonate per gram of trisodium citrate dihydrate as an indication of citrate's sequestration capacity for calcium in detergent solutions.

In addition to its water hardness sequestration performance, citrate also acts as a deflocculant for soil particles. Citrate may be used at levels of 10–25% in combination with surfactants and auxilliary ingredients to formulate liquid detergent concentrates. It has excellent solubility and solution stability characteristics. Citrate shows good compatibility with other detergent ingredients, is not degraded under wash conditions in the presence of detergent bleaches such as hypochlorite or perborate, and will not damage textile fibers when neutralized [7].

Citrate may be employed as a builder in heavy-duty laundry detergents at levels ranging from 3–30% as sodium citrate. Typically, citrate-built laundry detergents contain 5–10% sodium citrate. Ethoxylated alcohol nonionic surfactants perform better with citrate than anionic surfactants, such as linear alkyl benzene sulfonates or ethoxylated alcohol sulfates. However, nonionic and anionic surfactant combinations are frequently used in citrate-built detergents. Citrate also is used at low levels, from <1–3%, as a formulation aid in detergents to stabilize other ingredients such as enzymes [8], and to help maintain the clarity and color of liquid products during storage.

Other laundry and household cleaning products contain citrate as well. Phosphate-free liquid hard surface cleaners are an important product category for citrate use as a builder in the U.S. Phosphate-free dishwash detergents and rinse aids marketed in Europe contain citrate at high levels to prevent water spotting and film formation. Citrate also is used in laundry spot removers, bleach additives, fabric softeners, fine fabrics detergents, liquid abrasive cleansers, disinfectants and bar soaps.

Natural Occurrence of Citrate

The natural occurrence of citrate serves as the basis for the evaluation of the environmental acceptability of citrate in detergents. The U.S. FDA GRAS affirmation record for citric acid contains over 2300 references, and an equal number of references exist in the scientific literature reporting the presence and reactions of citric acid in living systems. The following discussion is a brief review of the avilable information on the occurrence of citrate in the environment.

Citric acid is involved in the intracellular metabolism of most organisms and is found in almost all living systems. It occurs as an intermediate in the tricarboxylic acid (TCA) or Krebs cycle, the primary pathway by which acetate from various biochemical pathways is converted into carbon dioxide and energy (as reduced pyridine nucleotides for synthesis of energy production). The TCA cycle is nearly ubiquitous and is known to occur in bacteria, fungi, plants and animals. The cycle usually progresses as shown in Fig. 1 [9], although most steps are reversible in some organisms, sometimes utilizing different enzymes. Many yeasts, molds and bacteria found in soil and water produce citric acid. Citric acid can be completely metabolized and serve as an energy source, furnishing 2.47 kcal/g [10].

Citric acid occurs naturally at the greatest levels in the juices of citrus fruits. Lemon juice contains 4 to 8% citric acid and limes up to 7%. Both were once used as a commercial source. Orange, grapefruit and tangerine juices contain citric acid at levels of about 1–2%. Noncitrus fruits, such as strawberries, pineapples, currants, blueberries, raspberries and gooseberries, also contain significant amounts of citric acid. Many vegetables, the seeds and juices of a variety of flowers and plants, all animal tissues and fluids, and even wine contain citrate either as the free acid or a salt [1]. The total circulating citric acid in

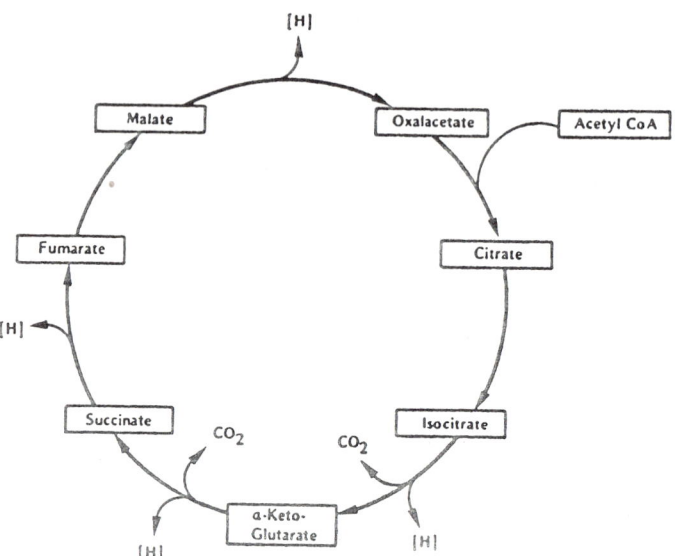

Fig. 1. Tricarboxylic Acid or Krebs Cycle as a supplier of reduced pyridine nucleotides

human serum is about 1 mg/kg body weight and the normal daily excretion of citric acid in the urine of humans is 0.2–1.0 g [1]. The human body contains about 80 g of citrate, mostly as a component of bone [10].

With the widespread occurrence of citrate in living oganisms, it is no surprise that citrate is found at low levels in soil and water. Even in the absence of commercial manufacture for industrial purposes, citrate will be introduced into the environment. Land-based and aquatic microorganisms and vegetation are major sources for the natural introduction of citrate into soil and aquatic systems. Sewage is an important source of citrate for the aquatic environment. The citrate contribution of animals, principally from excretions, is probably small.

Ambient citrate levels in natural waters range from 0.025 to 0.222 mg/L, but values as high as 8.7 mg/L have been reported [9, 11]. The average concentration of citrate in drinking water is 0.002–0.005 mg/L [11]. Sewage effluent has been shown to contain citrate at levels of <0.1 to 1.4 mg/L with a median level of 0.2 mg/L [11, 12]. The calculated possible intake of citrate from food sources for man is 4.5 g/day for the age group 2 to 65+ years old [10, 13].

Aquatic Chemistry of Citrate

The aquatic chemistry of citrate is dominated by its interaction with metal ions to form chelates. In natural waters with a composition similar to the U.S. Great Lakes at pH 8, it is predicted that citrate will form complexes with Fe(III) and Cu(II), predominantly, and may form complexes with Pb(II), Ni(II) and Zn(II) [14]. Metal-chelate stability constants of citric acid are given in Table 3 [2]. The relative importance of citrate to the aquatic behavior of these few metal ions may be considerably reduced by the presence of other complexing agents such as

Table 3. Metal-chelate stability constants of citric acid

Metal ion	Atomic weight	Log K value	Metal ion	Atomic weight	Log K value
Al^{+++}	26.98	7.0	Fe^{+++}	55.85	11.85
					25.0, K_1
Ba^{++}	137.36	2.84	Mg^{++}	24.32	3.20
Ca^{++}	40.08	3.15	Mn^{++}	54.93	3.54
Cd^{++}	112.41	4.22	Ni^{++}	58.71	4.4
Co^{++}	58.94	4.4	Pb^{++}	207.21	6.50
Cu^{++}	63.54	14.21			
		18.0, K_1	Sr^{++}	87.63	2.85
Fe^{++}	55.85	3.08	Zn^{++}	65.38	4.5

Citric = H_4L; values given as log K^M_{MHL}; log K_1 = log K^M_{ML}

naturally occurring humic acids, and by the competition that will exist between the various chemical reactions that can occur including complex formation, precipitation, redox and adsorption equilibria. The overall metal-binding affinity of citrate in aquatic systems is considered to be relatively weak [14].

Biodegradation of Citrate in Natural Waters and Soils

Because citrate is a natural product and a central metabolite in organisms, it is readily metabolized by the many aerobic microbes which possess the enzymes of the tricarboxylic acid cycle. This statement assumes that citrate can enter the microbial cells and reach the enzymes of this metabolic system. Fortunately, the number and variety of organisms that can elaborate citrate-specific transport mechanisms is so great that there is virtually no limit to the biodegradability of citrate in any environment where microbial life can exist. The availability of dissolved oxygen will not necessarily limit citrate degradation, because anaerobic microorganism growth is readily supported by citrate. Control mechanisms exist that enable organisms to increase degradation or decrease synthesis of citrate in the presence of exogenous citrate [9, 14].

The degradation of exogenous citrate down to ambient levels has been shown to be rapid and complete in natural waters, occurring with very short lag times and short half-lives. In die-away tests run with fresh river water spiked with citrate at levels as high as 15 mg/L, the time for disappearance of half the citrate ranged from 0.2 to 3.5 hours and citrate concentrations dropped to 0.15 mg/L (a 99% reduction) in 8 to 24 hours [9]. Substantial disappearance of exogenous citrate has been verified in soil as well [9]. It is not known if low temperatures affect the time required to establish microbial populations acclimated to citrate. Biodegradation of citrate occurs in the presence of metal ions, generally without significant inhibitory effects. When mixtures of 10 mg/L citrate with equimolar concentrations of metal ions were tested in semi-continuous activated sludge columns, only Cr(III) and Hg(III) seriously inhibited citrate degradation, presumably due to the toxicity of the metal ion. At lower concentrations of chromium, citrate degradation was not inhibited. Metals that either did not affect citrate degradation or, at worst, showed only a slight inhibitory effect included copper, nickel, zinc, vanadium, magnesium, sodium, aluminum, iron as ferric ion, lead and cadmium [14]. It has been concluded that citrate biodegradability should remain high under a range of environmental conditions, and should be affected in predictable ways by various environmental factors [14].

Biodegradation of Citrate in Sewage

Sewage normally contains some level of citrate. It has been estimated that raw sewage will contain 0.6–3.1 mg/L citrate from human excrement [9]. Sewage influent has been shown to contain citrate at 0.1–5.3 mg/L with an average of 1.5 mg/L in one study [15]. In another study sewage influent containd citrate at level of 0.4–2.9 mg/L with a median of 1.1 mg/L, and activated sludge treatment reduced the citrate to <0.1–1.4 mg/L with a median of 0.2 mg/L in the sewage effluent [12]. This corresponds to a citrate reduction of 82% based on the median values [12]. The measurements of citrate in sewage influent agree well with the estimated concentration, and the reduction of citrate after activated sludge treatment confirms the high biodegradability of citrate in sewage.

Assuming that a builder represents a maximum of 25% by weight of a detergent product, the concentration of a builder in waste-water influent in the U.S. has been estimated to be 12 mg/L [14]. It is extremely unlikely that citrate would be used exclusively as a builder at a 25% level in all detergent products, therefore, the estimated concentration of 12 mg/L represents the upper limit for the citrate content of raw sewage. Sewage influent from industrial sources may contain similar citrate levels. One study reported citrate levels of 11.3 and 27.6 mg/L in raw sewage from two food processing industry sources [12]. The citrate levels in this sewage dropped to 0.4 mg/L in both cases (a 96–99% reduction) after activated sludge treatment. Results from sewage seed acclimation studies [9] in Table 4 demonstrate that citrate concentrations up to 16.7 mg/L are rapidly degraded with 67 to 100% biodegradability in 5 days with or without prior acclimation of the seed. These findings, coupled with the observation that in natural waters citrate at 15 mg/L was 99% degraded within 24 hours, support the expectation that high citrate levels in sewage from commercial sources including detergents should remain highly degradable and will be reduced to ambient levels during treatment.

Table 4. Citrate biodegradability data*

Citrate conc.	Theoretical Oxygen uptake	7 Day seed Acclimation		15 Day seed Acclimation		No seed Acclimation	
		Oxygen uptake	% Degrad.	Oxygen uptake	% Degrad.	Oxygen uptake	% Degrad.
1.67	0.403	—	—	0.42	104.2	—	—
3.33	0.403	0.31	76.9	0.33	81.9	—	—
10.00	0.403	0.32	79.4	0.37	91.8	—	—
16.70	0.403	0.30	74.4	0.30	74.4	0.27	67.0

*Sodium citrate concentrations in seeded water given in mg/L. Oxyen uptake amounts given in mg oxygen/mg citrate. Theoretical oxygen uptake calculated for complete oxidation. Oxygen uptake measured after 5 days incubation. % Degradation = oxygen uptake after 5 days incubation/theoretical oxygen uptake × 100.

Citrate occurring in wastewater from detergent uses is expected to be removed principally by aerobic biological treatment processes. Removal of citrate in conjunction with floating and suspended solids during primary wastewater treatment is not known to occur and loss of citrate through sedimentation of particles or physicochemical treatment processes is not expected to be significant. Citrate degradation in aerobic biological treatment will occur without the need for prior acclimation, whether or not citrate from detergents is included in the wastewater. No difficulties are expected with the biological degradation of citrate in anaerobic sludge digestion [9, 14, 15]. Evidence supporting this statement was provided in a study with an anaerobic reactor where wastewater from a citric acid fermentation plant was successfully treated to remove 91% of the initial COD loading [16].

In the disposal of sludge by application to land or discharge at sea, citrate would be degraded. Citrate is not known to have any adverse effects on the biochemical oxygen demand or suspended solids removal efficiencies of wastewater treatment plants, and has not been known to cause transport of heavy metals through an activated sludge treatment system. Citrate will not interfere with phosphorous removal carried out during secondary treatment processes because it is rapidly degraded, but may cause an increase in the amount of treatment chemicals required for phosphorus removal in raw sewage. The theoretical oxygen demand for the citrate ion is 0.635 mg oxygen per mg citrate. If citrate levels in sewage were to increase from an average of 1.5 mg/L to 12 mg/L, sewage BOD would increase by 4.3 mg oxygen/L. This increase is less than 3% of the typical sewage 5-day BOD of 200 mg oxygen/L and would be inconsequential in terms of the effect on biological sewage treatment processes [9, 14, 15].

Because citrate is highly biodegradable under both aerobic and anaerobic conditions and its biodegradation is not significantly reduced in the presence of metal ions, citrate introduced into sewage through its use in detergent products may be successfully removed from wastewaters by conventional biological treatment processes. Sewage treatability studies have confirmed that nearly complete removal of citrate can be achieved without significant effects on sewage treament plant operation and performance, and that citrate will not cause an increase in heavy metal ion concentrations in sewage effluent [9, 14].

Safety of Citric Acid and Citrate Salts

The status of "generally recognized as safe" (GRAS) for use in foods has been accorded to food-grade forms of citric acid and its sodium, potassium, ammonium, calcium and ferric ammonium salts by the U.S. Food and Drug Administration [1, 17]. The safety of citric acid and citrate salts has been reviewed by the U.S. Food and Drug Adminisration [10, 13, 18–20], the Joint FAO/WHO

Expert Committee on Food Additives [21], and the Great Lakes Science Advisory Board of the International Joint Commission of Canada and the United States [11, 14]. The FAO/WHO Committee concluded that the acceptable daily intake for man for citrate is "not limited" [21], and the International Joint Commission concluded that "the use of citrates in detergents poses no hazard to man [11]".

The oral LD_{50} for citric acid is 5 g/kg in mice and 12 g/kg in rats [10]. The oral LD_{50} of sodium citrate in mice is 7.1 g/kg [10]. The signs of acute toxicity from citric acid in mice and rats are those of organic acidosis and calcium deficiency. Acute, subchronic and chronic toxicity tests of citrates show no significant oral effects. Citric acid can be a mild irritant to the eyes, nasal passages and skin. Mutagenicity and teratogenicity tests have shown no effects for citrates and no potential carcinogenicity is suspected for citric acid or its salts [11]. Citric acid and sodium citrate show low acute toxicity to a variety of aquatic organisms when pH is above 5 [11]. Increases in either pH or water hardness result in a decrease of aquatic toxicity for citrate. A summary of citrate

Table 5. Citrate Aquatic Toxicity Data

Test material	Test species	Dilution water	pH	Results*	Ref.
Sodium citrate	Rainbow trout (*Salmo gairdnerii*)	Soft	7.1	96 h LC_{50} = 833 mg/L	9, 14
	Bluegill (*L. macrochirus*)	Soft	7.1	96 h LC_{50} = 1516 mg/L	14
	Bluegill (*L. macrochirus*)	Soft	—	96 h LC_{50} = 1710 mg/L	14, 22, 23
	Snail (*Amnicola limosa* Say.)	Soft	—	96 h LC_{50} = 1415 mg/L	14, 22, 23
	Gammarus pulex	Soft	—	72 h LC_{50} = 1044 mg/L	14
	Gammarus pulex	Hard	—	72 h LC_{50} = 1750 mg/L	14
	Diatom (*Navicula seminulum* var. *hustedii* Patr.)	Soft	—	Growth reduction at 7 days, EC_{50} = 1200 mg/L	22, 23
	D. magna	Lake Erie	—	48 h concentration for immobilization, EC_{50} = 825 mg/L	14, 24, 25
Sodium citrate/ citric acid	*Polycelis nigra*	Distilled	7.2	48 h threshold toxicity = 2830 mg/L	26
Citric acid	Goldfish (*C. auratus*)	Hard	4.5	Not lethal, 100 h at 625 mg/L	27
	Goldfish (*C. auratus*)	Hard	4.0	Lethal, 4–28 h at 894 mg/L	27
	D. magna	Soft	5.9	No kill, 72 h, 80 mg/L	27
	D. magna	Soft	5.5	40% kill, 72 h, 120 mg/L	27
	D. magna	Soft	5.0	100% kill, 17 h, 185 mg/L	27
	D. magna	Lake Erie	—	Threshold concentration for immobilization, EC_{50} = 153 mg/L	25, 28

* EC_{50} = Concentration producing a given effect in 50% of the test organisms
  LC_{50} = Concentration lethal to 50% of the test organisms

aquatic toxicity data is provided in Table 5. Citrate use in detergents is not expected to pose an acute aquatic hazard, and citrate does not appear to affect the aquatic toxicity of metal ions [9]. Citrates may be safely introduced into soil and will not affect seed germination or plant growth [9].

Eutrophication Effects

Eutrophication is the process of nutrient enrichment of water bodies by chemical compounds or elements that can stimulate the growth of algae and, thereby, cause a depletion of dissolved oxygen when these algae decay. Citrate introduced into natural waters might directly affect algal growth by increasing the available inorganic carbon as a result of biodegradation or by providing an organic substrate for growth. It might also indirectly affect algal growth by chelating metal ions, either making a micronutrient more available or reducing the effect of a toxic metal ion, and by affecting organisms growing on the algal population.

The changes in the forms and amounts of trace metal ions that can occur in natural waters after the introduction of a chelating agent like citrate, will affect the availability of toxic free metal ions and the growth rates and forms of phytoplankton, resulting in shifts in algal speciation from tolerant to sensitive species. Many factors will influence the extent to which environmental quality may be affected, including the biodegradability of the chelating agent and the type and degree of metal deposition that occurs prior to the chelating agent introduction. The effects on an aquatic system will be site specific, and those caused by a biodegradable material should disappear as it is degraded. Because cirate is rapidly biodegraded, it is anticipated that the introduction of citrate to the aquatic environment from uses such as detergents will produce negligible effects on metal ion speciation and phytoplankton growth [14].

The International Joint Commission reviewed the possible eutrophication effects of citrate and concluded that "additions of citrate to most waters, at the concentrations expected (from use in detergents), will have no discernible effect on their algal populations, directly or indirectly [14]".

Conclusion

Citric acid and its sodium, potassium and calcium salts occur naturally, are normal human metabolites of carbohydrates and are GRAS substances suitable for use in foods with an unlimited acceptable daily intake. Citrate is readily biodegraded by many organisms under aerobic and anaerobic wastewater treatment conditions. The citrate concentrations expected in the environment from detergent uses are safely below those known to be toxic. The International Joint Commission has concluded that "the use of citrate as a detergent builder will not be detrimental to the environment" [14] and that "there is no evidence that ingestion of citrate in drinking water would pose a health hazard to humans [11]".

References

1. Citric acid (1979) In: Kirk-Othmer: Encyclopedia of chemical technology, vol 6, 3rd edn, John Wiley, New York
2. Pfizer organic chelating agents (1979) Data Sheet 681, Pfizer Inc., Chemical Division, New York
3. Comparison of properties: Citric acid vs other organic acids (1982) Information Sheet 2051, Pfizer Inc., Chemical Division, New York
4. The Merck index (1983) 10th edn, Merck & Co., Rahway, NJ, pp 330 and 1233
5. Mial LM, Rose AH (eds) (1978) Primary products of metabolism, Academic, New York
6. Galante DC, Clayton ET (1978) Liquid laundry detergents, Chemical Trends and Times, p 36
7. Muck DL, Gewanter HL (1972) The detergent building properties of trisodium citrate, Presented at the 1972 Joint Meeting of the American Oil Chemists' Society and the Japan Oil Chemists' Society, April 25, Los Angeles, CA
8. Milwidsky B (1988) Laundry liquids update, HAPPI, October, p 49
9. Environmental impact of Citrate (1988) Information Sheet 2120, Pfizer Inc., Chemical Division, New York
10. Evaluation of the health aspects of citric acid, sodium citrate, potassium citrate, calcium citrate, ammonium citrate, triethyl citrate, isopropyl citrate and stearyl citrate as food ingredients, (1977) prepared for the Bureau of Foods, Food and Drug Administration by the Federation of American Societies of Experimental Biology, Bethesda, MD, 1977 (U.S. Department of Commerce, National Technical Information Service No. PB-280-954)
11. Health implications of non-NTA detergent builders (1980) Report to the Great Lakes Science Advisory Board of the International Joint Commission, Windsor, Ontario, Canada, October, revised March 1981
12. Shannon EE, Schmidtke NW, Fowlie PJA (1977) Effect of citrate and carbonate based detergents on wastewater characteristics and treatment, Res. Rpt. No. 61, Project No. 73-3-7, Environmental Protection Service, Fisheries and Environment Canada, Ottawa, ON
13. Scientific literature reviews on generally recognized as safe (GRAS) food ingredients: Citric acid (1974) Prepared for the U.S. FDA by Tracor-Jitco Inc., Rockville, MD, (U.S. Department of Commerce, National Technical Information Service No. PB-241-967)
14. Ecological effects of non-phosphate detergent builders (1980) Report to the Great Lakes Science Advisory Board of the International Joint Commission, Windsor, ON
15. Payne AG (1973) Environmental testing of citrate: Bioassays for algal stimulation, Proc. 16th Conf. Great Lakes Res., p 100
16. Wu W et al (1987) Wat Res 21:789
17. Citric Acid—21 CFR 182.1033 and 182.6033; Sodium Citrate—21 CFR 182.1751 and 182.6751; Potassium Citrate—21 CFR 182.1625 and 182.6625; Calcium Citrate—21 CFR 182.1195, 182.5195, 182.6195 and 182.8195; Ferric Ammonium Citrate—21 CFR 184.1296; Ammonium Citrate—Included in the FDA's GRAS review program
18. Scientific literature reviews on generally recognized as safe (GRAS) food ingredients: Citrates (1973) Prepared for the FDA by the Food and Drug Research Laboratories, Inc., Maspeth, NY (U.S. Department of Commerce, National Technical Information Service No. PB-223-850)
19. Teratologic evaluation of FDA 71-54 (Citric Acid) (1973) Prepared for the U.S. FDA by Food and Drug Research Laboratories Inc., East Orange, NJ (U.S. Department of Commerce, National Technical Informaion Service No. PB-223-814).
20. Mutagenic evaluation of compound FDA 71-54 (Citric Acid) (1975) Prepared for the U.S. FDA

by Litton Bionetics Inc., Kensington, MD (U.S. Department of Commerce, National Technical Information Service No. PB-245-463)
21. FAO Nutrition Meetings Report Series No. 40 A,B,C, Food and Agricultural Organization of the United Nations' World Health Organization, New York, 1967
22. Schwartz AM, Davis AW (1974) The development of phosphate free heavy duty detergents, EPA-600/2-74-003
23. Gillette Company Research Institute, Interim Report, EPA Contract FWQA 14-12-875, January 7, 1972
24. EPA, Water Pollution Control Research Series No. 18050 GWV 05/71 (1971) Water Quality Criteria Data Book, vol 3, Effect of chemicals on aquatic life
25. Anderson BG (1946) Sew Wks Jour 18 : 82
26. Jones JR (1941) J Exp Biol 18:170
27. Ellis MM (1937) Bull Bur Fish 48:365
28. Anderson BG (1944) Sew Wks Jour 16:1156

EDTA—Ethylenediaminetetraacetic Acid

K. Wolf and P. A. Gilbert

Unilever Research, Quarry Road East, Bebington, Wirral, Merseyside, L63 3JW

Introduction 244
Analysis. 246
Environmental Fate of EDTA. 247
   Physico-Chemical Properties 247
   Biodegradability 248
   Photodegradation. 249
Levels in the Environment 251
   Estimate Based on Use in Detergents 251
   Levels Found by Environmental Monitoring 251
      Germany 251
      United Kingdom. 252
      Switzerland 252
      Austria. 252
Possible Effects on the Environment 252
   Effect on Waste Water Treatment. 252
   Toxicity to Aquatic Organisms 253
   Bioaccumulation. 253
   Stimulation of Algal Growth. 254
   Heavy Metal Mobilisation 254
Mammalian Toxicology 255
   Acute and Longer Term Toxicity 255
   Teratology. 257
   Acceptable Daily Intake for Man 257
References 257

Summary

EDTA is used at low levels in some fabric washing powders as a bleach stabiliser. It is also used in soaps as a stabiliser and in some liquid products to enhance the action of preservatives. EDTA has, however, many other industrial applications and its use in detergents accounts for only approximately 20% of the total use. Calculations based on the worst case estimates of sewage flows suggest that the concentrations in European sewage effluents of EDTA arising from detergents use should be in the range of 100–300 $\mu g\, L^{-1}$ and in receiving waters 10–30 $\mu g\, L^{-1}$. Where industrial discharges are also present, total concentrations of EDTA could of course be considerably higher. In many cases however, the concentrations of EDTA actually found in European rivers (0–60 $\mu g\, L^{-1}$) are well below the range predicted from the combined industrial and detergent uses.

Since EDTA has been shown to be subject to biodegradation under some conditions and the Fe III chelate also undergoes photodegradation, it should not persist indefinitely in receiving waters and soil. However, the rate of biodegradation seems to be too slow to ensure efficient removal in the normal two-stage sewage treatment processes. EDTA has a low toxicity to aquatic organisms and, in the concentration ranges found, should have no adverse effect on sewage treatment processes or aquatic life. It also has no potential for bioaccumulation. As a powerful sequestrant, EDTA has the potential to mobilise heavy metals and to keep them in aqueous solution. This is, however, unlikely to be significant at realistic concentrations on stoichiometric grounds alone. Natural variations in pH and the concentrations of competing ligands and precipitating anions are likely to have a more significant impact on aqueous concentrations of heavy metals in the practical situation.

EDTA has a low toxicity to mammals and is a permitted additive for foodstuffs in a number of countries. The WHO recommends that the acceptable daily intake of EDTA for man is 2.5 mg kg^{-1}. Even assuming up to 25 µg L^{-1} EDTA in drinking water, the daily intake from drinking 2 liters of this water would be well below this limit and a large margin of safety exists. There is therefore no risk to human health.

Introduction

EDTA has the chemical structure:

$$\text{HOCOCH}_2 \diagdown \qquad \diagup \text{CH}_2\text{COOH}$$
$$\text{NCH}_2\text{CH}_2\text{N}$$
$$\text{HOCOCH}_2 \diagup \qquad \diagdown \text{CH}_2\text{COOH}$$

Molecular Weight 292   Empirical Formula $C_{10}H_{16}N_2O_8$
CAS Registry Number [60-00-4]
Name (IUPAC Nomenclature) N,N'Ethane bis-[N-(carboxylmethyl)] glycine
Synonyms— Ethylenediaminetetraacetic acid
          EDETIC acid

EDTA is produced industrially by the reaction of ethylene diamine with sodium cyanide and formaldehyde under alkaline conditions at elevated temperature. A partial vacuum is used to remove ammonia which is a by-product. The tetrasodium EDTA salt produced by this reaction is easily converted to the tri-, di- or monosodium salt, or the free acid, by treatment with sulphuric or hydrochloric acid. In the United States, total production was 22 700 tonnes per annum in 1979 [1]. More recent estimates [2] put the production of aminopolycarboxylates (excluding Nitrilotriacetic acid—NTA) at 50 000 tonnes in the United States and more than 20 000 tonnes in Western Europe. The European Amino-Carboxylate Producers Committee (EAC) a sector group of CEFIC, has estimated that about 28 000 tonnes per annum of EDTA, calculated as H_4 EDTA (or 37 000 + pa calculated as Na_4 EDTA) were used in 1987 and 1988 in Western Europe [82]. The quantities produced in Western Europe are believed to be similar.

EDTA is a powerful hexadentate chelating ligand and is the most widely used sequestering agent. It is incorporated into some fabric washing powders at relatively low levels (0.1–0.5%) as a bleach stabiliser. By chelating traces of metal ions, such as iron, copper and manganese which may be present in the wash system, EDTA prevents the wasteful metal-catalysed decomposition of peroxide bleaches. It is also incorporated into soap bars to prevent metal-catalysed oxidative degradation and into some liquid detergent products (0.1–0.5%) as a preservative enhancer since, at those concentrations, it can sensitise micro-organisms to the action of biocides used as preservatives. It has, however, many other important applications (Table 1) most of which will also result in its discharge to the aquatic environment.

Historically, the first major use of EDTA was in the textiles industry, where it is used to remove trace metal impurities prior to processing and dyeing of natural and synthetic fibres. In industrial cleaners it prevents precipitation of calcium, magnesium and heavy metal salts which can cause deposits on both the cleaning equipment and the material to be cleaned. In alkaline degreasing baths it stabilises phosphates and prevents flocculation of lime soaps, thus enabling the baths to be used longer. It also enhances their cleaning action and prevents tarnishing of non-ferrous metals. After acid pickling it is used to ensure complete removal of iron oxide hydrates that are water and acid insoluble. It is also used in electroplating and chemical reduction baths in place of cyanide as a complexing agent and as a regenerating salt for masking impurities. Cooling water can be softened with EDTA, with the added advantage that the neutral or weakly alkaline medium is non-corrosive, unlike the normal acid treatment.

Natural latex is washed with EDTA prior to polymerisation in order to remove the rubber poisons copper and manganese. EDTA is also added to the catalyst system as a complexing agent for divalent iron.

In the photographic industry, developers which are to be prepared with hard water are protected by the addition of EDTA to prevent precipitation of, for example, sulphite and carbonates onto the photosensitive layer and developing equipment. In colour photography $FeNH_4$ EDTA is a popular oxidising agent of the silver to be subsequently removed. All stages of pulp and paper processing are facilitated by removal of metal ions by EDTA.

In several countries, EDTA is an approved additive for pharmaceuticals, cosmetics and food. Its main function is to prevent trace metal catalysed oxidation which causes rancidity.

It is also used specifically as an antidote for heavy metal poisoning and radioactive contamination. In plant nutrition, EDTA complexes of copper, iron,

Table 1. The major uses of EDTA

Industrial and institutional cleaning	30%
Household detergents	25%
Photographic industry	10%
Textiles	10%
Agriculture, pulp and paper, metal plating	10%
Other uses	15%

Estimates made by EAC [82]

manganese and zinc are used to rectify deficiencies, as the chelated forms cannot be precipitated as biologically inactive insoluble salts.

Analysis

Most of the methods currently employed for analysis of mixtures containing EDTA are based on gas chromatography (GC) or high performance liquid chromatography (HPLC). The high sensitivity of GC enables low level determinations to be carried out. The EDTA must firstly be isolated from its aqueous matrix in a preconcentration step. This is generally achieved by ion exchange—solid phase extraction. After elution, the EDTA is derivatised to produce a volatile species for GC determination. The tetramethyl [4, 5] and tetraethyl [8] and tetrabutyl [7] esters have been used, though the tetrabutyl derivative is reported to offer superior stability in analysis.

Separation is carried out on a glass or fused silica capillary column, or a megabore column, with a chemically bonded non-polar stationary phase. Either a flame ionization detector (FID) or a nitrogen-specific detector (NPD) can be used: the latter offers the advantage of selctivity. Quantitation is achieved by the addition of an internal standard so that a peak area ratio calibration graph can be generated. Substances used as internal standards include octadecanonitrile [7], heptadecanonitrile [8], 1,6-hexamethylenediamine tetraacetic acid (HDTA) [4].

Quantitation limits of 25 μg L^{-1} (tetramethyl ester, FID, 50 ml sample) [4], 1 μg L^{-1} (tetrabutyl ester, NPD, 25–100 mL sample) [8], 10 μg L^{-1} (tetramethyl ester, FID) ca 10 mL sample) [5] and 15 μg L^{-1} (tetraethyl ester, FID, 25 mL sample) [6] have been reported.

The commonest approach to the analysis of EDTA by HPLC is its spectrophotometric determination in the form of a metal complex. One species of metal ion must be chelated by all the EDTA in a sample for quantitative analysis to be possible, and the resulting complex must be stable over the time required for analysis. The use of iron III [9, 10] and of copper II [11, 12] as the complexing metal has been reported. It is claimed [12] that the methods using the copper II EDTA complex are superior because the iron III EDTA complex is photochemically unstable.

The basis of each method is that an excess of the metal ion is added to the filtered samples. Either a reversed-phase column or an anion exchanger can be used to effect separation of the metal-EDTA complex from the matrix. The UV absorbance of the column elluant is monitored at a suitable wavelength (254 nm for copper II EDTA; 258 nm for iron III EDTA). Quantitation is achieved by the addition of an internal standard such as NTA, or by an external standard method. A limit of detection of around 0.2 mg L^{-1} is claimed for iron III EDTA using an anion exchange column [10], and a minimum quantifiable level of ca. 2.8 ng (equivalent to 1.5 mg L^{-1}) is claimed for copper II EDTA using an octadecyl reversed-phase column [11].

An alternative to spectrophotometric detection of the copper II EDTA complex is provided by atomic absorption spectrometry [13]. A limit of detection of 0.2 mg L^{-1} is reported for this species in the presence of copper complexes of other chelating ligands separated on a weak anion exchange column with AAS detection.

Aminopolycarboxylic acids can also be quantified directly using an amperometric detector [14]. A minimum detectable quantity of 0.15 mg L^{-1} is claimed for EDTA separated on a reversed-phase column.

In selecting a method for determination of EDTA in environmental samples, the main factors influencing choice are the levels to be measured, and the nature of possible interfering substances in the matrix.

Environmental Fate of EDTA

In order to predict the fate of EDTA in the environment and to estimate the concentrations likely to occur at key points it is necessary to take account of:

Physico-Chemical Properties

EDTA is insoluble in organic solvents and soluble in water (1000 mg L^{-1} at 25 °C) [1]. It is neutralised by alkali metal hydroxides to form a series of highly water-soluble salts containing one to four alkali metal cations.

Several chemical and physico-chemical processes can be used to eliminate EDTA from waste waters containing high levels of the chelator (ca. 1 g L^{-1}). EDTA is degraded by ozonisation, UV-irradiation [15, 16, 17], and reaction with hypochlorous acid [18]. Heavy metal EDTA complexes can be dealt with by separation of the metal, for example by alkaline precipitation or electrodialysis, followed by acidification of the solution to precipitate the EDTA. The latter can then be recovered [19, 20, 21]. Thermal decomposition followed by hydrolysis of the EDTA is also possible and reverse osmosis may be applied to remove EDTA from highly polluted waste waters.

These methods cannot, however, be applied to sewage containing sub-mg L^{-1} levels of EDTA. The water solubility of the acid and its chelates renders unlikely any reduction of their levels by adsorption or precipitation during settling of raw sewage or biological treatment. This was confirmed by Gardiner [19] who reported negligible adsorption of EDTA onto suspended solids (such as silica, kaolin, river sediment and sewage plant effluent solids). It was also demonstrated by Cheng et al. [20] that EDTA is readily leached from soils.

Examination of the drinking water in the city of Essen revealed EDTA concentrations similar to those of the River Ruhr, from which the water is abstracted [21]. This suggested that the filtration and ozonisation processes used to prepare the water did not eliminate low levels of EDTA. More recent investigations by Brauch et al. [33] show that EDTA can be eliminated by ozonisation (up to 80%) but efficiency depends on the stage of the drinking water preparation process. Results from a Swiss study showed that in sewage plants employing tertiary treatment, EDTA elimination in the range 16–86% was achieved. Results from several other studies indicate that in general EDTA removal in normal two-stage sewage treatment plants is not substantial.

Biodegradability

In the simple laboratory tests for biodegradability, EDTA generally displays low levels of biodegradation. No decrease of EDTA (5–20 mg L^{-1}) was found in a four week test after three successive subcultures of the inoculum of sewage bacteria [22]. Using the effluent of a municipal sewage treatment plant as inoculum in the AFNOR test, no Dissolved Organic Carbon (DOC) removal occurred within 42 days. With pre-adaptation of the inoculum, 10% carbon dioxide evolution and 22% DOC removal were observed in the Sturm Test. After 19 days incubation in the OECD Screening Test a 10% DOC removal was found and in the Closed Bottle Test the oxygen consumption was 3% of the theoretical BOD [23]. A somewhat higher level of DOC removal (37%) in the Zahn-Wellens test was reported [23].

In a study of the mineralisation of EDTA (1–100 μg L^{-1}) in buffer media (EDTA present predominantly as the free acid or the calcium, magnesium, iron or aluminum complex) no degradation of any EDTA species by sewage bacteria was observed over the 72 day test period [24]. Incubation of (2-^{14}C) EDTA for four weeks plus two weeks post incubation in a screening test system inoculated with bacteria and microalgae gave no evidence of biodegradation [25].

In contrast, (2-^{14}C) and (ethylene ^{14}C) radiolabelled FeNH$_4$ EDTA was found to degrade considerably when the test mixture was inoculated with microorganisms from an aerated lagoon receiving EDTA-containing industrial effluents [26]. ^{14}CO$_2$ evolution of ca. 30% resulted within five days incubation in the dark. The mineralisation rate was markedly enhanced in the presence of organic compounds such as unlabelled FeNH$_4$ EDTA, NTA or ethylene diamine, but was reduced after addition of certain sugars and amino acids. Since the ^{14}CO$_2$ evolution was stopped by heat treatment or the addition of disinfectant, a biological mechanism for EDTA degradation is indicated. In the same test system unlabelled NaFe EDTA gave 89% removal and a 63% DOC decrease within five days. Gas chromatographic analysis showed the main transient metabolites to be ethylene diamine triacetic acid and iminodiacetic acid, plus a possible further minor intermediate.

In a model system which simulated the purification of river water on an activated carbon filter, EDTA (22 and 78 mg L^{-1}, the only carbon source) was eliminated quantitatively after an adaptation time of about three weeks [27].

In a study of a Swedish river and lake continuously supplied with waste water from a paper factory, the EDTA levels found in the river (detention time about two weeks) were as expected, while those found in the lake (detention time about thirty weeks) were considerably lower, which indicated that the EDTA was being eliminated. The relative contributions of biodegradation and photodegradation could not be assessed [28, 29].

A study of the biodegradability of EDTA in aquatic sediments and agricultural soils demonstrated slow but steady mineralisation of all the EDTA carbon [30, 31]. Optimum carbon dioxide evolution occurred in the physiological temeprature range (30–40 °C) and mineralisation was inhibited in sterilised soils, indicating that the degradation was due to microbial action. The biodegradation rates of free EDTA and its copper, cadmium, zinc, manganese, calcium and iron chelates were comparable, while the nickel complex showed lower biodegradability. Since intermediates of EDTA biodegradation could not be detected, the initial degradative step appeared to be the rate determining step for mineralisation. No significant biodegradation was observed under anaerobic conditions.

The results show that EDTA is susceptible to biodegradation under aerobic conditions [32]. However, the rate of biodegradation is too slow to ensure removal during the biological treatment of sewage, but EDTA should not persist indefinitely in receving waters and soils.

Photodegradation

In the absence of evidence for the rapid biodegradation of EDTA [32], other mechanisms for its breakdown in the environment have been sought, such as photodegradation, i.e. degradation resulting from the absorption of sunlight [34–38].

The susceptibility of a species to photodegradation depends on two factors, its absorption spectrum and the quantum efficiency of the photochemical reaction [39]. The absorption spectra of metal chelates reflect three types of electronic transitions:

1. transitions typical of the central metal ion, modified by the ligand field of the chelating molecule,
2. charge transfer transitions between metal ion and ligands,
3. transitions typical of the ligand.

Of these, the charge transfer transitions are most likely to induce photochemical reactions, and typically involve absorption of radiation in the UV-visible wavelength range. Therefore, those species that exhibit strong absorbance bands in the region of the solar spectrum (wavelengths greater then 295 nm) are likely to be more susceptible to photodegradation.

At pH 7, the ferric complex of EDTA shows a strong absorption maximum at 254 nm, which tails into the solar spectral region with the molar absorptivity at 300 nm (6500 L mol^{-1} cm^{-1}) being more than fifty percent of that at 254 nm (12 000 L mol^{-1} cm^{-1}). This has been interpreted as a ligand-to-metal charge transfer band [36]. The absorption spectrum of the copper II complex of EDTA contains a weak metal ion absorption band in the visible part of the spectrum, and a strong band, most likely due to a charge transfer transition, in the near UV. At its maximum (260–265 nm), the molar absorptivity is 4400 L mol^{-1} cm^{-1} and at 300 nm it is 1800 L mol^{-1} cm^{-1} so the complex does absorb in the solar region [39].

The zinc II EDTA complex shows no strong absorbance above 250 nm, and only a small maximum at 200 nm (possibly an artifact resulting from the onset of oxygen absorbtion) plus a very weak absorbance at 350 nm. This could only contribute to photodegradation in the event of a very high quantum efficiency [39]. The nickel II complex shows weak ligand field transitions in the visible and near-UV ranges, plus a strong, diffuse absorption below 250 nm which may be due to the ligand alone [39].

The quantum efficiency of a photochemical reaction is defined as the number of molecules (or moles) to react per photon (or Einstein) absorbed [40]. A proposed dependence of quantum efficiency on the wavelength of irradiation was not supported by experimental results. However, the iron III EDTA complex was shown to possess a high quantum efficiency (10^{-2} initially) for degradation at pH 7. None of the EDTA complexes of copper II, nickel II or zinc II showed any measureable degradation at pH 7 under the prevailing experimental conditions, and so the quantum efficiencies of these complexes are evidently very small [39].

The practical consequences of these observations are, firstly, that iron III EDTA photodegrades rapidly because of its substantial quantum efficiency and relatively strong UV absorption. The suggested decay mechanism is photoreductive decarboxylation initiated by intramolecular electron transfer [35, 36]. Photodegradation is therefore an effective sink for iron III EDTA under natural conditions from Spring to Autumn, and degradation also proceeds moderately fast in Winter. The calculated half life of the iron III EDTA is less than one day in shallow rivers and the surface layers of lakes, under average climatic conditions [39].

The nickel II, copper II and zinc II complexes of EDTA were observed to be inert with respect to photodegradation. Significant photodegradation of EDTA in surface waters will only therefore occur where the iron (III) chelate is present in a significant proportion. Analysis of samples from Lake Geneva showed that the EDTA concentration was independent of depth [41, 42]. The authors concluded that degradation was unlikely to be taking place in these waters.

Levels in the Environment

Estimate Based on Use in Detergents

An assessment was made in 1987 [3] of the quantities of EDTA then used in detergent products. Since that time, it is likely that such use of EDTA will have declined in some countries. However, using these estimates it is possible to calculate the resulting contribution of detergent products to the levels of EDTA likely to be present in the aquatic environment (Table 2).

Since, as we have seen, the use of EDTA in detergents accounts for approximately only one fifth of the total use it is likely that in some areas, where there are industrial discharges, these levels could be exceeded.

Levels of EDTA in domestic sewage are expected to be typically in the range of 0.14–0.31 mg L^{-1}. In rivers immediately below a sewage outfall levels will typically be an order of magnitude lower, being generally less than 30 μg L^{-1}.

Levels Found by Environmental Monitoring

Germany

Dietz [64] reported that over a ten month period the median concentration of EDTA in the River Ruhr was 23 μg L^{-1} with a 95 percentile value of 56 μg L^{-1}. Linckens and Reichert [65] reported that levels of EDTA in the Main, Rhein, Mosel and Sieg were generally in the range 0–20 μg L^{-1}. The highest concentrations were found in the Main, where occasional peak values of up to 50 μg L^{-1} were reported.

Table 2. Estimated levels of EDTA in the environment

Country	France	Germany	Italy	U.K.
Population ($\times 10^6$)	55	61	57	58
Water Consumption (L/inhabitant/day)	220	225	225	200
Consumption of EDTA in Detergents (Tonnes/Year)	1400	1000	900	600
Levels (mg L^{-1}) *of EDTA in*				
Sewage Effluents (assuming no removal)	0.31	0.20	0.19	0.14
Rivers Receiving Sewage (assuming 1:10 dilution)	0.031	0.020	0.019	0.014

United Kingdom

Information obtained from the Thames Water Authority on the rivers Thames and Lee shows that prior to 1980, fairly high concentrations of EDTA (260–4200 $\mu g\,L^{-1}$) were found. After elimination of a large industrial discharge, over a five year period the EDTA levels were frequently below 15 $\mu g\,L^{-1}$ (the limit of detection of the analytical method) with occasional higher concentrations (up to 60 $\mu g\,L^{-1}$) reported. The sewage effluent content of these rivers is typically in the range 50 to 60%, and up to 70% at times of very low flow. Assuming a typical EDTA concentration of 140 $\mu g\,L^{-1}$ in U.K. sewage effluents (Table 2) substantial degradation of EDTA in these rivers is indicated.

Switzerland

Concentrations of EDTA in Swiss rivers have been found [66] to reach a maximum of 54 $\mu g\,L^{-1}$ (River Glatt at Rheinsfelden), but most contain EDTA at levels below 20 $\mu g\,L^{-1}$, and in many cases below 10 $\mu g\,L^{-1}$. In Lakes Geneva and Bienne, concentrations of EDTA are ca. 1 $\mu g\,L^{-1}$. Concentrations in groundwater of 3.7–43 $\mu g\,L^{-1}$ are reported from the Glattal area, whereas in the Rhine basin at Basle the range is 2.4–10 $\mu g\,L^{-1}$.

Austria

The Internationale Gewasserschutzkommission fur den Bodensee reported that concentrations of EDTA in Lake Constance were in the range 2–4 $\mu g\,L^{-1}$. Rivers flowing into the lake contained 17–43 $\mu g\,L^{-1}$, and sewage treatment plant effluents contained 93–215 $\mu g\,L^{-1}$ EDTA.

Possible Effects on the Environment

Effect on Waste Water Treatment

The settling properties of activated sludge from a plant receiving mainly industrial sewage have been shown to deteriorate gradually in the presence of increasing EDTA [43] in the concentration range 10–500 mg L^{-1}. Some inhibition of microbial activity was also observed at 10 mg L^{-1} EDTA. In the presence of 50 mg L^{-1} EDTA chemical coagulation with ferric chloride prior to dewatering required increased iron salts and resulted in a higher moisture content of the filter cake.

However, inhibition of microorganisms at environmentally-realistic EDTA concentrations seems unlikely, especially in view of its low toxicity to aquatic life (see later) [44, 45]. The efficiency of phosphorus removal by tertiary treatment of sewage waters has been shown [3] not to be impaired by the presence of EDTA.

Toxicity to Aquatic Organisms

The observed toxicity of EDTA is influenced by the form in which it is presented to the test organisms.

These differences (Table 3) are probably due to the influence of the supplied form on pH and, more importantly, on the free calcium ion concentration of the test medium. This is supported by the results obtained for the 96 hour LC_{50} to Bluegill Sunfish of the EDTA tetrasodium salt in very soft water (61.2 mg L^{-1}), medium hard water (401.7 mg L^{-1}) and very hard water (807.3 mg L^{-1}). EDTA concentrations of 95 and 190 mg L^{-1} in soft water (30–32 mg L^{-1} as $CaCO_3$) were shown to cause a significant decrease in the absorption and an increase in the excretion of calcium by goldfish (*Carassius auratus*) [47], such that the calcium requirements of the fish were not satisfied. Under practical environmental conditions, there will always be a large stoichiometric excess of calcium ions, so EDTA will not exert toxicity by affecting the calcium balance of aquatic organisms. Information on the chronic toxicity of EDTA to such organisms is lacking, though it is reported that the 'no observed effect concentrations' derived from such studies are usually higher than one tenth and almost always higher than one hundredth of the corresponding LC_{50} [48]. Actual concentrations of EDTA in surface waters are therefore many orders of magnitude below those at which any adverse effects on aquatic life should occur.

Bioaccumulation

A highly polar, water soluble compound such as EDTA would not be expected to bioaccumulate by partitioning into the lipid component of aquatic organisms. A whole body bioconcentration factor of 1, with a half life for depuration of

Table 3. 96 Hour LC_{50} of several forms of EDTA to bluegill sunfish (*Lepomis Macrochirus Rafinesque*)

Test compound	LC_{50}/mg L^{-1}	pH
EDTA—Acid	159	3.7
EDTA—Tetrasodium Salt	486	8.9
EDTA—Calcium disodium Salt	2340	7.4

128–242 hours, was observed for Bluegill Sunfish exposed for 28 days to radiolabelled EDTA (0.76 and 0.08 mg L^{-1}) [49]. EDTA will not therefore bioaccumulate in the aquatic food chain and the concentrations of EDTA in fish eaten by man or other terrestrial animals will be no higher than that of the water from which they were taken.

Stimulation of Algal Growth

The stimulating effects of EDTA on the growth of algae and other organisms under laboratory conditions are well documented [50–54]. EDTA is generally added to nutrient solutions for the purpose of keeping iron and calcium phosphates in solution. A possible mechanism for the stimulation of algal growth is photodegradation of the ferric complex to form a less stable ferrous complex, resulting in temporary enhanced availability of iron which stimulates growth. Other trace elements are also kept in solution by complexation with EDTA. These complexes are not sensitive to photodegradation and have lower complex constants than iron III EDTA, so the availability of the metals may be increased [50]. Reduction by EDTA of the toxicity of heavy metals may also stimulate growth [52].

When EDTA is used to stimulate growth of higher plants in, for example, hydrocultures or forests, the iron III EDTA complex is not photodegraded until it is taken up by the plants. In this case EDTA may help to prevent adsorption of metals onto the soil and complexation with humic acids. It also increases the availability of iron in alkaline soils allowing successful growth of calcifuge plants in lime-containing soils.

In theory, therefore, EDTA may extract trace levels of essential metals from sludges and humic acids, making them more available for algae and other plants. In practice, the stimulation effect is negligible without additional supplies of nitrogen and phosphorus [55–57]. In surface waters the levels of metals are far lower than in nutrient solutions, the level of EDTA is much lower, and the levels of nitrogen and/or phosphorus are growth-limiting. The effects of EDTA may be overshadowed by comparable effects of natural humic substances. It is concluded that the possible effects of EDTA on algal growth in surface waters are very small compared with the influence of natural variables.

Heavy Metal Mobilisation

The stability constants for a number of heavy metal EDTA chelates are given in Table 4.

EDTA has the potential to take into aqueous solution heavy metals which are present as precipitates or bound to solid surfaces. The extent to which EDTA

Table 4. Stability constants for EDTA—metal chelates

Cation	pK mL
Ca	10.6
Cd	16.3
Co	15.9
Cu	18.6
Fe	25.0
Hg	21.2
Mg	7.8
Ni	19.4
Pb	18.7
Zn	16.9

is able to mobilise heavy metals in the environment is influenced not only by the concentrations of EDTA and the metal ions but also by the presence of competing ligands and precipitating anions, and by the pH and redox potential of the water.

Many studies have been undertaken of the mobilisation potential of EDTA in model systems [19, 58–62]. The results do not confirm any significant degree of mobilisation. No conclusive experimental evidence of the heavy metal mobilisation potential of EDTA at environmentally realistic concentrations is available [63]. Nonetheless, stoichiometric considerations show that the maximum concentrations of heavy metals that can possibly be mobilised by EDTA in surface waters are strictly limited; 20 μg L^{-1} of EDTA can at most complex 4 μg L^{-1} of copper [3]. Normal variations in pH will have a larger effect on the aqueous concentrations of heavy metal ions.

Mammalian Toxicology

In the aquatic environment, a large stoichiometric excess of calcium ions is likely to be present in surface waters. The calcium salt of EDTA should therefore be used for the safety assessment. Exposure to any EDTA in drinking water is primarily via the oral route, though washing will also involve some dermal exposure [3].

Acute and Longer Term Toxicity

The published data, which are summarised in Table 5, indicate that conversion from the tetrasodium salt to the calcium disodium salt greatly reduced toxicity.

Table 5. Toxicity of EDTA to mammals

Form	Route	Species	LD_{50} (mg Kg^{-1})	Ref.
CaNa$_2$ EDTA	Oral	Rat	10000	[67]
		Rabbit	7000	
		Dog	12000	
Na$_2$ EDTA		Rat	2000–2200	[68]
		Rabbit	2300	[69]
Na$_4$ EDTA		Rat	2400	[70]
Oral doses above 250 mg/animal caused diarrhoea in rats				
Irritancy: a dose of 1900 µg of Na$_4$ EDTA is irritant to rabbit eyes				[71]

Subchronic toxicity studies on CaNa$_2$EDTA

Species	Route	Dose level	Duration	Observations	Ref.
Rat	Oral	0.5% and 1% in diet	205 days	No differences versus control	[72]
Dog	Oral	50, 100, 250 mg kg^{-1} in diet	12 months	No differences versus control	[67]

Chronic toxicity studies on CaNa$_2$EDTA

| Rat | Oral | 50, 125, 250 mg kg^{-1} in diet | 2 years | No differences vs control. Death and tumour incidence not related to dose levels | [67] |
| Dog | Oral | 50, 125, 250 mg kg^{-1} in diet | 1 year | No differences from normal found in urine and blood chemistry, state of bones, organ weights or histopathology | [67] |

Teratology

Where studies involving EDTA have caused congenital abnormalities, there is support for the hypothesis that EDTA-induced zinc deficiency is the cause. 3% EDTA in the diet produced 100% malformations, whereas no congenital abnormalities were found in the young of rats given a similar diet containing an additional 1000 ppm zinc. It is also possible that this excess zinc complexed the EDTA so that the latter was no longer available to chelate another essential element [73].

From a study involving subcutaneous administration of Ca EDTA, Zn EDTA and a mixture of the two to rats, from day 11 to day 15 of gestation, it was concluded that Ca EDTA is teratogenic in rats at concentrations which produce no discernible toxicity in the dam other than reduced weight gain. These concentrations are comparable to the recommended therapeutic dose in humans (1500 mg m^{-2}/day, corresponding to 4 mmol m^{-2}/day). Protection is afforded by incorporating zinc in the chelate [74].

Acceptable Daily Intake for Man

Reflecting its low toxicity to man, EDTA is a permitted food additive in the United States [81], the Netherlands [76, 77, 78], the United Kingdom [80] and Denmark, [79] and its addition to a range of foodstuffs is permitted.

The level causing no toxicological effect in the rat was determined to be 5000 ppm in diet, equivalent to 250 mg kg^{-1} [75]. Based on this, the WHO level of acceptable daily intake for man is 0 to 2.5 mg kg^{-1} calculated as CaNa$_2$ EDTA (1/100 of no-effect level) [75].

The levels of EDTA in rivers used for the preparation of drinking water are usually below 25 μg L^{-1} and pose no hazard to human health. Absorption through the skin is essentially zero. Drinking two litres of water per day would only involve ingestion of up to 50 μg of EDTA, which is well below any effect level and also well below the WHO value of 2.5 mg kg^{-1} for acceptable daily intake.

References

1. Kirk-Othmer Encyclopaedia of Chemical Technology, 3rd edn, John Wiley, (1979)
2. SRI International—Chemical Economics Handbook (1987)
3. An assessment of the implications of the use of EDTA in detergent products: Report of the AIS Environmental Safety Working Group (1987)
4. Ribick MA, Jemal M, Cohen AJ (1987) J Pharm Biomed Anal 5:687
5. Rudling L (1972) Water Res 6:871

6. Gardiner J (1977) Analyst (London) 102:120
7. Schaffner C, Giger W (1984) J Chrom 312:413
8. Rottiers A, Procter and Gamble (personal communication)
9. Yagamuchi A, Rajput AR, Ohzeki K, Kambara T (1983) Bull Chem Soc Jpn 56:2621
10. Harmsen J, van den Toorn A (1982) J Chromatogr 249:379
11. Hall L, Takahashi L (1988) J Pharm Sci 77:247
12. Bauer J, Heathcote D, Krogh S (1986) J Chromatogr 389:422
13. Jones DR IV, Manahan SE (1976) Anal Chem 48:502
14. Dai J, Helz GR (1988) Anal Chem 60:301
15. Bober TW, Dagon TJ (1975) IWPCF 47:2114
16. Macur GJ, Alpaugh WA, Sharkness JE (1981) Proc Ind Waste Conf 1980 35:809
17. Chlem R, Hodgson AT (1978) Anal Chem 50:102
18. BASF (personal communication)
19. Gardiner J (1976) Water Research 10:507
20. Cheng SM, Thomas RL, Elrich DE (1972) Can J Soil Sci 52:337
21. Data from Dr Koppe, Ruhrverband Essen
22. Bunch RL, Ettinger MB (1962) Proc Purdue Univ Engineering Bul Exr Ser 129:393
23. Gerike P, Fischer WK (1979) Ecotoxicology and Environmental Safety 3:159
24. Madsen EL, Alexander M (1985) Appl Environm Microbiol 50:342
25. Steber J (1985) BMFT—Forschungsbericht 037278
26. Belly RT, Lauff JJ, Goodhue CT (1975) Appl Microbiol 29:787
27. Kern G (1986) Poster Presentation Fachgruppe Wasserchemie Bad Wilbad
28. Foundation for Water and Air Pollution Research of the Swedish Forest Industries (SSVL): DTPA and other complexing agents in the bleaching of mechanical wood pulp process technology and environmental aspects. Stockholm 1973
29. Foundation for Water and Air Pollution Research of the Swedish Forest Industries (SSVL): Technical importance and environmental effects of heavy complexing agents used in forest industry Appendix 2. Stockholm 1974-04-05
30. Tiedje JM (1975) Appl Microbiol 30:327
31. Tiedje JM (1977) J Environm Qual 6:21
32. Means JL, Kucak T, Crerar DA (1980) Environm Pollution (Series B) 1:45
33. Brauch HJ, Schullener S (1987) Vom Wasser 69:155
34. Lockhart HB Jnr, Blakely RV (1975) Environ Sci Technol 9:1035
35. Natarajan P, Edicott JF (1973) J Phys Chem 77:2049
36. Carey JH, Langford CH (1973) Can J Chem 51:3665
37. Lockhart HB Jnr, Blakely RV (1975) Environ Letters 9:19
38. Carey JH, Langford CH (1975) Can J Chem 53:2436
39. Battelle Institute Report to the AIS Environmental Safety Working Group (1988)
40. Freed VH, Chiou CT, Haque R (1977) Environ Health Persp 20:55
41. EAWAG Jahresbericht, Dubendorf (1987)
42. Egli T (1988) Microbiol Sci 5:36
43. Potos C (1965) JWPCF 37:1247
44. Yonezawa Y, Urushigawa Y, Hirai M, Tanaka M (1977) Kogai Shigen Kenkyusho Iho 6:15
45. Tomlinson TG, Boon AG, Trotman CNA (1966) J Appl Bact 29:226
46. Batchelder TL, Alexander HC, McCarty WM (1980) Bull Environm Contam Toxicol 24:543
47. Berg M (1970) 1st Ital Idrobiol 26:257
48. The EEC Sixth Amendment. A Guide to Risk Evaluation for Effects in the Environment. Technical Report No 13, ECETOC (1984)
49. Bishop WE, Maki AW (1980) Proc 3rd Ann Symp on Aquat Tox ASTM STP 707
50. Tevlin MP (1978) Water Res 12:1027
51. Victor DM, Martin DF (1977) Water Res 11:447
52. Nusch EA (1977) ZF Wasser und Abwasser-Forschung 10:49
53. Segot H, Codomier L (1981) Bot Mar 24:63
54. Bender ME, Matson WR, Jordan RA (1979) Environ Sci Technol 4:520
55. Sakamoto MT (1981) Verh Int Verein Limno 21:256
56. Hamilton RD (1972) Symposia I 217–221, Fishery Res Board Canada, Winnipeg
57. Stockner JG, Evans DE (1974) Fish Res Board of Canada Technical Report 416
58. Muller G, Forstner U (1976) Z Wasser und Abwasser 9:150
59. Frimmel F (1976) Z Wasser und Abwasser Forschung 9:167
60. Barcia J, Stainton MP, Hamilton AL (1973) Water Res 7:1791

61. Schmidt M: Berichte aus der Dortmunder Stadtwerke AG No 176
62. Chubin RG, Street JJ (1981) J Environmental Quality 10:225
63. Salomons R: report to AIS Standing Committee
64. Dietz F (1985) Bericht Korrespondenz Abwasser 988–989
65. Linckens AHM, Reichert JK (1982) Vom Wasser 58:27
66. Sprecher C, Giger W, Schaffner C: Determination of NTA and EDTA in Waste Waters and Natural Waters in Switzerland; 16th Annual Symposium on the Analytical Chemistry of Pollutants, Lausanne, March 1986
67. Oser BL, Oser M, Spencer HC (1963) Toxicol Appl Pharmacol 5:142
68. Yang SS (1964) Fd Cosmet Toxicol 2:763
69. Shibata S (1956) Folio Pharmacol, Japan 52:113 (Cited in Ref 74)
70. Procter and Gamble (unpublished data)
71. Niosh Reg Tox Effect Chem Subst 1979 VI 31
72. Chan MS (1964) Fd Cosmet Toxicol 2:765
73. Swenerton H, Harley LA (1971) Science 173:62
74. Brownie CF, Brownie C, Noden O, Krook L, Haluska M, Aronsen AL (1986) Toxicol Appl Pharmacol 82:426
75. Toxicological evaluation of some food additives including anticaking agents, antimicrobials, antioxidants, emulsifiers and thickening agents. WHO Food Additive Series No 5, WHO, Geneva (1974)
76. The Mayonnaise and Salad Cream Decree, Netherlands, 1971
77. The Preserved Vegetables Decree, Netherlands, 1978
78. The Soaked Pulses Decree, Netherlands, 1978
79. The Permitted Additives List, Denmark, 1984
80. The Miscellaneous Additives in Food Regulations, United Kingdom, 1980 as last amended 1982
81. U.S. Code of Federal Regulations. Title 21. Part 172. 120
82. Dr Opgenorth, BASF (personal communication)

Environmental Properties and Safety Assessment of Organic Phosphonates Used for Detergent and Water Treatment Applications

William E. Gledhill[1] *and Tom C. J. Feijtel*[2]

[1] Monsanto Company—U4G, 800 N. Lindbergh Blvd., St. Louis, MO. 63167, USA
[2] Procter and Gamble Company, B 1853 Strombeek-Bever, Belgium

Introduction—Structures, Uses and Production 261
Chemical Structure 263
Environmental Fate 264
    Environmental Entry and Predicted Environmental Levels 264
    Physico-chemical Properties 265
    Environmental Partitioning/Groundwater 265
    Persistence 267
        Microbial Degradation—Pure Cultures 267
        Microbial Degradation—Natural Populations 268
        Ready Biodegradation Tests 268
        Fate in Sewage Treatment Systems 268
        Surface Waters 269
        Soil 271
        Photochemical Degradation 271
    Metal Mobilization 273
    Metal Complexation 274
    Bioconcentration 278
    Drinking Water Purification 278
Environmental Effects 279
    Toxicity to Plants 282
Environmental Safety Assessment 282
Acknowledgements 284
References 284

Summary

This review article will focus on the environmental safety assessment of organic phosphonates used in water treatment and detergent application. The review will summarize use, environmental fate, persistence and aquatic toxicity.

Introduction—Structures, Uses and Production

Phosphonate compounds containing more than one phosphonate group are effective sequestrants and possess other useful properties such as high water solubility, chemical stability, threshold effect, effective dispersion/flocculation of

Table 1. Major uses and production of phosphonate compounds

	ATMP		HEDP		EDTMP		HDTMP		DTPMP	
	U.S.A.	Europe	U.S.A.	Europe	U.S.A.	Europe	U.S.A.	Europe	U.S.A.	Europe
Volume MT**/yr	3820	3360	5270	2010	0	1090	180	90	275	5270
Use, % of total										
Ind. boiler/cooling	70	100	85	17	0	0	95	95	20	2
Ind. cleaning	10	0	0	38	0	4	0	0	—	7
Laundry detergents	0	0	2	38	0	96	0	0	0	85
*Other	20	0	13	7	0	0	5	5	80	6

Phosphonate (Active basis)

* Other includes use in oil field drilling, textile and pulp/paper industries
** MT = Metric Tonnes

solids and peroxy bleach stabilization. They find wide applicability in water treatment and cleaning applications. Because of regulated detergent phosphate reduction, phosphonates, particularly EDTMP and DTPMP, are finding specialty uses at low concentration (<1%) in combination with other specialty products in detergents [1]. Their ability to prevent precipitation of calcium salts at substoichiometric concentrations (threshold effect), stabilize peroxy bleaches by inactivating metals ions that catalyze peroxide decomposition and prevent scale build-up makes them particularly attractive for such applications. These materials are also used as scale or corrosion inhibitors at sub-stoichiometric concentrations (1–10 ppm) in cooling towers, oil fields, boilers and other water treatment applications.

Phosphonates are characterized by the presence of one or more $-C-PO_3^-H_2$ groups (see structures below). Numerous chemicals containing this structure exist, but only a few are cost effective for the applications discussed above. Most are synthesized from phosphorous acid (made from PCl_3 and H_2O) by reaction with formaldehyde and either ammonia (ATMP) or amines (EDTMP, HDTMP, DTPMP). HEDP is formed from PCl_3 and acetic acid. They are usually sold as solutions of the acids or sodium salts in water at concentrations ranging from 25% to 60%. Approximate production volumes and uses are shown in Table 1.

Chemical Structures

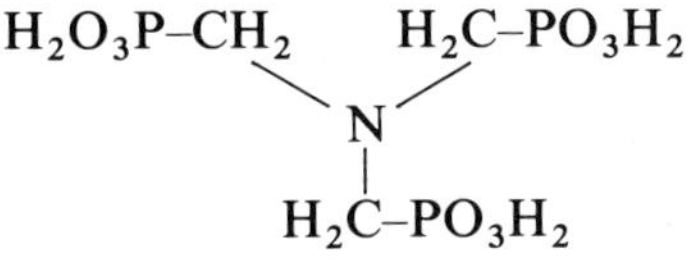

Product	Common abbreviation
1. Amino tris(methylenephosphonic acid) CAS No. 6419-19018 CAS No. 2235-43-0: Penta Sodium Salt	ATMP

$$H_2O_3P-CH_2 \diagdown \quad \diagup H_2C-PO_3H_2$$
$$N$$
$$|$$
$$H_2C-PO_3H_2$$

2. 1-Hydroxy-ethane diphosphonic acid CAS No 2809-21-4 CAS No. 2666-14-0: Tetra Sodium Salt	HEDP

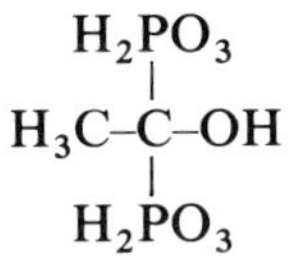

$$H_2PO_3$$
$$|$$
$$H_3C-C-OH$$
$$|$$
$$H_2PO_3$$

3. Ethylenediamine tetra(methylenephosphonic acid)   EDTMP
   CAS No. 1492-50-1
   CAS No. 22036-77-7: Penta Sodium Salt

$$\begin{array}{c} H_2O_3P-CH_2 \\ \diagdown \\ H_2O_3P-CH_2 \end{array} NCH_2CH_2N \begin{array}{c} H_2C-PO_3H_2 \\ \diagup \\ H_2C-PO_3H_2 \end{array}$$

4. Hexamethylenediamine tetra(methylenephosphonic acid)   HDTMP
   CAS No. 23605-74-5
   CAS No. 56744-47-9: Sodium Salt

$$\begin{array}{c} H_2O_3P-CH_2 \\ \diagdown \\ H_2O_3P-CH_2 \end{array} N(CH_2)_6N \begin{array}{c} H_2C-PO_3H_2 \\ \diagup \\ H_2C-PO_3H_2 \end{array}$$

5. Diethylenetriamine penta(methylenephosphonic acid)   DTPMP
   CAS No. 15827-60-8
   CAS No. 22042-96-2: Sodium Salt

$$\begin{array}{c} H_2O_3P-CH_2 \\ \diagdown \\ H_2O_3P-CH_2 \end{array} N-CH_2CH_2\underset{|}{N}CH_2CH_2N \begin{array}{c} H_2C-PO_3H_2 \\ \diagup \\ H_2C-PO_3H_2 \end{array}$$
$$H_2O_3P-CH_2$$

Environmental Fate

Environmental Entry and Predicted Environmental Levels

ATMP and HEDP are the major phosphonates used in the U.S.A. primarily in industrial cooling and boiling water applications (Table 1). Smaller amounts find use in oil field drilling, metal finishing, paper mill processes and industrial cleaning applications. Consequently, entry to the environment from these sources occurs only on an intermittent basis. Phosphonate manufacturing facilities provide additional point source locations for release to the environment. ATMP and HEDP use levels (5–15 mg/L) in cooling towers and boiler waters are such that even with minimum dilution, intermittent concentrations in receiving waters would be well below 1 ppm. Concentration or bioavailability is further decreased by the partitioning and degradation processes discussed below. In contrast to the U.S.A., a significant percentage of European phosphonate use occurs in detergents; consequently, environmental release occurs on a more continual basis. It would be desirable to have data on environmental levels of phosphonates; however, at present sufficiently sensitive analytical methods are unavailable. One must rely on modelling to develop estimates of environmental concentrations. Modelling that has been done for western European countries [2a] based on annual detergent use of 6300 MT (Table 1), predicts maximum raw sewage levels of phosphonates to range from 170 to 290 ppb. Removal primarily via partitioning would result in average effluent levels ranging from 90 to 235 ppb and these levels would decrease by an order of magnitude upon dilution in receiving waters. Therefore, maximum phosphonate environmental levels would be in the low µg/L (<30) range. If partitioning to sediments (100:1) and limited photo/bio-degradation are accounted for, average environmental levels in European streams would be on the order of 0.25 µg/L. Limited use in U.S.A. detergents coupled with higher per capita water consumption would suggest U.S. environmental levels to be significantly lower than those in Europe. However, because of intermittent discharge of cooling tower water, concentrations of ATMP and HEDP in the vicinity of the discharge might approach a low µg/L level depending on the nature of the discharge and its treatment.

Physico-Chemical Properties

Phosphonate salts are highly water soluble and some are formulated as aqueous solutions containing up to 70% active material. Free acids are less soluble than the salts; however, aqueous solutions in excess of 1% active material can be prepared (Table 2). Phosphonates are non volatile, having vapor pressures in the region of 1×10^{-7} Torr. An estimate of Henry's Constant (vapor pressure/molar water solubility) indicate that the Henry's constant would be on the order of 10^{-5} or less.

Environmental Partitioning/Groundwater

In general, extremely water soluble chemicals don't readily adsorb to sediments or soil; however, chelation properties of phosphonates cause these materials to

Table 2. Physico-chemical properties of phosphonates

	ATMP	HEDP	EDTMP	HDTMP	DTPMP
Molecular weight	299	206	436	492	573
Water solubility (g/L)	inf.	inf.	21	15	inf.
Molar water solubility	inf.	inf.	0.048	0.030	inf.
Vapor pressure (Torr)	10^{-7}	10^{-7}	10^{-7}	10^{-7}	10^{-7}
Henry's constant	—	—	$<10^{-5}$	$<10^{-5}$	—
K_{ow}	−3.53	−3.49	−4.10	−4.43	−3.40

have high affinity for the mineral portion of soils and sediments. Sediment/water partition coefficients for radiolabelled phosphonates were developed using a National Bureau of Standards river sediment (standard reference material 1645) and are summarized in Table 3 [3a]. Sediment properties were: pH 7.3, organic carbon 11.8%, cation exchange capacity 4.73 meq/100 g, sand 74%, silt 18%, clay 8%. The data indicate that anticipated environmental levels of phosphonates (<1 mg/L) will have high affinity for sediments. In general, little difference was noted between partition coefficients developed in soft and hard water. At levels up to 1 mg/L, HDTMP had the greatest affinity for the sediment.

Water hardness and phosphonate concentration were found to control sorptive behavior. However, irrespective of the mechanism and extent of this control, effective partitioning to the sediments can be reported in all cases (Table 3). Sorption coefficients for ATMP, EDTMP and DTPMP decreased with increasing phosphonate concentration. Highest partitioning would occur at realistic environmental phosphonate concentrations. Sorption coefficients for ATMP, EDTMP and DTPMP were respectively 1300, 1100 and 1100 at phosphonate concentrations between 50 and 100 μg/L. Sorption experiments with HEDP indicated a partition coefficient of 1300 at 50 μg/L in hard water and 2400 at 5 mg/L HEDP in soft water.

Mobility of phosphonates in soils is, therefore very low, thereby reducing the risk of groundwater contamination. Freundlich isotherm constants for HEDP and ATMP in different soils ranged from 20–190 and 32–237, respectively [4, 5]. Soil partition coefficients for these and other phosphonates were also summarized in Table 3 [3e]. According to EPA, substances with these affinities for soil are classified as moderately to slightly mobile in soils. The different adsorption characteristics for different soils may be due to a variety of things including mineral composition, surface area, organic matter and iron oxides. Leaching experiments with three standard soils simulating a 12–20 month rainfall period in Germany showed elution of only 0.4–1.7% of the 1 μg ^{14}C radiolabeled HEDP adsorbed to 30 cm soil columns [4]. These findings also

Table 3. Phosphonates sediment/water partition coefficients

| Phosphonate | Initial water concentration | | | | | | | |
| | 0.05 mg/L | | 0.10 mg/L | | 1.0 mg/L | | 5.0 mg/L | |
	S	H	S	H	S	H	S	H
ATMP	1100	1300	1200	1300	360	1000	170	230
HEDP	920	1300	1100	1100	940	1200	2400	680
EDTMP	250	920	300	1100	170	320	660	90
HDTMP	3900	3900	3900	7900	2600	3700	80	220
DTPMP	720	720	810	1100	280	400	52	130

Adsorption measured by placing 20 g sediment in 800 mL of soft (S) Milli-Q water or hard (H) (211 mg/L as $CaCO_3$) water containing the various concentrations of radiolabelled phosphonates. After 24 h of mixing, samples were centrifuged and radioactivity remaining in the water column was determined by liquid scintillation counting suggest HEDP to be tightly bound to soil. In another study [3e], more than 95% of the radiolabeled phosphonates were tightly bound to the sediment after 38–50 days and only 5% of the radioactivity could be extracted by ultrasonic homogenization with a 0.25 N HCl-xylene solvent.

Persistence

Microbial Degradation—Pure Cultures

A variety of natural [6] and synthetic chemicals contain the C–P bond. The C–P bond provides the molecule stability and relative high resistance to chemical, photolytic and thermal decomposition. The N–P, S–P and O–P linkages are more reactive than C–P bonds [7, 8]. Nevertheless, several laboratory studies report phosphonate degradation by pure microbial cultures when supplied as the sole phosphorus source [8–17]. Indeed, orthophosphate has been found to supress phosphonate utilization in many microorganisms [7, 15–17]. Organisms perferentially use inorganic phosphate, which may explain low biodegradability of phosphonates in synthetic test media and natural sewage systems.

Workers at Cornell [7] were the first to isolate a pure culture of *Pseudomonas testosteroni* and elucidate the metabolic pathway for its growth on several alkyl phosphonates as the sole P source. Products included the corresponding alkane and inorganic phosphate. Subsequently, a variety of other organisms have been reported to have C-P lyase capabilities [8, 9, 12, 18, 19]. Species include *Bacillus, Pseudomonas, Arthrobacter, Escherichia, Actinomyces, Rhizobium, Acinetobacter, Alcaligenes* and *Klebsiella* and are similar in their preferential use of P_i over the organic phosphonate. The *Arthrobacter* sp. [11, 20] is apparently unique in that it cleaves only a single phosphonate group from ATMP, EDTMP, HEDP and DTPMP, perhaps due to inhibition caused by P_i. Because of the diversity of taxonomic classes possessing C–P lyase activity, it can be expected that these materials would undergo biodegradation in phosphate limited environments. Indeed, a recent study [11] has demonstrated that phosphonate degraders are ubiqitous in nature and that isolates can completely mineralize HEDP and EDTMP. Recently, batch and continuous culture experiments with a bacterial strain endea MMM101a and methylphosphonate as a model compound revealed that phosphates did not supress phosphonate utilization by this species [21]. Phosphonates clearly contributed to cell growth even in the presence of phosphate. Phosphonates were not the favored P-sources, however. The affinity for orthophosphate ($Ks = 0.17 \mu M$) was more than two orders of magnitude higher than for methylphosphonate ($Ks = 66 \mu M$) [21]. The above suggests the presence of phosphates in environmental samples does not necessarily exclude phosphonate biodegradation. The effect of high phosphate levels (>0.5–1 mM) requires further investigation.

Microbial Degradation—Natural Populations

Ready Biodegradation Tests

Phosphonates seem to be recognized by bacteria only as a possible P-source, which may explain poor biodegradation results in standard biodegradation tests. Numerous studies have shown that little, if any, primary or ultimate biodegradation occurs for any phosphonate product in biodegradation tests such as the OECD Screening Test, BOD_{20} Test, Sapromat test and Closed Bottle Test [4, 5, 22–24]. When some removal has been noted, it is generally attributable to adsorption or, as will be discuss below, photolysis. Based on the work with pure cultures, this lack of biological degradation may result from the preferential use of or enzymatic repression by P_i in these test systems. However, it is expected that phosphonates serve as a carbon source at very high phosphonate concentrations. The DOC removal of 23–33% in a Zahn-Wellens test for HEDP and ATMP is related to the typical high test concentrations [4, 5].

Phosphonate degrading bacteria are found in polluted and unpolluted samples; however, environmental conditions, bioavailability or numbers of phosphonate degrading bacteria cause the limiting factors in the degradation of phosphonates. One of the environmental parameters is the orthophosphate concentration. The Pho-regulon (group of genes) decodes, among other enzymes, for the C-P lyase enzyme, which is induced under P stress. However, the presence of low phosphate concentrations does not supress the C-P lyase activity once this enzyme system is induced. Phosphonate utilization to an extent of 94 and 97% HEDP and EDTMP, respectively, provided evidence that complete breakdown of these compounds with respect to the C–P bond can be achieved by some bacteria [11]. The pentaphosphonate DTPMP was somewhat less susceptible with only about 60% P_n conversion.

Fate in Sewage Treatment Plants

Several studies have examined phosphonate biodegradation in sewage treatment systems. The removal of HEDP and ATMP was only moderate, while removal of EDTMP ranged between 75–90% at day 126. However, HEDP and ATMP removal increased significantly when the pH was buffered to about 7. Removals increased to 100% and 90%, respectively, after 26 days [22]. Acidification of non-buffered sludge and extended time period may explain the limited phosphonate removal. However, the authors used DOC as a measure of removal and could not, therefore, discriminate between adsorption and biodegradation. A two month study with ATMP, HEDP and EDTMP showed no indication of degradation as judged by loss of chelant titer and orthophosphate measurement [3c]. However, phosphonate levels up to 160 mg/L had no inhibitory effect on COD or MBAS removal in the test units. Biodegradation of radiolabelled ATMP, HEDP, EDTMP, HDTMP and DTPMP at 5 mg/L in

SCAS systems resulted in 0.5 to 10.2% conversion to $^{14}CO_2$ over a 210 day test period [3b]. These extents of degradation are indicative of recalcitrance and may be attributed to minor impurities in the radiolabelled products or a combination of slow photochemical degradation followed by biodegradation.

Muller et al. [25] reported an HEDP elimination of 50–70% in the primary stage of a sewage treatment plant. Field studies in municipal sewage treatment plant (2-stage STP) showed that phosphonates are eliminated to the extent of 50% [24]. Adsorption studies with radiolabelled HEDP and sludges from 3 different municipal sewage treatment plants showed more than 90% adsorption within 24 h [4]. Similar experiments with radiolabelled ATMP showed comparable results [5]. Adsorption of ^{14}C labelled EDTMP and DTPMP at an expected environmental concentration of 115 µg/L on two activated sludges from Belgia sewage treatment plants varied between 23–70% for EDTMP and 80–90% for DTPMP [22]. In summary, it is concluded that phosphonate removal is high in sewage treatment plants even though biological degradation may be low.

Reports of anaerobic biodegradability are sparse. Steber and Wierich [4, 5] reported little anaerobic degradation (<4% conversion to $^{14}CO_2$ and $^{14}CH_4$) for HEDP and ATMP in a model digestor. The materials did not inhibit digestor performance even at high loading.

Concerning water treatment, phosphate precipitation by low concentrations of $FeCl_3$ may be affected by phosphonate levels. Horstmann and Grohmann [26] showed that ATMP and EDTMP inhibited phosphate removal in the quaternary treatment of waters by flocculation filtration. Effect levels were noted in the 0.2 to 1.0 mg/L concentration. Such results are not unexpected because of the ability of these materials to complex Fe^{+++}. Addition of excess $FeCl_3$ overcame this inhibition. Contrary to these findings, HEDP in the 2–2.4 mg/L range had no adverse effect on tertiary treatment of sewage [25]. A study conducted under the conditions of ASTM procedure D-2035-74 as measured by scintillation counting of ^{14}C labelled materials indicated that removal of phosphonates by chemical coagulation-flocculation is strongly dependant on the precipitation reagent used. Addition of ferrous sulfate flocculates about 60% EDTMP and DTPMP, while a mixture of ferrous sulfate and aluminum sulfate removed both phosphonates almost completely from raw sewage [2b].

Surface Waters

A river water die-away study was conducted using Meramec River water supplemented with 2 mg/L ^{14}C-labelled ATMP, HEDP, EDTMP, HDTMP and DTPMP and results are summarized in Table 4 [3d]. Degradation was followed over a 60 day period. Abiotic degradation seemed responsible for 3–6% when all biological activity was inhibited. This was due mainly to photolysis. In biologically active water, ultimate degradation increased up to 17% for HEDP

Table 4. Phosphonate mineralization in Meramec river water and eutrophic lake water*

Phosphonate	Average % $^{14}CO_2$ evolution, 60 days			
	Sterile natural water		Active natural water	
	Dark	Sunlight	Dark	Sunlight
ATMP	0.1	6.2	12.3	13.6
HEDP	0.2	2.7	2.0	17.2
EDTMP	0.2	4.1	5.2	14.4
HDTMP	0.3	4.2	6.4	9.4
DTPMP	1.4	3.0	4.8	14.3

* Degradation rates similar in both river and lake water; therefore, data combined (duplicate flasks)

Table 5. Phosphonate mineralization in lake water/sediment and river water/sediment microcosms

Phosphonate	Average % $^{14}CO_2$ evolution			
	Lake water[1]		River water[2]	
	Sterile	Active	Sterile	Active
ATMP	1.6–8.6	4.7–12.3	—	—
HEDP	0.2–5.2	4.6–9.6	—	—
EDTMP	3.8	9.6	0.1–0.6	10.4–25.2
HDTMP	—	—	0.2–1.8	38.9–41.0
DTPMP	0.1–9.2	14.8–16.9	0.2–1.0	28.6–30.7

[1] Pristine lake water/sediment, 50 day study, 1000 ppb phosphonate
[2] Missouri River water/sediment, 38 day study, 300–500 ppb phosphonate

in the light. Thus, enhancement of CO_2 evolution rates was due to microbial activity, sunlight and their interaction [3d].

In lake and river water/sediment microcosms (Ecocores [27]), slower, but somewhat more extensive, biodegradation than in the previous study occurred over 38–50 days [3e]. Microcosms were incubated at ambient temperature with a 16 h/8 h light/dark cycle. Results are summarized in Table 5 and indicate somewhat less extensive mineralization in the lake water than the river water. For the former, $^{14}CO_2$ evolution ranged from 5–17%, while in the latter 10–41%. It was also noted from these studies that phosphonates were rapidly removed from the water column. From 66–80% of the radioactivity was lost from the water column after 11 days and by the end of the study >95% of the radioactivity was tightly bound to the sediment (only 5% ^{14}C activity extractable via ultrasonic homogenization with 0.25 N HCl-xylene solvent).

In summary, data from persistence studies in natural water both with and without sediment indicated that mineralization of phosphonates to carbon dioxide can occur at a slow to moderate rate. Degradation was probably a combination of both biological and photochemical processes. Extrapolation of these data to nature would suggest that phosphonate entering such environments would initially be suceptible to both processes and as material partitioned to sediment, continued biological degradation would be expected, especially in phosphate limited environments.

Soil

Schowanek and Verstraete [11] have shown that phosphonate utilizing bacteria are ubiquitous. Phosphonate degrading bacteria were not only present in sludge, but also in soils and peat. Phosphonate degrading microorganisms were readily isolated from soils and sediments having no previous contact with these materials. Reported soil biodegradation studies are limited and difficult to compare. Steber and Wierich [4] found that in two soils, mineralization of HEDP to $^{14}CO_2$ amounted to only 0.4 to 4.8% over an 80 day period. Linearity of kinetics suggested slow abiotic degradation. Corresponding values for ATMP in three soils ranged from 2 to 53% over ten weeks [5]. A more extensive study reported similar degradation for HEDP and ATMP and significantly more extensive mineralization of EDTMP, HDTMP and DTPMP [3d, 3f]. The studies involved four different soils supplemented with 10 μg/gm ^{14}C-labelled phosphonates. Soils were adjusted to 60% of their moisture holding capacity and $^{14}CO_2$ evolution was followed over a 119 to 148 day period. Results are summarized in Table 6 below. Substantial mineralization of all phosphonates except ATMP was shown to occur. The extent of HEDP mineralization varied from low to moderate depending on soil type, but EDTMP, HDTMP and DTPMP displayed extensive mineralization in all soils. Half-lives of these latter three materials ranged from 40 to 50 days. Mineralization in sterile controls also indicated abiotic processes were involved; however, the substantial difference between sterile and biologically active systems indicated biodegradation to be the major degradation pathway.

Photochemical Degradation

Results from control flasks in the various biodegradation studies indicated that photochemical processes were important in the degradation of various phosphonates. Steber and Wierich [4] reported substantial conversion of HEDP to

Table 6. Biodegradation of radiolabelled phosphonates in soils

Phosphonate	Average % $^{14}CO_2$ Evolution, 119–148 days		T1/2 Days
	Sterile*	Active	
ATMP	0.3–0.8	0.6–14.6	—
HEDP	3.8–22.6	3.3–47.5	>148
EDTMP	3.1–18.1	68.7–72.6	40–50
HDTMP	5.6–32.4	60.9–76.2	40–50
DTPMP	3.7–19.2	62.6–64.0	40–50

*Sodium azide added at 0.1%

$^{14}CO_2$ (up to 70%) in the AAP medium after 3 to 7 days. Exposure of sterile AAP medium to sunlight followed by addition of active bacterial and algal inoculum and dark exposure indicated the initial photochemical process resulted in degradation products that were biologically degradable. For HEDP, P_i and acetate were the primary photochemical reaction products. Similar studies conducted with ATMP and EDTMP [5] showed an analogous process. However, in contrast to 95% release of P_i from HEDP, very little was noted from ATMP even though half-lives for primary degradation were on the order of 6 h. Consequently, more slowly degrading intermediates were detected in the test medium. Of the three primary reaction intermediates formed under sterile conditions, aminomethylene phosphonic acid (AMP) was biologically stable. Evidence was presented that release of inorganic phosphate from the parent ATMP and EDTMP may slow the biodegradation of the phosphonate intermediates. This inhibition could be circumvented if additional carbon sources such as glucose or glutamic acid were added to the AAP medium.

An additional study demonstrated little direct photolysis of phosphonate products in purified water, but substantial sensitized photolysis in the presence of some metal ions (particularly Fe^{+++}) [3f]. Solutions consisting of 10 mg/L phosphonate active acid and equimolar concentrations of metal ions were prepared in quartz tubes at three pH values. Samples were exposed to natural sunlight (46–82% of that possible at temperatures ranging from 9–28 °C) for a 17 day period. Dark controls were also included. Degradation was assessed by following release of P_i and is summarized in Table 7. All products underwent substantial photolysis in the presence of ferric ions (44–91% conversion to P_i in 17 days). With the exception of HEDP, other metals were not as effective and degradation was generally not seen in the dark controls. HEDP underwent

Table 7. Phosphonate degradation in the presence of sunlight and metal ions.

Phosphonate Solution	pH	% Transformation to orthophosphate, 17 days				
		ATMP	HEDP	EDTMP	HDTMP	DTPMP
Phosphonate	4	0	44	4	9	4
	7	2	7	11	12	14
	10	0	3	4	4	4
Phosphonate/Ferric Nitrate	4	45	55	64	64	91
	7	44	78	56	82	70
	10	38	78	25	19	47
Phosphonate/Chromic Nitrate	4	3	16	4	15	7
	7	1	6	3	14	11
	10	3	6	6	14	10
Phosphonate/Zinc Nitrate	4	0	41	2	13	5
	7	1	6	5	15	8
	10	3	6	4	13	9
Phosphonate/Cupric Nitrate	4	2	77	1	7	4
	7	2	70	1	11	9
	10	3	67	4	8	7
Dark Controls	4,7,10	<2	<37*	<2	<3	<2

*Phosphonate/Cupric Nitrate, pH 7 = 37, all others = <10

67–77% conversion to P_i in the presence of cupric ions. HEDP also displayed a wide range (16–77%) of degradation both with and without metal ions at pH 4. In an additional study designed to simulate more closely a cooling tower situation (HEDP or ATMP/Fe^{+++} ratio of 10:1, 10 mg/L phosphonate), there was not a significant difference between the extent of degradation in solution with and without Fe^{+++} over a 17 day period [3g, 3h]. However, extents of degradation were lower and ranged from 2–6% for ATMP and 22–27% for HEDP.

A more detailed study into the mechanism of EDTMP photodegradation in the presence of ferric ions was presented by Matthijs et al. [28]. The reaction mechanism was very similar to that reported for EDTA and NTA [29–31] occurring at a very rapid rate (a few minutes to a few hours) with up to 75% of the organic phosphonate being converted to P_i. A non-chelating, photostable product was identified as N-methylaminomethylenephosphonic acid. The authors [28] predicted a half-life in the upper layer of surface waters under central European conditions to be on the order of 26 hours, and estimated half-lives in a German river of a few minutes to a few hours. It was concluded that photodegradation could provide a major degradation pathway in nature, but that the extent depends on the ammount of sunlight, species of metal ions present and their competition for EDTMP and the interaction of EDTMP with particulates.

Metal Mobilization

The introduction of chelating agents into the environment may affect the distribution and partitioning of metals in soils, sediments and sludges. Complexing agents potentially can cause active desorption of trace metals from particulates or interfere with natural adsorption processes. Concern arises when heavy metals are mobilized, thereby making them more bioavailable either to aquatic species or to man. As was shown in the previous section, phosphonates strongly partition to sediments. Thus, it would be expected that at environmental levels ($< 10 \mu g/L$) metal mobilization by phosphonates would be minimal. This is in fact is what most studies have shown.

Metal mobilization was studied in the same system described above for the soil partitioning studies using the same standard river sediment and phosphonate concentrations in hard and soft water [3b]. Metals examined were Cd, Cr, Cu, Fe, Ni, Pb, Zn and Hg. Cd, Cu, Ni, Hg and Pb were not mobilized in either hard or soft water by any phosphonate at all concentrations studied (0.05, 0.1, 1.0 and 5.0 mg/L). Zn, Fe and Cr were mobilized to varying degrees with concentrations of metals peaking by 24 h and then declining. In general, the extent of mobilization in hard and soft waters was less than 0.5 and 0.2 mg/L, respectively per metal. In soft water, Zn was slightly mobilized ($< 30 \mu g/L$) at 0.05 and 0.10 mg/L ATMP and HEDP and not mobilized by 0.05 mg/L

EDTMP, HDTMP and DTPMP. Fe and Cr were solubilized to concentrations of <250 and <100 μg/L, respectively by all phosphonates at 0.05 mg/L.

Another study was conducted to examine the concentrations of EDTMP and DTPMP that cause release of metals or interference with adsorption of metals in natural European sediments [2]. Studies were conducted with radioactive Cd, Ni and Zn added to sediments and with natural unamended sediment/water systems. Results similar to the study above using the standard reference sediment were obtained. Metal mobilization did not occur until phosphonate concentrations of 0.3 to 1.0 mg/L were reached, levels significantly above those anticipated in nature. Muller et al. [25] found that concentrations of HEDP in the range of 2 to 2.4 mg/L did not mobilize Cu, Cd, Cr, Ni, Pb or Zn, in fact 50 to 70% of the HEDP was removed via adsorption to solids in municipal sewage and adsorptive removal was >90% through tertiary treatment. Gunther et al. [32] also found no solubilization of thallium or cadmium from natural river sediments by HEDP at concentrations of 2.0 and 100 mg/L. Borzetto [33] studied mobilization of a range of metals by HEDP, EDTMP and DTPMP and found no mobilization at concentrations below 10 mg/L.

Metal Complexation

Phosphonates function primarily to complex metal ions, thus holding them in solution. This process is dependant not only on the metal ions present, but also the pH of the solution, presence of cations and the extent of phosphonate adsorption to particulates in the system. Phosphonates generally have the greatest affinity for metal ions when present as the fully dissociated species [alkaline pHs] in purified water. Consequently, most stability constant measurements are made under such conditions. Acid/base titrations are used to establish the protonation or dissociation constants [i.e. the pH at which protons are

Table 8. Acid dissociation constants for phosphonates, EDTA and NTA

Compound	pK^1 1	2	3	4	5	6	7	8	Ref.
ATMP	12.3	6.7	5.5	4.3	<2	<2	—	—	34
HEDP	11.4	7.0	2.5	1.7	—	—	—	—	34
EDTMP	13.1*	9.9*	7.9*	6.4	5.1	2.9	—	—	3h, i
HDTMP	13.3*	13.1*	6.8	6.1	5.3	4.6	<2	—	3h, i
DTPMP	13.4*	12.3*	9.1	7.6	6.5	5.5	4.2	<2	3h, i
NTA	9.7	2.5	1.9	—	—	—	—	—	34
EDTA	10.3	6.2	2.7	2.0	—	—	—	—	34

[1] 0.001 M compound in 0.1 M KNO_3
* Estimated from measured values at 2.0 M KNO_3

Table 9. Phosphonate, EDTA and NTA metal stability constants

Compound	Log K stability constants									
					Metal ion ($^{+2}$ Valence state)					
	Ca	Cd	Co	Cu	Hg	Mg	Ni	Pb	Zn	Ref.
ATMP	7.6	12.7	18.4	17.0	21.7	6.7	15.5	16.4	14.1	3j
HEDP	6.8	15.8	17.3	18.7	16.9	6.2	15.8	Insol.	16.7	3j
EDTMP	9.6	21.4	15.8	24.3	31.8	10.0	20.2	23.0	21.1	3j
HDTMP	5.4	13.3	18.9	19.8	21.9	6.0	20.6	17.0	19.5	3j
DTPMP	6.7	9.7	17.3	19.5	22.6	6.6	19.0	8.6	19.1	3j
NTA	6.4	9.8	10.4	13.0	14.6	5.5	11.5	11.3	10.7	35
EDTA	10.6	16.6	16.2	18.8	21.8	8.7	18.6	17.9	16.5	34

added to the fully dissociated compound]. Table 8 summarizes dissociation constants [pK's] for various phosphonates as well as for NTA and EDTA.

Several reports have presented metal ion stability constants for various phosphonates [28, 34]. Log K values for various metal ions are summarized in Table 9. For comparative purposes NTA and EDTA values are included. The data indicate phosphonates to have greater affinity for metal ions than NTA. ATMP and HEDP are similar to EDTA, while EDTMP, HDTMP and DTPMP generally bind metal ions more strongly than EDTA.

Table 10 summarizes the logarithm of the apparent or conditional stability constants for various phosphonate-metal ion complexes as a function of pH which were developed in a separate study [3h, 3i]. Phosphonates generally possess greater affinity for metal ions at higher pH values. At typical environmental pHs (6–8) substantial complexation ability exists; however, with exception of Hg^{+2}, apparent stability constants are generally several orders of magnitude lower than for the fully ionized anions. Thus, because of many factors (e.g. pH, excess Ca and Mg, adsorption, etc.), ability to complex metals in the environment is much less than that indicated in Table 10.

An additional study [3k] is summarized in Table 11. In contrast to the previous study above, stability constants for many of the mono-protonated phosphonates were greater than the fully unprotonated acids.

Models are currently available [35] which allow prediction of the types of metal-ligand complexes formed under simulated environmental conditions. In general, as ligand (phosphonate) concentration decreases, the ligand becomes increasingly associated with the metal ions for which it has the greatest stability constant. For example, with NTA [35] even though it has comparatively low affinity for Ca and Mg, environmental concentrations above 100 μg/L will exist primarily as the Ca or Mg complex simply because of the high concentrations of these cations in nature. Only as NTA levels decrease to below 25 μg/L would it exist primarily as the Cu (60%) and Ni (40%) complex. A similar computer model has also been run for EDTMP using simulated Rhine river water ingredients [2]. Table 12 indicates that under the conditions established for the model, as the concentration of EDTMP decreases in the river water it becomes

Table 10. Logarithm of the conditional stability constants for metal ion-phosphonate complexes at different pHs

Compound/Metal	Log. metal stability constant pH									
	3	4	5	6	7	8	9	10	11	12
ATMP										
Ca	—	—	0.4	2.2	3.1	3.9	4.5	5.4	6.3	7.1
Cd	—	1.9	4.0	5.8	7.5	9.5	11.4	10.0	9.5	7.8
Co	1.6	5.5	8.8	11.3	12.9	14.1	15.1	16.1	17.1	17.9
Cu	1.9	4.6	8.0	11.5	13.8	15.4	13.7	14.7	15.7	16.5
Hg	4.4	6.9	8.2	8.7	8.3	7.5	6.5	5.5	4.5	3.3
Mg	—	—	—	1.4	2.2	2.8	3.2	4.4	5.4	6.2
Ni	—	2.6	5.9	8.4	10.0	11.2	12.1	12.5	12.6	15.0
Pb	1.2	4.1	6.8	9.2	10.9	11.9	13.0	14.0	14.9	15.8
Zn	—	1.2	4.5	6.9	8.6	9.8	10.8	11.8	12.8	13.6
HEDP										
Ca	—	—	1.0	2.5	4.0	5.0	5.9	6.8	—	—
Cd	2.6	3.0	4.5	6.1	8.2	10.5	12.1	10.7	—	—
Co	—	1.7	3.7	5.7	7.5	8.9	9.9	10.8	11.2	—
Cu	2.4	3.6	5.3	8.2	11.2	13.1	14.1	10.4	9.8	—
Hg	1.3	2.0	2.0	2.0	1.9	1.2	0.3	—	—	—
Mg	—	—	—	1.1	2.2	2.9	3.5	2.2	—	—
Ni	—	0.1	2.1	4.1	5.9	7.3	8.3	8.5	8.0	—
Pb	Insoluble at all pHs									
Zn	—	1.0	3.0	5.0	6.9	8.2	9.3	10.1	10.5	—
EDTMP										
Ca	—	—	—	—	2.1	3.9	5.5	6.8	7.7	8.4
Cd	—	2.2	5.0	8.3	11.8	13.9	15.4	16.5	16.7	16.4
Co	—	—	2.5	5.9	8.8	10.9	12.6	13.8	14.5	—
Cu	0.6	3.8	7.5	11.4	14.7	17.1	18.6	19.4	19.6	19.3
Hg	3.7	7.5	10.4	12.6	14.0	14.9	15.0	14.7	13.9	12.6
Mg	—	—	—	0.2	1.9	3.2	4.4	5.1	8.0	8.7
Ni	—	—	2.7	6.9	10.4	13.2	15.2	16.3	16.6	—
Pb	0.6	3.2	6.3	9.7	12.6	14.9	16.5	15.1	14.8	12.3
Zn	—	3.6	7.8	11.3	14.1	16.2	17.7	16.7	14.4	—
HDTMP										
Ca	—	—	—	—	—	—	0.9	1.9	3.2	4.8
Cd	—	—	—	2.3	3.8	4.9	6.5	7.7	10.8	12.5
Co	—	—	0.9	5.2	8.3	10.5	12.2	13.5	14.2	13.9
Cu	—	—	3.2	6.6	9.9	11.4	13.1	13.7	14.6	15.3
Hg	—	—	0.1	2.3	3.4	3.7	3.7	3.6	3.5	3.2
Mg	—	—	—	—	—	—	0.7	1.8	2.9	3.9
Ni	—	—	2.7	6.9	10.0	12.3	14.2	15.5	16.5	—
Pb	—	—	2.7	5.2	6.9	8.5	10.0	11.7	14.5	16.2
Zn	—	—	1.6	5.8	8.9	11.2	13.0	12.7	11.6	10.2
DTPMP										
Ca	—	—	—	—	—	1.4	2.6	3.6	4.6	5.2
Cd	—	—	—	—	—	4.8	6.0	6.9	7.9	7.4
Co	—	—	—	3.2	7.6	10.7	12.7	13.8	14.0	12.8
Cu	—	2.1	6.5	10.9	14.5	17.3	18.0	18.9	19.2	18.9
Hg	—	—	—	2.6	4.9	6.2	6.5	6.3	5.6	4.4
Mg	—	—	—	—	—	1.5	2.4	3.4	4.1	4.4
Ni	—	—	—	4.9	9.3	12.5	14.7	15.9	16.3	18.7
Pb	—	—	—	—	2.1	3.96	4.9	5.7	6.2	6.1
Zn	—	—	—	5.0	9.4	12.6	14.7	14.3	12.6	10.3

Table 11. Phosphonate metal stability constants for the un-, mono- and di-protonated phosphonates

Metal	LIGAND, pH*								
	EDTMP			HDTMP			DTPMP		
	13.1	9.9	7.9	13.3	13.1	6.8	13.4	12.3	9.1
Mg	9.3	10.1	8.5	7.3	11.8	11.0	10.8	12.0	9.1
Ca	10.2	9.5	7.9	6.6	12.1	11.2	10.7	11.8	9.1
Ba	8.0	10.1	8.3	5.8	11.4	8.3	8.2	11.8	9.1
Mn	14.6	8.5	7.0	12.0	10.8	8.6	17.3	10.4	7.9
Cu	21.7	7.7	6.2	19.9	9.5	6.1	25.3	9.2	7.3
Zn	19.2	8.0	5.9	15.7	9.7	6.9	20.1	11.1	7.5

* pH values represent Log K_{ML}, Log K_{MLH} and Log K_{MLH2}, respectively

Table 12. EDTMP metal ion speciation (%) under simulated Rhine river water conditions

Complex	EDTMP Concentration, $\mu g/L^1$					
	1	10	50	100	250	500
Ca—EDTMP	0	0	0.9	2.9	17.4	37.2
Mg—EDTMP	0	0	0.5	1.6	9.2	19.7
Cu—EDTMP	100	100	68.7	34.4	13.7	6.9
Zn—EDTMP	0	0	29.0	58.4	49.2	26.2
Ni—EDTMP	0	0	0.9	2.7	10.4	9.8
H—EDTMP	0	0	0	0	0.1	0.3
Complex	pH value^2					
	6.5	7.0	7.5	8.0	8.5	
Ca—EDTMP	16.4	9.4	4.8	1.3	0.4	
Mg—EDTMP	10.2	5.1	2.1	0.9	0.5	
Cu—EDTMP	34.4	34.4	34.4	34.4	34.4	
Zn—EDTMP	33.1	46.6	55.6	61.2	63.1	
Ni—EDTMP	4.5	4.4	3.1	2.2	1.7	
H—EDTMP	1.4	0.3	0	0	0	

Cation conc. for simulated Rhine river water — Cu^{++} 5 μg/L, Zn^{++} 20 μg/L, Cd^{++} 0.2 μg/L, Pb^{++} 3 μg/L, Ni^{++} 9 μg/L, Fe^{+++} 130 μg/L, Ca^{++} 80 mg/L, Mg^{++} 12 mg/L
^{1} pH 7.7
^{2} 100 μg/L EDTMP

predominantly associated with the Cu^{++} ion. As the pH decreases, Ca^{++} and Mg^{++} play more of a role in EDTMP complexation; however, Cu^{++} and Zn^{++} remain the primary complexed metal ions. In contrast to NTA, soluble metal ion complexes of EDTMP would not be expected to remain in solution, but based upon previously described sorption experiments, would be expected to partition to sediments. Thus, the likelihood of heavy metal ion mobilization is minimized.

Bioconcentration

Octanol/water partition coefficients were measured for radiolabelled samples of ATMP, HEDP, EDTMP, HDTMP and DTPMP [3m]. As expected for highly water soluble compounds, log P values were extremely low (-3.5, -3.5, -4.1, -4.4 and -3.4, respectively). Calculated fish bioconcentration from these values is well below 1. Actual bioconcentration studies with Zebra fish have also been conducted [4, 5] for radiolabelled samples of ATMP (1 μg/L) and HEDP (50 μg/L). For ATMP, a steady state bioconcentration factor of 18 to 24 was determined after a 4 week uptake phase and radiolabel was rapidly reduced when fish were placed in clean water for a 10 day period. Corresponding BCF for HEDP was also 18 after a 6 week exposure phase and clearance was rapid and nearly complete during a 14 day period in clean water. Thus, for these chemicals, and presumably other phosphonates, bioconcentration in fish would be insignificant. The likelihood of bioconcentration is further reduced by the tendency of phosphonates to partition to sediments.

Drinking Water Purification

Surface waters used for local community water supply are usually treated prior to use as potable water. Treatment usually involves a flocculation/precipitation step followed by chlorination or in many European countries, ozonation. Because of their adsorptive capacity, phosphonates would be expected to be removed in the flocculation/precipitation step of water purification. In studies with ATMP, HEDP and DTPMP at concentrations ranging from 0.1 to 5.0 mg/L, alum, ferrous sulfate and ferric sulfate were very effective in phosphonate removal from water [3h]. Overall, ferrous sulfate proved the most effective resulting in phosphonate removal in excess of 95% at all phosphonate concentrations in both hard and soft water. Lime and alum were also very effective ($>95\%$) in removal of phosphonates at the 0.1 mg/L concentration; however, removals at the higher concentrations were lower (20–67% for lime and 44–98% for alum). As discussed previously, concentrations anticipated in receiving waters would be in the low ppb range. Thus, typical flocculation/precipitation procedures in water treatment plants should significantly reduce these low levels by another 1–2 orders of magnitude.

Chlorination studied [3h] also indicate phosphonates are subject to reaction/decomposition in the presence of chlorine. Using ASTM method D-1291-75, 1.0 mg/L of each of the five phosphonates and chlorine levels ranging from 0.4 to 8.8 times the theoretical oxygen demand, breakdown and removal of detergent phosphonates from drinking water was demonstrated. Identities and stability of chlorination products has not been determined.

Environmental Effects

Aquatic Toxicity

Acute and chronic toxicity data of phosphonates to algae, invertebrates and fish are summarized in Table 13 [3n]. For fish and invertebrates, acute toxicity generally occurs at concentrations exceeding 100 ppm, well above any anticipated environmental concentration. Similar values were reported by Huber [36] for carp for ATMP and HEDP (48 h LC_{50} values of 263 and 223 mg/L, respectively). Another summary report [24] reported fish and *Daphnia* acute toxicities for ATMP, HEDP and EDTMP in the same range (LC 0 and EC 0 ranging from > 160 to 510 mg/L). The oyster, as judged by the shell deposition test, was the most sensitive species, with effects noted in the 67 to 212 ppm range. This effect was not entirely unexpected since shell growth depends on Ca^{++} metabolism and precipitation, and phosphonates are effective Ca^{++} chelators. For fish, longer term toxicity studies (14 days) [2o] provided data that were not significantly different from the 96 h data, indicating these materials not to be accumulatively toxic. Both the Rainbow Trout 4 and 14 day data suggest toxicity to be inversely related to molecular weight, the larger molecules being less toxic. Fish embryo/larval studies with ATMP and DTPMP indicated lowest observable effect levels in the 20 to 50 ppm range [3p]. These values are similar to the no effect levels in the 14 day studies and further confirm the comparatively non toxic properties of phosphonates and their low potential to bioaccumulate. Chronic toxicity to *Daphnia* survival and reproduction for ATMP and HEDP occurred in the 10 to 50 ppm range [3q]. Phosphonates, being effective chelants, also offer protection to aquatic organisms from the toxic effects of heavy metals. Table 14 indicates the acute toxicity (LC-50) of Cd^{++}, Cu^{++}, Pb^{++} and Zn^{++} to be greatly reduced in the presence of phosphonates [3r].

Phosphonate interaction in algal assay studies is a more complex issue. The algal assay medium contains a precise level of trace nutrients (essential metals, phosphorus, etc.) held in solution by a hydrolytically and biologically stable chelator, EDTA. Thus, the introduction of an additional chelator, such as a phosphonate, which has both the ability to bind essential metals and to photodegrade to release additional phosphorus can profoundly effect (inhibit or stimulate) algal response. A 14 day algal growth study with *Selenastrum capricornutum* is also summarized in Table 13 [3s]. The study involved addition of 10,000 cells/mL to algal growth medium, continuous exposure to 400 ft. candles of light and cell counts at 4 and 14 days. Cell counts at 4 days indicated effect levels (growth inhibition) for various phosphonates in the 0.4 to 20 ppm range. EDTMP was the most effective phosphonate in algal growth inhibition with effects occurring at concentrations below 1 ppm. By 14 days, growth inhibition observed at the lower concentrations for HEDP, EDTMP and

Table 13. Aquatic toxicity of phosphonates

		Toxicity value mg/L									
		ATMP		HEDP		EDTMP		HDTMP		DTPMP	
Species	Test	EC	NOEC	EC	NOEC	EC	NOEC	EC	NOEC	EC	NOEC
Fish											
Bluegill Sunfish	96 h LC-50	>330	330	868	529	>164	164	>273	273	758	576
Channel Catfish	96 h LC-50	1212	924	695	529	967	522	>2400	2400	657	432
Sheepshead Minnow	96 h LC-50	8132	4831	2180	104	1513	605	>954<1670	954	5377	2125
Rainbow Trout	96 h LC-50	>330	330	368	151	>164	164	>273	273	>180<252	180
Rainbow Trout	14 day LC-50	160	—	200	—	250	—	440	—	573	—
	60 day chronic	150	47	180	60	250	35	440	74	>262	139
Rainbow Trout		<47	>23	—	—	—	—	—	—	<34	>26
Invertebrates											
Chironomus	48 h EC-50	11000	7040	8910	3925	7320	1956	4660	1803	9910	7589
Grass Shrimp	96 h LC-50	7870	4575	1770	104	1436	605	942	537	4849	2125
Eastern Oyster	96 h EC-50	201	95	89	<52	67	55	212	<161	156	56
Daphnia magna	48 h EC-50	297	125	527	400	510	250	574	125	242	125
Daphnia magna	28 day chronic	<54	>25	<25	>12	—	—	—	—	—	—
Algae											
Selenastrum	96 h EC-50	19.6	7.4	3.0	1.3	0.42	0.09	28	10.2	1.9	5.2
	14 day EC-50	19.6	7.4	39.1	13.2	27.1	9.3	27	10.2	8.7	5.2

Table 14. Effect of phosphonates on acute toxicity of heavy metals to Bluegill Sunfish

LC_{50}, mg/L Metals	Metals alone	ATMP	LC_{50} Metals plus phosphonate,[1] mg/L			
			HEDP	EDTMP	HDTMP	DTPMP
Cd^{++}	4.8	147	>870<1000	243	124	254
Cu^{++}	0.81	326	101	121	104	317
Pb^{++}	73	>1000	>1000	536	475	>1000
Zn^{++}	2.2	>1000	549	216	150	611
$Zn^{++\,2}$	—	11	9.8	55	62	149
$Zn^{++\,3}$	—	61	75	31	46	38

[1] Molar ratio of 1/1 metal/phosphonate used
[2] Molar ratio of 1/0.2 metal/phosphonate used
[3] Molar ratio of 1/5 metal/phosphonate used. For this set, conc. of phospohonate used exceeded its toxicity

Table 15. Effects of phosphonates on growth of 3 species of algae

Compound	Conc., mg/L	Growth of species[1]		
		Selenastrum	Anabaena	Chlorella
Control	—	143	21	380
ATMP	0.1	345	29	592
	1.0	230	29	928
	10.0	320	13	535
	100.0	1	0	8
HEDP	0.1	328	16	488
	1.0	200	6	658
	10.0	81	19	765
	100.0	1	0	8
EDTMP	0.1	126	19	665
	1.0	41	42	839
	10.0	1	8	570
	100.0	1	2	43
HDTMP	0.1	52	31	746
	1.0	56	39	832
	10.0	1	46	552
	100.0	1	24	716

[1] Growth measured after 8 days via chlorophyll-a fluorescence

DTPMP was not noted and higher concentrations were required before effects were observed. Thus, initial concentrations of phosphonates may have chelated some essential micronutrients for algal growth. By four days, HEDP, EDTMP and DTPMP may have photodegraded to release these nutrients plus additional phosphorus which resulted in the observed growth stimulation. In an earlier study, similar results were observed [3t] (Table 15). At day five, concentrations of all four phosphonates ranging from 0.1 to 100 ppm were inhibitory to growth of species of algae. However, at day eight, lower concentrations were generally stimulatory to algal growth, while higher concentrations were inhibitory. Exceptions were noted for HDTMP with *Selenastrum* and *Chlorella*. Stimulation may have resulted from the photolytic release of phosphorus from the phosphonates. In an additional algal assay with *Chlorella*, an EC-50 of 10 mg/L was established for EDTMP; however, when tested as the Ca^{++} salt, no toxicity [EC-50 > 500 mg/L] was noted [37]. Nyholm and Kallqvist [38] reported that *Selenastrum* and *Scenedesmus* species to be more sensitive than *Chlorella* and also pointed out extreme variability in interlaboratory results probably due to the algal assay medium. The EC draft guideline for the OECD Guideline 201 method even specifies that chelator content should be below 1 μmol/L. Thus this particular assay is not necessarily appropriate for chelating agents.

Toxicity to Plants

ATMP and HEDP have been screened for both pre-emergent and post-emergent phytotoxicity to a wide variety of weeds and crops [3i]. ATMP was essentially non phytotoxic in these screening studies, while HEDP showed only minimum toxicity at application rates ranging from 3 to 30 kg/ha. Similar results for tomato, cucumber and radish plants were noted by Unilever Research [37]. They examined concentrations ranging from 1–100 mg/L and found no significant effects on seed germination at any concentration.

Environmental Safety Assessment

The basic concepts of safety assessment [39–41] focus on exposure [predictable from environmental fate data] and ecological effects of phosphonates. Quantitative safety assessment is based on the ratio [safety margin] of the highest observed no-effect level for the most sensitive aquatic species and the aqueous environmental exposure concentration. The greater this ratio the more confidence there is that production, use and disposal of phosphonates will not adversely impact the aquatic environment. Additional considerations including potential to mobilize metals in the environment, bioaccumulation to higher life forms and long term environmental sinks should also be considered.

The studies summarized here permit an initial safety assessment for phosphonates used in water treatment and detergent applications. Phosphonates, primarily ATMP and HEDP, are used for industrial cooling and heating water applications in the U.S.A. resulting in intermittent discharge to the environment. In Europe, these products are used for similar purposes; however, HEDP, EDTMP and DTPMP also find significant use in detergent applications. Consequently, discharge to the environment occurs on a more continual basis. One of the major limitations to the environmental safety assessment of phosphonates is the lack of a suitable analytical method to establish environmental levels. As a result, modelling has to be used to estimate exposure concentrations. In Europe, average surface water concentrations of 0.25 µg/L are predicted. General concentrations in U.S.A. waters are expected to be lower because of lack of detergent use. However, concentrations in the vicinity of some industrial water discharges might intermittently approach a few ppb depending on the amount of discharge, its treatment and dilution in receiving waters.

Low environmental exposure levels result from relatively low use volumes, biotic and abiotic degradation processes and partitioning to sediments. Photolysis in the presence of ferric ions is rapid for all phosphonates. Biological degradation is slow in systems containing excess P_i; however, several key genera of microorganisms can grow on phosphonates as their sole carbon and phosphorus source. Slow to moderate biodegradation rates occur in natural sediment/water and soil systems. In addition to finite photo- and biodegradation rates, partitioning to sediments is rapid and significantly reduces bioavailability of phosphonates. Drinking water purification processes further reduce natural water concentration of these materials by another two orders of magnitude. Even though phosphonates are extremely effective metal sequestrants, concentrations found in the environment (<50 ppb) do not result in metal solubilization from sediments. Bioconcentration studies have also shown phosphonates not to significantly accumulate in fish tissue.

From both aquatic and terrestrial toxicity standpoints, phosphonates can be classified as non toxic to slightly toxic to a wide variety of species. Acute toxicity values to invertebrates and fish generally exceed 100 mg/L. Oysters (shell deposition) were the most sensitive species with effect concentrations ranging from 67 to 212 mg/L. This result is not entirely surprising because of the sequestration properties of phosphonates. Chronic toxicity studies with both Rainbow Trout and *Daphnia magna* provided similar results with no-effect/effect levels (MATC values) ranging from >12 to <54 mg/L. Algal growth effect studies are a little more complex to interpret due to the chelation and photodegradation properties of phosphonates. Nevertheless, no-effect concentrations in excess of 5 mg/L were demonstrated for all phosphonates in a 14-day algal assay study.

To summarize, using a safety margin approach (i.e. maximum no-effect conc./exposure conc.) safety margins of 20,000 to 50,000 are anticipated for most European waters (worst case). Safety margins would be greater in U.S.A. waters since phosphonates are not used in detergents in the U.S.A. Even in cases where intermittent discharge of phosphonates occurs directly to a receiving water, significant safety margins would still exist. This safety assessment supports the conclusion that current use/disposal practices for this class of water treatment and detergent chemicals does not constitute a hazard to the functioning or well being of our natural ecosystems. This conclusion will be further strengthened with development of suitable analytical tools to establish actual exposure concentrations in nature.

Acknowledgements: We acknowledge the critical review of this manuscript by Dr. Davide Calamari, Professor, Institute Agricultural Entomology University of Milan, Italy and input from Dr. Josef Steber, Henkel KGaA, Dusseldorf, Germany.

References

1. May HB, Nijs H, Godecharles V (1986) HAPPI 23:50, 52, 89
2. Procter and Gamble internal report (unpublished)*
3. Monsanto internal reports (unpublished)*
4. Steber J, Wierich P (1986) Properties of hydroxyethane diphosphonate affecting its envrionmental fate: Degradability, sludge adsorption, mobility in soils and bioconcentration. Chemosphere 15:929
5. Steber J, Wierich P (1987) Properties of aminotris(methylenephosphonate) affecting its environmental fate: Degradability, sludge adsorption, mobility in soils and bioconcentration. Chemosphere 16:1323
6. Kittredge JS, Roberts E (1969) A carbon-phosphorus bond in nature. Science 164:37
7. Daughton CG, Cook AM, Alexander M (1979) Bacterial conversion of alkylphosphonates to natural products via carbonphosphorus bond cleavage. J Agric Food Chem. 27:1375
8. Cook AM, Daughton CG, Alexander M (1978) Phosphonate utilization by bacteria. J Bacteriol 133:85
9. Quinn JP, Peden JMM, Dick RE (1989) Carbon-phosphorus bond cleavage by Gram-positive and Gram-negative bacteria. Appl Microbiol Biotechnol 31:283
10. Egli T (1988) (An)aerobic breakdown of chelating agents used in household detergents. Microbiol Sci 5:36
11. Schowanek D, Verstraete W (1990) Phosphonate utilization by bacterial cultures and enrichments from environmental samples. Appl Environ Microbiol 56:895
12. Pipke R, Amrhein N (1988) Degradation of the phosphonate herbicide glyphosphate by *Arthrobacter atrocyaneus* ATCC 13752. Appl Environ Microbiol 54:1293
13. Rosenberg H, LaNauze JM (1967) The metabolism of phosphonates by microorganisms. The transport of aminomethylphosphonic acid in *Bacillus cereus*. Biochim Biophys Acta 141:79
14. Fitzgibbon J, Braymer HD (1988) Phosphate starvation induces uptake of glyphosate by *Pseudomonas* sp. strain PG2982. Appl Environ Microbiol 54:1886
15. Wackett LP, Shames SL, Vendetti CP, Walsh CT (1987) Bacterial carbon-phosphorus lyase: products, rates, and regulation of phosphonic and phosphinic acid metabolism. J Bacteriol 169:710
16. Wackett LP, Wanner BL, Vendetti CP, Walsh CT (1987) Involvement of the phosphate regulon and psiD locus in carbon-phosphorus lyase activity of *Escherichia coli* K-12. J Bacteriol 169:1753
17. Pipke RN, Amrhein N, Jacob GS, Schaefer J, Kishore GM (1987) Metabolism of glyphosate in an *Arthrobacter* sp. GLP-1. Eur J Biochem 165:276
18. Shinabarger DL, Schmitt EK, Braymer HD, Larson AD (1984) Phosphonate utilization by the glyphosate-degrading *Pseudomonas* sp. strain PG2982. Appl Environ Microbiol 48:1049
19. La Nauze JM, Rosenberg H, Shaw DC (1970) The enzymatic cleavage of the carbon-phosphorus bond: purification and properties of phosphonase. Biochim Biophys Acta 212:322
20. Weissenfels W (1987) M.S. Thesis Bochum University, FRG
21. Schowanek D, Verstraete W (1990) Phosphonate utilization by bacteria in the presence of alternative phosphorus sources. Biodegradation (in press)
22. Horstmann B, Grohmann A (1988) Investigations into the biodegradability of phosphonates. Vom Wasser 70:163
23. Huber L (1975) Studies on the biodegradability and fish toxicity of 2 organic complexing agents based on phosphonic acid (ATMP and HEDP). Tenside Detergents 12:316
24. Schöberl P, Huber L (1988) Okologische relevante daten von nichttensidischen inhaltsstoffen in wash- und reinigungsmitteln. Tenside 25:99
25. Muller G, Steber G, Waldhoff H (1984) The effect of hydroxyethane diphosphonic acid on phosphate elumination with $FeCl_3$ and remobilization of heavy metals. Vom Wasser 63:63

26. Horstmann B, Grohmann A (1986) The influence of phosphonates or phosphate elimination. Z Wasser-Abwasser-Forsch 19:236
27. Bourquin AW, Hood MA, Garnas RT (1977) An artificial microbial ecosystem for determining effects and fate of toxicants in a salt-marsh environment. Develop Indust Microbiol 18:185
28. Matthijs E, deOude NT, Bolte M, Lemaire J (1989) Photodegradation of ferric ethylenediaminetetra(methylenephosphonic acid (EDTMP) in aqueous solution. Water Res 27:845
29. Svenson A, Kaj L, Bjorndal (1989) Aqueous photolysis of iron (III) complexes of NTA, EDTA and DTPA. Chemosphere 18:1805
30. Frank R, Rau H (1990) Photochemical transformation in aqueous solution and possible environmental fate of ethylenediaminetetraacetic acid (EDTA). Ecotox Environ Safety 19:55
31. Lockhart HB, Blakeley RV (1975) Aerobic photodegradation of Fe(III)-(ethylenedinitrilo)tetraacetate. Environ Sci Technol 9:1035
32. Gunther K, Henze W, Umland F (1987) Mobilization behavior of thallium and cadmium in a river sediment. Fresen Z Anal Chem 327:301
33. Borzetto P (1985) Unpublished data, University of Modena, Institute of Hygiene
34. Martell AE, Sillen LG (ed) (1968) Stability Constants—Supplement No. 1 Special publication 25. The London Chemical Society. Alden Press, London
35. Great Lakes Research Advisory Board (1978) Ecological effects of non phosphate detergent builders. Final report on NTA
36. Huber L, Studies on the biodegradability and fish toxicity of two organic complexing agents based on phosphonic acid (ATMP and HEDP). Tenside Detergents 12:316
37. Unilever Research. Internal reports (unpublished)
38. Nyholm N, Kallqvist T (1989) Methods for growth inhibition toxicity tests with freshwater algae. Environ Toxicol Chem 8:689
39. Kimerle RA, Gledhill WE, Levinskas GJ (1978) In: Cairns J Jr, Dickson KL, Maki AW (eds) Estimating the hazard of chemical substances to aquatic life, American Society for Testing and Materials, STP 657, p 132
40. Stern AM, Walker CR (1978) In: Cairns J Jr, Dickson KL, Maki AW (eds) Estimating the hazard of chemical substances to aquatic life, American Society for Testing and Materials, STP 657, p 81
41. Duthie JR (1977) In: Mayer FL, Hamelink FL (eds) Aquatic toxicology and hazard evaluation, American Society for Testing and Materials, STP 643, p 17

---

* Reference 2 and 3. Letters reference specific Procter and Gamble and Monsanto internal reports, respectively

Perborate

Karen Raymond and Lucy Butterwick
Environmental Resources Ltd., 106 Gloucester Place, London W1H 3DB, England

Introduction 288
Chemical and Physical Characteristics of Perborate 288
Manufacture of Perborate 290
Use of Perborate 291
Discharge of Perborate to the Environment 292
Analytical Methods 293
Ecotoxicology of Boron: The Aquatic Environment 294
    Boron Toxicity to Fish 294
    Boron Toxicity to Aquatic Life Other than Fish 300
    Factors Affecting the Toxicity of Boron 304
    Regulations Protecting Aquatic Life 304
Ecotoxicology of Boron: The Terrestrial Environment 305
    Boron Toxicity to Higher Plants 305
    Boron Toxicity to Grazing Animals 310
Safety of Boron in Aquatic and Terrestrial Environments 312
Clinical Toxicology 312
    Evidence of Industrial Intoxication 312
    Chronic Studies 313
    Safety of Boron in the Diet 314
References 315

The Handbook of Environmental Chemistry,
Volume 3 Part F, Ed. O. Hutzinger
© Springer-Verlag Berlin Heidelberg 1992

Introduction

Sodium perborate is used as a bleach in domestic and industrial cleaning products and is released into the environment during manufacture and end use. This chapter assesses the environmental implications of such releases.

The chemical and physical properties that make perborates suitable compounds for use as bleaches are first summarised; this is followed by a description of how perborates are manufactured and a discussion of the present and future market for this product.

The environmental implications of perborate manufacture and end use are then assessed by examining the manner and form in which perborate is released into the environment, and by reviewing the evidence for boron toxicity to terrestrial and aquatic ecosystems and its effects on human health. The safety of perborate release to the environment is assessed by comparing critical levels that have been identified for boron toxicity with environmental concentrations of boron that have been observed. Present regulations that apply to perborate in this context are also discussed.

Chemical and Physical Characteristics of Perborate

The basic structural form of perborate is shown below.

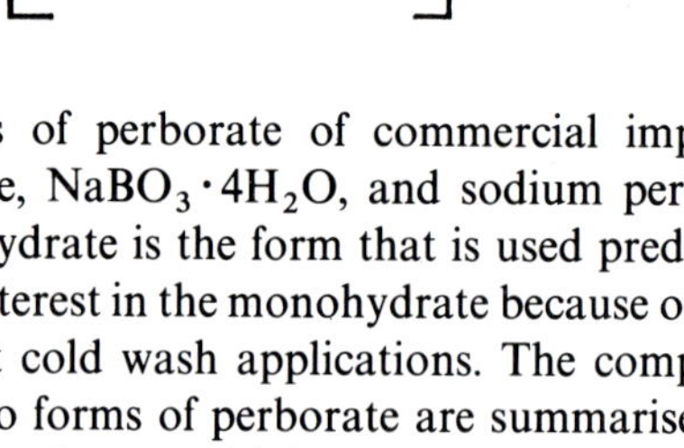

$$\left[\begin{array}{c} HO \quad O-O \quad OH \\ \diagdown B \quad B \diagup \\ \diagup \quad \diagdown \\ HO \quad O-O \quad OH \end{array} \right]^{2-} \quad 2Na^+ \; xH_2O$$

The two main forms of perborate of commercial importance are sodium perborate tetrahydrate, $NaBO_3 \cdot 4H_2O$, and sodium perborate monohydrate, $NaBO_3 \cdot H_2O$. Tetrahydrate is the form that is used predominantly at present, but there is growing interest in the monohydrate because of its rapid solubility in water and consequent cold wash applications. The composition and physical properties of these two forms of perborate are summarised in Table 1.

The properties of perborate which, according to manufacturers, explain its value in bleaches and detergents include the following:

— water solubility;
— effective bleaching;

Table 1. Composition and physical properties of perborates [1]

		Sodium Perborate Tetrahydrate [$NaBO_3 \cdot 4H_2O$]		Sodium Perborate Monohydrate [$NaBO_3 \cdot H_2O$]	
Composition		Sodium perborate tetrahydrate	min. 96% wt.	Sodium perborate monohydrate	min. 96% wt.
		Available oxygen	min. 10% wt.	Available oxygen	min. 15% wt.
		B_2O_3	min. 22.5% wt.	B_2O_3	min. 34% wt.
		Na_2O	min. 20.0% wt.	Na_2O	min. 30% wt.
Physical properties	Appearance		white, crystalline, free-flowing odourless powder		white, crystalline, free-flowing odourless powder
	Bulk density		740–820 kg m^{-3}		500–600 kg m^{-3}
	Solubility in water at 20 °C		approx. 23 g l^{-1}		approx. 15 g l^{-1}
	at 30 °C		approx. 37 g l^{-1}		approx. 24 g l^{-1}
	pH (1% solution)		approx. 10.4		approx. 10.4
	Particle size (typical)		>0.85 mm <1%		>0.85 mm <1%
			>0.15 mm >80%		>0.15 mm >80%
	Free moisture (standard grade) Special grades with very low moisture content also available				max. 1% wt.

— a bactericidal action;
— mild alkalinity;
— an odour-free additive, able to eliminate some unpleasant odours;
— capability of blending with many other chemicals;
— good storage stability.

Manufacture of Perborate

Sodium perborates can be produced by both chemical and electrochemical methods. The chemical production of sodium perborate is now the dominant method of manufacture and involves the treatment of borax (and in most cases the pentahydrate form, $Na_2B_4O_7 \cdot 5H_2O$) with sodium hydroxide and then hydrogen peroxide:

$$Na_2B_4O_7 + 2NaOH + 4H_2O_2 + 11H_2O \longrightarrow 4NaBO_3 \cdot 4H_2O$$

The process is illustrated schematically in Fig. 1. After treatment the tetrahydrate crystallises out and can be easily separated from the parent liquid by standard filtration or centrifugal methods. The wet product obtained usually contains about 10% free water which can be removed by drying in a stream of hot air to yield the tetrahydrate. Further controlled drying will yield the monohydrate [2, 3].

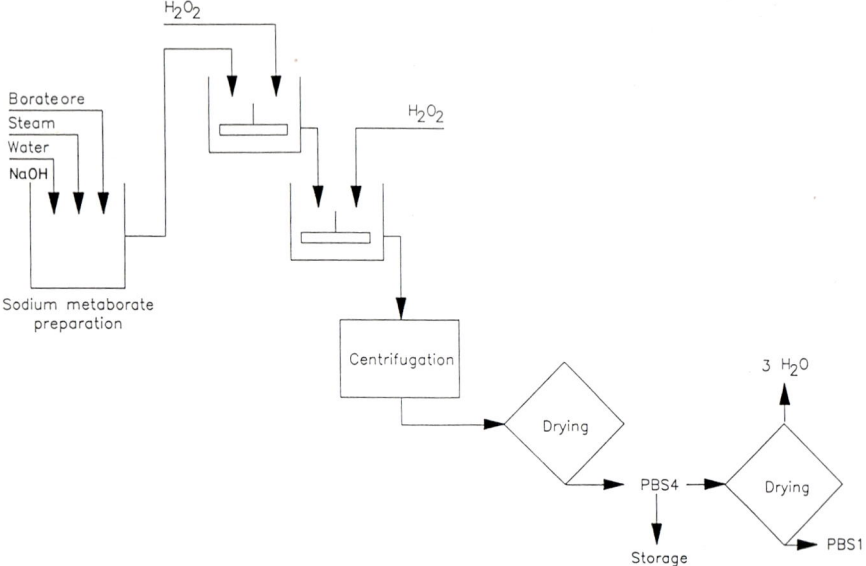

Fig. 1. Manufacture of perborate

Use of Perborate

In Western Europe, sodium perborate is used as a bleaching agent in detergent formulations and some household cleaners. The use of sodium perborate as a bleach in North America has been limited by the extensive use of cold wash machines. Perborate loses much of its function as a bleach below about 55 °C, and hence chlorine-based bleaches are more commonly used in North America. Hot water washing is more usual in Europe although it is decreasing.

Interest in sodium perborate by U.S. bleach and detergent producers declined in the 1970s. However, interest in sodium perborate in detergents has recently increased and test marketing has been taking place for several years. The interest appears to be due to activators (e.g. tetraacetyl-ethylenediamine (TAED), alkyloxybenzene sulfonate (AOBS) or pentaacetylglucose (PAG)) that can be added in the detergent along with the perborate, and by the use of the more active sodium perborate monohydrate, which is more readily soluble than the tetrahydrate. The activators interact with the perborate to form a peracid; for example with TAED, the sodium perborate and TAED react to form peracetic acid (CH_3CO_3H), which oxidises stains at temperatures as low as 40 °C. The use of activator enhanced sodium perborate monohydrate bleaches might thus substitute for chlorine-based bleaches in the U.S. bleach and detergents market to some extent. As a result of this development, the U.S. market for perborates could grow by as much as 4% per year—sodium perborate is currently being imported from Western Europe to meet this demand.

In Western Europe the total heavy duty detergent market has been increasing at an average rate of 2% per annum and with it the use of sodium perborate. The European production capacity for sodium perborate in 1985 was 790 000 tonnes per annum (see Table 2).

Table 2 [4]. Western European producers of sodium perborate—annual capacity 1985

Country	Sodium perborate ('000 tonnes)
Austria	17
Belgium	110
France	123
Germany	160
Italy	135
Portugal	10
Spain	65
Sweden	10
Switzerland	35
U.K.	125
	790

Table 3 [4]. Estimates of use in Western Europe of boron in cleaning products

Year	'000 tonnes boron pa
1965	40
1970	52
1975	62
1980	50
1983	53
1984	52

Estimates for the use of boron in cleaning products in Western Europe is shown in Table 3. The situation has not changed significantly over the last few years.

Most of the perborate capacity listed is used to make sodium perborate tetrahydrate, although there is an important trend towards the production of the monohydrate.

There is currently a trend in Western Europe to use lower wash temperatures. This is leading to a growing interest in the use of sodium perborate monohydrate as a replacement for sodium perborate tetrahydrate in heavy duty detergents. In addition, the introduction of bleach activators and liquid detergents in Western Europe is expected to result in a decrease in the quantity of sodium perborate used in detergents. Demand for sodium perborate is expected to remain fairly constant in Western Europe in the near future.

Discharge of Perborate to the Environment

Perborate can be discharged to the environment during production and end use of bleaches and detergents. Perborate is non-volatile and negligible amounts of solid waste are generated during manufacture; the main discharge route is therefore through the liquid phase. Perborate discharges to water are minimal during manufacture with recycling of crystallisation liquors common practice. After use, however, all the boron contained in bleaches and detergents enters the environment directly via the sewage system and sewage treatment facilities or, in the case of some industrial processes, directly to receiving waters.

Little or no boron is removed during conventional sewage treatment, and therefore that which is present in sewage will usually be discharged to the receiving water in treated effluent [5].

Borates are inorganic salts and therefore non-biodegradable. Boron discharged through use of bleaches and detergents eventually ends up in the sea, which has a natural level of boron of about 5 mg l^{-1}; this remains unchanged by any additions of boron from anthropogenic sources.

The predominant species of boron in most natural freshwater systems (pH 6–9), regardless of its initial formulation, is undissociated boric acid [6]. Therefore in assessing the toxicology of perborate in the following sections a more general review is made of the toxicology of boron and boric acid, as these are the substances to which humans, and aquatic and terrestrial environments are exposed, as a result of use of perborate in bleaches and detergents.

Analytical Methods

Methods for the water quality analysis of boron are described in Table 4. The curcumin and titration methods are based on well established traditional principles but suffer particularly in that they are time consuming. Methods based on the colour reaction with azomethine-H are relatively new, but are simpler and quicker than earlier methods; the automated method is particularly useful for a large number of samples. The azomethine-H methods are carried out in aqueous

Table 4. Comparison of methods for determining boron in solution [8]

Method	Curcumin	Azomethine-H (automated)	Azomethine-H (manual)	Titration
Description	(1)	(2)	(3)	(4)
Type of sample	non-saline	all	all	all
Range in mg l^{-1}	0–5	0–4.5	0–10	1–1000
Limit of detection in mg l^{-1}	0.004	0.04	0.04	0.5
Interferences	nitrate, chloride, and high level of fluoride	high levels of fluoride	high levels of fluoride	slight interference from very high levels of phosphate, fluoride and buffer anions
Number of analyses achievable per day	15	100	80	6

(1) Boric acid reacts quantitatively with curcumin under anhydrous conditions in a sulphuric acid-acetic acid medium to produce the boron-curcumin complex rosocyanine which is coloured red. The rosocyanine is considered to be formed via protonation of the curcumin by acid which then reacts with boric acid to form the complex. Excess curcumin in its protonated form is also coloured red. This is indistinguishable from rosocyanine until the reaction mixture is diluted with alcohol, whereupon the excess protonated curcumin reverts to its normal yellow form dissolving together with rosocyanine to produce an orange to red solution
(2), (3) Azomethine-H, which is a condensation product of H-acid 8-aminonaphth-1-ol-3,6-disulphonic acid and salicylaldehyde, reacts in aqueous solution with dissolved forms of boron at a pH of about 5. A yellow complex is formed, the absorbance of which is measured at 420 nm and is related to the boron concentration by means of a calibration curve. Possible interfering cations are masked by the use of ethylenediaminetetraacetic acid
(4) Following treatment with an anion exchange resin to remove possible interfering cations, the solution is neutralized to pH 7.0. Mannitol is added to promote ionization of boric acid followed by titration with a standard alkali solution of pH 7.0. The mannitol-boric acid complex titrates as a monoacid provided the solution is saturated with mannitol

solution, avoiding the use of strong mineral acids. Azomethine-H is the preferred colorimetric method at the present time.

More recently inductively-coupled plasma atomic emission spectrometry has been used for the analysis of boron in surface and wastewater. Although the capital cost is greater than the colorimetry methods described above, it has the major advantage of being able to measure boron more accurately over a much broader range of concentrations, 0.05–20 mgBl^{-1}. It also provides rapid multi-element analysis and can be used to analyse solutions with high salt concentrations. [7].

Ecotoxicology of Boron: The Aquatic Environment

Toxicological effects of boron on aquatic life have been assessed primarily in laboratory studies. A number of early investigations were concerned with determining the acute effects on fish. More recent work has addressed chronic effects on the invertebrate, *Daphnia magna*, and several species of fish, algae and amphibian early-life stages. Results from these investigations are summarised below.

Boron Toxicity to Fish

Available data on the effects of boron on freshwater and saltwater fish are summarised in Tables 5 and 6.

Of all the species and life stages investigated in aquatic toxicity studies, the early life stages of rainbow trout appear to be most sensitive to boron with a consistent dose-response related lowest observable effect concentration (LOEC) of 0.1 mgBl^{-1} in reconstituted water as the experimental medium [16].

Whilst most laboratory toxicity studies are based on reconstituted water as the experimental medium, recent work by Procter and Gamble [18] has found that when trout embryo-larval stages were exposed to boron in natural water courses, it was found to be substantially less toxic.

Birge et al. [20] as well as Procter and Gamble [18] have recorded LOEC under natural water exposures of 1.0 mgBl^{-1} (boron amendment to natural waters), and found that 0.750 mgBl^{-1} boron (natural background boron) did not affect rainbow trout early life stages. Another more recent early life stage rainbow trout study by Procter and Gamble [18] (90 days starting with green eggs) conducted in natural water, indicated no impairment to rainbow trout early life stages at 17.0 mgBl^{-1}, the highest test concentration used. It therefore seems that the low level effects observed in reconstituted laboratory water are not predictive of the much higher first effect levels under natural water exposure

Table 5. Summary of boron toxicity data for freshwater fish

Test organism	Boron compound tested	Type of test	Water quality characteristics	Test response(s) reported	Boron concentration (mg l^{-1})	Ref.
Bluegill sunfish (*Lepomis macrochirus*) Average size 7 cm, 5 g	Sodium tetraborate $Na_2B_4O_7 \cdot 10H_2O$	Static	Dechlorinated Philadelphia tap water; 20 °C; pH 6.9–7.5; alkalinity 33–81 mg l^{-1}; hardness 84–163 mg l^{-1}	24-h TLm	4.6	9
Bluegill sunfish (*Lepomis macrochirus*) Average size 7 cm, 5 g	Boron trifluoride BF_3	Static	Dechlorinated Philadelphia tap water; 20 °C; alkalinity 1750 mg l^{-1}	24-h TLm	2389	9
Coho salmon (*Oncorhynchus kisutch*) alevins 0.19–0.7 g	Sodium metaborate $Na_2B_2O_4 \cdot 8H_2O$	Static-renewal (daily)	Well water, 11.0 °C; hardness 47 mg l^{-1}	283-h LC$_{50}$	113	10
Fathead minnow (*Pimeohales promelas*) eggs and fry	Boric acid H_3BO_3	Flow-through early life stage	Well water; 25 °C; pH 7.1–7.9; alkalinity 33–38 mg l^{-1}; hardness 38–46 mg l^{-1}	30-d NOEC-LOEC (reduction in growth) 60-d NOEC-LOEC (reduction in fry survival)	14–24 24–88	11
Minnow	Boric acid H_3BO_3	—	20 °C; distilled H_2O	6-h minimum lethal dose	3145–3319	12
Minnow	Borax $Na_2B_4O_7 \cdot 10H_2O$	—	20 °C; hard H_2O 19 °C; distilled H_2O 17 °C; hard H_2O	6-h minimum lethal dose Minimum lethal dose Minimum lethal dose	3319–3407 340–374 793–850	13 13
Mosquito fish (*Gambusia affinis*) adult females	Boric acid H_3BO_3	Static	20–23 °C; pH 5.4–7.3; turbidity 210–250 mg l^{-1}	24-h TLm 48-h TLm 96-h TLm No mortalities in 96-h	3145 1834 978 <314	14
Mosquito fish (*Gambusia affinis*) adult females	Borax $Na_2B_4O_7 \cdot 10H_2O$	Static	22–26 °C; pH 8.6–9.1; turbidity 410–650 mg l^{-1}	24-h TLm 48-h TLm 96-h TLm 144-h TLm No mortalities	1360 929 408 215 <204	14
Rainbow trout (*Salmo gairdneri*)	Boric acid H_3BO_3	—	—	Darkening of skin immobilisation and loss of equilibrium	874 13 976	15

Table 5. (Continued)

Test organism	Boron compound tested	Type of test	Water quality characteristics	Test response(s) reported	Boron concentration (mg B l^{-1})	Ref.
Trout, fingerling	Sodium perborate NaBO$_3 \cdot$ 4H$_2$O	—	—	80% mortality	23.7	13
Channel catfish (*Ictalurus punctatus*) embryo-larval stages	Boric acid H$_3$BO$_3$	Flow-through	Reconstituted water; hardness 50 mg CaCO$_3$ l^{-1}	9-d NOEC-LOEC	1.01–5.42	16
Channel catfish (*Ictalurus punctatus*) embryo-larval stages	Boric acid H$_3$BO$_3$	Flow-through	Reconstituted water; hardness 200 mg CaCO$_3$ l^{-1}	9-d NOEC-LOEC	0.75–1.0	16
Channel catfish (*Ictalurus punctatus*) embryo-larval stages	Borax Na$_2$B$_4$O$_7 \cdot$ 10H$_2$O	Flow-through	Reconstituted water; hardness 50 mg CaCO$_3$ l^{-1}	9-d NOEC-LOEC	9.0–25.9	16
Channel catfish (*Ictalurus punctatus*) embryo-larval stages	Borax Na$_2$B$_4$O$_7 \cdot$ 10H$_2$O	Flow-through	Reconstituted water; hardness 200 mg CaCO$_3$ l^{-1}	9-d NOEC-LOEC	0.49–1.04	16
Goldfish (*Carassius auratus*) embryo-larval stages	Boric acid H$_3$BO$_3$	Flow-through	Reconstituted water; hardness 50 mg CaCO$_3$ l^{-1}	7-d NOEC-LOEC	9.2–22.5	16
Goldfish (*Carassius auratus*) embryo-larval stages	Boric acid H$_3$BO$_3$	Flow-through	Reconstituted water; hardness 200 mg CaCO$_3$ l^{-1}	7-d NOEC-LOEC	6.8–8.33	16
Goldfish (*Carassius auratus*) embryo-larval stages	Borax Na$_2$B$_4$O$_7 \cdot$ 10H$_2$O	Flow-through	Reconstituted water; hardness 50 mg CaCO$_3$ l^{-1}	7-d NOEC-LOEC	26.50–48.75	16
Goldfish (*Carassius auratus*) embryo-larval stages	Borax Na$_2$B$_4$O$_7 \cdot$ 10H$_2$O	Flow-through	Reconstituted water; hardness 200 mg CaCO$_3$ l^{-1}	7-d NOEC-LOEC	8.53–27.33	16
Rainbow trout (*Salmo gairdneri*) embryo-larval stages	Boric acid H$_3$BO$_3$	Flow-through	Reconstituted water; hardness 50 mg CaCO$_3$ l^{-1}	28-d NOEC-LOEC	0.11–1.00	16
Rainbow trout (*Salmo gairdneri*) embryo-larval stages	Boric acid H$_3$BO$_3$	Flow-through	Reconstituted water; hardness 200 mg CaCO$_3$ l^{-1}	28-d NOEC-LOEC	0.001–0.01	16

Species	Compound	Test type	Test water	Duration	Concentration	Ref.
Rainbow trout (Salmo gairdneri) embryo-larval stages	Borax, $Na_2B_4O_7 \cdot 10H_2O$	Flow-through	Reconstituted water; hardness 50 mg $CaCO_3$ l^{-1}	28-d NOEC-LOEC	0.96–9.70	16
Rainbow trout (Salmo gairdneri) embryo-larval stages	Borax $Na_2B_4O_7 \cdot 10H_2O$	Flow-through	Reconstituted water; hardness 200 mg $CaCO_3$ l^{-1}	28-d NOEC-LOEC	9.63–49.70	16
Large mouth bass (Micropterus salmoides) freshly fertilised eggs	Boric acid H_3BO_3	Flow-through	Reconstituted hard water 200 mg $CaCO_3$ l^{-1}	11-d NOEC-LOEC	1.39–12.17	17
Rainbow trout (Salmo gairdneri) freshly fertilised eggs	Boric acid H_3BO_3	Flow-through	Reconstituted hard water 200 mg $CaCO_3$ l^{-1}	32-d NOEC-LOEC	0.01–0.1	17
Rainbow trout (Salmo gairdneri) early life stages	Boric acid H_3BO_3	Flow-through	Natural water exposures	36-d NOEC-LOEC	0.75–1.0	18
Rainbow trout (Salmo gairdneri) early life stages	Boric acid H_3BO_3	Flow-through	Well water; water hardness 27 mg $CaCO_3$ l^{-1}; pH 6.5–7.5	60-d LOEC	>17	18

Table 6. Summary of boron toxicity data for marine fish

Test organism	Boron compound tested	Type of test	Water quality characteristics	Test response(s) Reported	Boron concentration (mgB l^{-1})	Ref.
Coho salmon (*Oncorhynchus kisutch*) under yearlings, 1.8–3.8 g	Na$_2$B$_4$O$_7 \cdot$8H$_2$O	Static-renewal (daily)	Seawater; 8 °C; salinity 28‰	283-h LC$_{50}$	12.2	10
Coho salmon (*Oncorhynchus kisutch*) under yearlings, 1.8–3.8 g	Na$_2$B$_4$O$_7 \cdot$8H$_2$O	Static-renewal (daily)	Seawater; 8 °C; salinity 28‰	96-h LC$_{50}$	40.0	10
Dab (*Limanda limanda*)	Na$_2$B$_4$O$_7$		Seawater; salinity 34.8‰	24-h LC$_{50}$	88.3	19
Dab (*Limanda limanda*)	Na$_2$B$_4$O$_7$		Seawater; salinity 34.8‰	72-h LC$_{50}$	75.7	19
Dab (*Limanda limanda*)	Na$_2$B$_4$O$_7$		Seawater; salinity 34.8‰	96-h LC$_{50}$	74.0	19

conditions. A large part of the first effect variability within and between water types is also associated with the very flat dose-response curve shown by boron for aquatic species. Consistent across all the trout toxicity tests is the observation that it takes >1.0 mgBl^{-1} boron to get a 10% increase in the control adjusted mortality (see Fig. 2).

Support for the higher first effect values comes from the observation that in the wild, healthy rainbow trout have been found in surface waters containing up to 13 mgBl^{-1} [21]. Waters supplying trout hatcheries in California have also been found to range between 0.02 to 1 mgBl^{-1} with no apparent observable effects to rainbow trout early development.

Early life stages of non-salmonid fish species appear relatively resistant to aqueous exposure to boron. As shown in Table 5, tests on the fathead minnow egg-fry indicate a 30-day LOEC (reduction in growth) of 24 mgBl^{-1} and a 60-day LOEC (reduction in fry survival) of 88 mgBl^{-1}. The LOECs for embryo-larval stages of the channel catfish and goldfish ranged from 1.04 to 25.9 mgBl^{-1} and 8.33 to 48.75 mgBl^{-1}, respectively. Tests with largemouth bass indicated a LOEC of about 12 mgBl^{-1}.

At low concentrations boron has in fact been found to be beneficial to some freshwater organisms. The addition of 0.4 mgBl^{-1} to ponds used for raising carp increased production by 7.6% [22]. In an examination of the use of growth promoting substances in aquaculture, Sen and Chatterjee [23] reported that the addition of 1 mgB fish^{-1}d^{-1} increased the survival rate of rohu (*Labeo rohita*) fry relative to experimental controls receiving no additional boron.

Data on the toxicity of boron to marine fish are very sparse. The results of two studies so far published, on a U.K. estuarine species the dab (*Limanda limanda*) and a North American salmonid species coho salmon (*Oncorchynchus kisutch*) are summarised in Table 6.

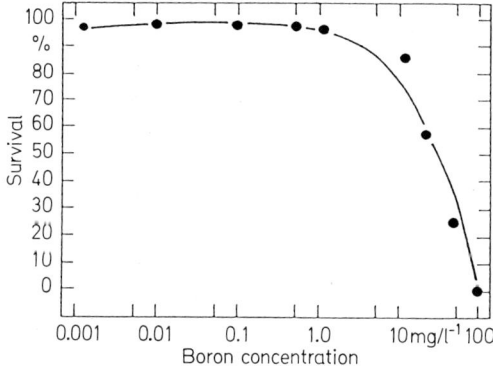

Fig. 2. Control adjusted survival of rainbow trout embryo-larval stages exposed to borax in soft reconstituted water (50 mg CaCO$_3$ l^{-1}) (Teratic larvae were counted as mortalities) [18]

The toxicity of boron in seawater is similar to that in freshwater. The range of LC_{50} values quoted in Tables 5 and 6 is approximately 5–3000 mgB l^{-1} for freshwater and 12–90 mgB l^{-1} for seawater. Only one test on a single species has shown boron to be more toxic in seawater than in freshwater. This species *Oncorchynchus kisutch* (coho salmon) was tested at the yearling stage, at which age this particular salmon still lives in freshwater [10].

Boron Toxicity to Aquatic Life Other than Fish

Table 7 summarises the results of work so far carried out on the toxicological effects of boron on aquatic life other than fish.

Studies have found amphibians to respond to boron at similar concentrations as fish, requiring 9.6 mgB l^{-1} for detectable effects. In the case of the toad *Bufo fowleri* no effects occurred on its embryos until a boron concentration of 53 mgB l^{-1} was applied [16].

With regard to invertebrates, in tests with *Daphnia magna*, investigations have included assessing effects resulting from both acute and chronic exposure to boron. NOEC and LOEC values for this species are seen to be about 6 and 13 mgB l^{-1}. McKee and Wolf [13] have presented additional data indicating a higher effect for boron in the form of sodium perborate. However, as stated previously, the formulation of boron should not significantly influence its aquatic effects at naturally occurring pH levels, as the predominant boron species under these conditions is undissociated boric acid.

Bringmann [26] determined that cell replication of the protozoan, *Entosiphon sulcatum*, was reduced by 5% when exposed to 1.0 mgB l^{-1}, and identified this as the boron "effects threshold" for this species.

Some further work has been conducted to assess the effects of boron on other aquatic organisms. In studies with phytoplankton, a boron concentration of 30 mgB l^{-1} was found to reduce photosynthesis in half the species tested [30] and 50 mgB l^{-1} decreased growth rates of five of the nineteen species exposed [31]. This study concluded that concentrations of boron of up to 10 mgB l^{-1} would be readily tolerated by marine phytoplankton and would cause little destabilisation of natural communities.

Stanley [32] has shown that a concentration of 40.3 mgB l^{-1} added as a tetraborate salt leads to a 50% inhibition of root growth in the freshwater plant *Myriophyllum spicatum* after thirty-two days treatment.

Some work has also been done on the toxic effects of boron on micro organisms in sewage treatment plants; 20 mgB l^{-1} has been shown to have no effect on activated sewage treatment [34], and no significant inhibition of anaerobic sludge digestion occurs at boron levels below 200 mgB l^{-1} [35].

There is no evidence that aquatic organisms bioaccumulate boron.

Table 7. Summary of boron toxicity data for aquatic life other than fish

Test organism	Boron compound tested	Type of test	Water quality characteristics	Test response(s) reported	Boron concentration (mgB l^{-1})	Ref.
Invertebrates						
Daphnia magna Straus	Boric acid H_3BO_3	Static	Lake Huron water; 20 °C; pH 6.7–8.1; hardness 150 mg l^{-1}	48-h LC$_{50}$	133	24
Daphnia magna Straus	Boric Acid H_3BO_3	Static renewal (3 times weekly)	Lake Huron water; 20 °C; pH 7.3–8.0; hardness 150 mg l^{-1}	21-d LC$_{50}$ 21-d NOEC-LOEC	52.2 6.4–13.6	24
Daphnia magna Straus	Boric acid H_3BO_3	Static	Carbon filtered well water; 19.2 °C pH 7.1–8.7; hardness 166 mg l^{-1}	48-h LC$_{50}$	226	25
Daphnia magna Straus	Boric acid H_3BO_3	Static (3 times weekly)	Carbon filtered well water; 19.2 °C; pH 7.1–8.7; hardness 166 mg l^{-1}	21-d LC$_{50}$ 21-d NOEC-LOEC	53.2 6–13	25
Daphnia magna	Sodium perborate $Na_2BO_3 \cdot 4H_2O$	—	Lake Erie H_2O; 25 °C	Threshold concentration for immobilisation	≪0.38 estimated to be 0.19 (≪5.2; estimated to be 2.6)	13
Daphnia magna	Sodium tetraborate $Na_2B_4O_7 \cdot 10H_2O$	—		Threshold concentration for immobilisation	≪27.2 estimated to be 13.6 (≪240; estimated to be 120)	13
Protozoan (*Entosiphon sulcatum*)	Sodium tetraborate $Na_2B_4O_7 \cdot 10H_2O$	Static	Culture medium adjusted to pH 6.9; 25 °C	Toxicity threshold (measured as a 5% reduction in cell replication after 72 h)	1.0	26
Mosquito larvae (*Anopheles quadrimaculatus*)	Boric acid H_3BO_3	—	—	100% mortality after 25 hours 92% mortality after 48 hours	125 25	27

Table 7. (Continued)

Test organism	Boron compound tested	Type of test	Water quality characteristics	Test response(s) reported	Boron concentration (mgB l^{-1})	Ref.
Sea urchin *Anthocidaris crassispina* (embryos)	—	—	—	Normal development Fatal concentration	37 75	28
Amphibians Toad (*Bufo vulgaris formosus*)	Boric acid H_3BO_3	Static; embryos exposed to B for 2 h at various embryonic stages and then cultured in tap water until 14 d past fertilization	Boric acid solutions prepared in tap water	Teratogenic defects and reduced survival	1747 (1% soln.)	29
Fowler's toad (*Bufo fowleri*) embryo-larval stages	Boric acid H_3BO_3	Flow-through	Reconstituted water, hardness 50 mg $CaCO_3$ l^{-1}	7-d NOEC-LOEC	48.7–96.0	16
Fowler's toad (*Bufo fowleri*) embryo-larval stages	Boric acid H_3BO_3	Flow-through	Reconstituted water, hardness 200 mg $CaCO_3$ l^{-1}	7-d NOEC-LOEC	22.3–53.5	16
Leopard frog (*Rana pipiens*) embryo-larval stages	Boric acid H_3BO_3	Flow-through	Reconstituted water, hardness 50 mg $CaCO_3$ l^{-1}	7-d NOEC-LOEC	32.5–47.50	16
Leopard frog (*Rana pipiens*) embryo-larval stages	Boric acid H_3BO_3	Flow-through	Reconstituted water, hardness 200 mg $CaCO_3$ l^{-1}	7-d NOEC-LOEC	45.7–86.0	16
Leopard frog (*Rana pipiens*) embryo-larval stages	Borax $Na_2B_4O_7 \cdot 10H_2O$	Flow-through	Reconstituted water, hardness 50 mg $CaCO_3$ l^{-1}	7-d NOEC-LOEC	7.04–9.60	16
Leopard frog (*Rana pipiens*) embryo-larval stages	Borax $Na_2B_4O_7 \cdot 10H_2O$	Flow-through	Reconstituted water, hardness 200 mg $CaCO_3$ l^{-1}	7-d NOEC-LOEC	7.04–10.50	16

Perborate

Plants						
Marine phytoplankton (10 species)	Boric acid H_3BO_3	Unialgal cultures	Seawater	Reduction in photosynthesis for 50% of species after 5 days	30	30
Marine phytoplanktor (19 species)	Boric acid H_3BO_3	Axenic cultures	Seawater; pH 7.6–8.0	Reduction in growth rate for 26% of species	50	31
Spiked or Eurasian watermillfoil (*Myriophyllum spicatum*)	Tetraborate salt $B_4O_7^{2-}$	—	Freshwater	No effect on growth 50% inhibition of roots weight after 32 days	10 40.3	31 32
Chlorella algae (*C. vulgaris*)	—	—	—	Toxic	50	33
Chlorella algae (*C. protothicoides* and *C. emersanii*)	—	—	—	Toxic	100	33

Factors Affecting the Toxicity of Boron

The toxicity of many trace elements depends on the chemical speciation of the dissolved species. Factors such as the formation of polymeric species and the formation of complexes with inorganic and organic moieties have been shown to be of importance [36].

Borate forms ion pairs with the major and minor cations present in natural waters; with stability constants decreasing in the order $Mg^{2+} > Ca^{2+} > Sr^{2+} > Li^+ > Na^+$ [37]. In hard waters enriched in calcium, it might be expected that the formation of ion pairs between Ca and $B(OH)_4$ would affect the speciation of boron [38].

Birge and Black [16] investigated the effect that water hardness (at 50 mg $CaCO_3 \, l^{-1}$ and 200 mg $CaCO_3 \, l^{-1}$) has on the toxicity of boron administered to aquatic organisms. In addition, a study of the comparative toxicity of boron administered as borax or boric acid was conducted. When LC_1 and LC_{50} data for all species were combined, no statistically significant differences between boric acid and borax toxicity could be demonstrated regardless of water hardness.

The hardness of the water in the test medium did however appear to exert an effect on the toxicity of boron to fish, but the effect was not found to be consistent. The tests showed that borax exhibited greater toxicity in hard water to embryos and larvae of both goldfish and channel catfish, unlike the embryos and larvae of rainbow trout which displayed a greater toxic reaction to borax in soft water. In contrast, boric acid was most toxic to rainbow trout life-stages in hard water. This pattern of toxicity was also exhibited in the reaction of channel catfish embryos and larvae to boric acid exposure. No pattern emerged in the toxic responses of goldfish embryos and larvae to boric acid in test waters of different hardness. None of the data from other reported tests allows any comparison of boron toxicity variation with hardness.

Regulations Protecting Aquatic Life

No legislation exists relating specifically to perborate; however, regulations are developing for boron which, as discussed above, are relevant to the discharge of perborate containing bleaches and detergents. Under the European Community Council Directive 76/464/EEC on Pollution by Certain Dangerous Substances discharged into the Aquatic Environment [39], which is designed to protect aquatic life, Member States are required to reduce pollution of both fresh and salt waters resulting from the discharge of List II substances, of which boron is one. All discharges liable to contain a List II substance require prior authorisation with emission standards being laid down. These emission standards are to be based on water quality objectives.

Table 8. Environmental quality standards (EQSs) for boron recommended in the U.K. for protection of aquatic life [40]

Nature of protection	EQSs (mgB l^{-1})
Protection of freshwater fish	2.0
Protection of other freshwater life and associated organisms	5.0
Protection of saltwater fish and shellfish	7.0
Protection of other saltwater life and associated non-aquatic organisms	7.0

Each Member State is required to submit its proposal for water quality objectives to the European Commission; EC-wide objectives will then be developed on the basis of these submissions. As an example of objectives being developed, the recommendations recently finalised in the U.K. for environmental quality standards for boron are summarised in Table 8. Other Member States are thought not to be any further advanced.

In the U.S.A., attention has been placed mainly on the development of boron water quality standards for the protection of agricultural crops; examples of irrigation water quality limits established are provided in Table 11, later in the chapter. A number of states have, however, set general surface water quality standards for boron; examples of these are provided in Table 9.

Ecotoxicology of Boron: The Terrestrial Environment

Boron was shown to be an essential nutrient for vascular plants sixty years ago [42], but there is no evidence that it is required by other terrestrial organisms. Conversely, the toxic effects of boron have been demonstrated in animals as well as in plants.

Boron Toxicity to Higher Plants

In terms of its impact on higher plants, boron has two roles: it is an essential nutrient for plant growth, but it is also toxic at higher levels (see Fig. 3). The literature contains much more information on boron deficiency than on boron toxicity, suggesting that boron deficiency has been the more serious problem [43]. Borate is often added to irrigation water or used as a borated fertiliser, or as a foliar spray, to improve the yield of crops where boron deficiency symptoms appear.

Table 9. General surface water quality standards for boron in the USA [41]

State	Date of Standard	Criteria (mgB l^{-1})
Illinois	1972	1.0
Puerto Rico	1976	1.0
North Dakota	1977	0.5
Oregon	1977	0.5
Utah	1978	0.75

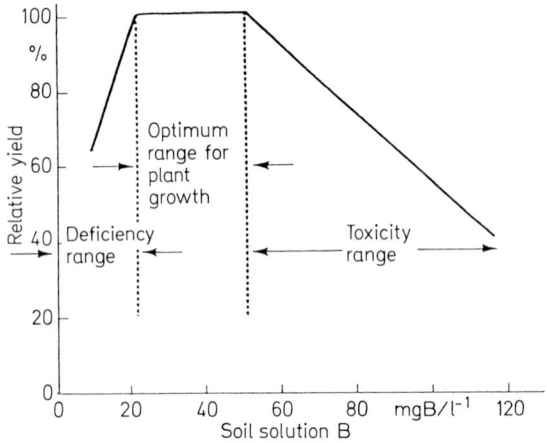

Fig. 3. Relative yield as influenced by soil solution boron (Idealised) [43]

In contrast to its effect on other living organisms, the range of concentrations within which boron is essential to some plants overlaps with that within which it is toxic to others. It is the small margin between the essential and the toxic concentration which makes a detailed understanding of the interactions of boron with higher plants so necessary.

Toxicity symptoms typically show first on older tissue, with yellowing of the leaf tip and chlorosis which subsequently progresses along the leaf margin and then spreads into the blade. Necrosis of the chlorotic tissue follows, after which leaf abscission takes place [44].

The most comprehensive summary of boron tolerance data based upon plant response to soluble boron (concentration in nutrient solution, solution used to irrigate sand culture, or in soil solution) is that of Maas [45] (and cited by Keren and Bingham [46] (Table 10)). The threshold concentration range identified is the maximum concentration that a given plant species tolerates without manifesting visual injury symptoms and/or a decrease in yield. The crop species are listed in order of increasing tolerance to boron, within each of three tolerance classes.

Table 10. Threshold concentration range of boron in soil water, field capacity basis, according to crop species[a] [46]

Crop species	Threshold concentration range (mol B m^{-3})	Ref.
Sensitive crops		
Lemon (*Citrus limon*)	0.028 to 0.046	[47, 48]
Blackberry (*Rubus* sp.)	(0.30–0.50 mgB l^{-1})	[47]
Avocado (*Persea americana*)		[48]
Orange (*Citrus sinensis*)		[48]
Grapefruit (*Citrus paradise*)		[48]
Apricot (*Prunus armeniaca*)		[49]
Peach (*Prunus persica*)		[47, 48]
Cherry (*Prunus Avium*)		[47]
Plum (*Prunus domestica*)		[49]
Persimmon (*Diosypysos kaki*)	0.046 to 0.074	[47]
Fig (*Ficus carica*)	(0.50–0.80 mgB l^{-1})	[47]
Grape (*Vitis vinifera*)		[47]
Walnut (*Juglans regia*)		[48]
Pecan (*Carya illinoensis*)		[48]
Cowpea (*Vigna sinensis*)		[47]
Onion (*Allium cepa*)		[47]
Garlic (*Allium sativum*)		[50]
Sweet potato (*Ipomoea batatas*)		[47]
Wheat (*Triticum aestivum*)		[51]
Mung bean (*Phaseolus aureux*)		[52]
Strawberry (*Fragaria* sp.)	0.074 to 0.093	[47]
Kidney bean (*Phaseolus vulgaris*)	(0.80–1.01 mgB l^{-1})	[47]
Lima bean (*Phaseolus lunatus*)		[47]
Semitolerant crops		
Sesame (*Sesamum indicum*)		[52, 53]
Red pepper (*Capsicum annum*)		[47]
Pea (*Pisum sativum*)		[47]
Carrot (*Daucus carota*)	0.093 to 0.19	[47]
Radish (*Raphanus sativus*)	(1.01–2.05 mgB l^{-1})	[47]
Potato (*Solanum tuberosum*)		[47]
Cucumber (*Cucumis sativus*)		[54]
Lettuce (*Lactuca sativa*)		[47]
Cabbage (*Brassica oleracea*)		[47]
Celery (*Cepium graveolens*)		[47]
Turnip (*Brassica rapa*)		[47]
Barley (*Hordeum vulgare*)		[51]
Corn (*Zea mays*)	0.19 to 0.37	[47, 54]
	(2.05–4.00 mgB l^{-1})	
Artichoke (*Cynara scolymus*)		[47]
Tobacco (*Nicotiana tabaacum*)		[47]
Sweet clover (*Melilotus indica*)		[47]
Squash (*Cucurbita pepo*)		[54]
Muskmelon (*Cucumis melo*)		[47, 54]
Tolerant crops		
Sorghum (*Sorghum bicolor*)		[51]
Alfalfa (*Medicago sativa*)		[47]
Purple vetch (*Vicia benghalensis*)		[47]
Oat (*Avena vulgare*)		[55]
Parsley (*Petroselium crispum*)	0.37 to 0.56	[47]
Red beet (*Beta vulgaris*)	(4.00–6.05 mgB l^{-1})	[47]

Table 10. (*Continued*)

Crop species	Threshold concentration range (mol B m^{-3})	Ref.
Tomato (*Lycopersicon lycopersicum*)		[47, 56]
Sugarbeet (*Beta vulgaris*)		[47, 57]
Cotton (*Gossypium hirsutum*)	0.56 to 0.93 (6.05–10.05 mgB l^{-1})	[47, 51]
Asparagus (*Asparagus officinalis*)	0.93 to 1.39 (10.05–15.03 mgB l^{-1})	[47]

ª After Mass (1984)

Much of the data reported by Maas [45] is drawn from studies undertaken by Eaton in 1944 [47, 58] whose work still represents the most comprehensive attempt to quantitatively define limits for boron toxicity for particular crop species. As a result of his work, Eaton proposed a scheme of critical concentrations of boron in irrigation water based on each species' relative sensitivity to boron. The original values suggested were 0.5–1.0 mgB l^{-1} for sensitive plants and 1.0–2.0 mgB l^{-1} for semi-tolerant ones.

Because these original guidelines were based on sand culture they do not take into account the ability of soils to adsorb and remove boron from the soil solution. This seems to be cofirmed by evidence in the field; no reported effects on crop growth or yield have been found for various crops exposed to boron concentrations of up to 2 mgB l^{-1} in irrigated wastewater (Table 11).

Later guidelines that have been developed are summarised in Table 12. The great limitation which lies in any attempt to set down limits for boron in irrigation water is the complexities which arise as a result of the specific climatic and soil conditions in a particular area. These factors compound the difficulties which arise from differences in individual species tolerance.

Studies by Eaton refer to conditions in California, where there is very limited rainfall—winter only—and irrigation is the primary source of water. The stated maximum permitted tolerances to boron are generally not relevant to European conditions, where there is adequate rainfall during the whole year.

In the U.K. irrigation guidelines have been developed which specifically take into account total boron loadings to soils [89]. The quantities of boron considered to be safe for application each year have been identified as 2.0, 3.0 and 4.0 kg/ha B for sensitive, intermediate and tolerant crops, respectively. The safe quantities of irrigation water related to its boron content have been calculated and these are given in Table 13. For example, if 200 mm of irrigation water is to be applied to plums, it should not contain more than 1.0 mgB l^{-1}.

The amount of boron taken up by plants varies considerably from one species to another and also depends on the stage of growth. There are also a number of soil and environmental factors which affect the uptake of boron by

Table 11. Levels of boron used on crops under wastewater irrigation with no reported effect on crop growth or yield

Crop	Soil type	Boron concentration (mgB l^{-1}) effluent	Location	Ref.
Barley, wheat, corn	Sandy loam	0.44	Mess, AR	[59]
Bermuda grass native vegetation	Fine loam	0.45–0.85	Salt River, AR	[60, 61]
Tomatoes, broccoli, spinach	Clay loam	0.85	Camarillo, CA	[62]
Orchards, green areas	—	0.5–0.7	Goleta, CA	[63]
Grasses	Gravelly sand and loam	1.4	Hollister, CA	[64]
Citrus	—	0.5–1.7	Santa Ana R. Valley, CA	[65]
Grain	—	0.53	Santa Rosa, CA	[63]
Desert chaparral	Sandy loam	0.2–0.3	San Bernadino	[66]
Citrus	—	0.15–0.8	Florida	[67]
Rye	Sandy clay loam	1.2	Athens, GA	[68]
Red pine[a]	Loamy clay	0.9	Michigan	[69, 70]
Rye, corn, sorghum, perennial grasses and legumes	Hard clay	0.3	Lansing, MI	[71]
Alfalfa	—	0.5–2.1	Carson City, NV	[72]
Corn, alfalfa	Loam	0.34	Roswell, New Mexico	[73]
Forest, corn, wheat grass	Silt/clay loam and sand loam	0.21–0.42	State College, PA	[74]
Red pine[a]	—		State University, PA	[75]
Various truck crops	Clay loam, sandy clay loam, clay	1.5–2.8	Mexico City, Mexico	[76, 77]
Various	Sandy loam	1.3–2.8	Tula R, Valley, Mexico	[78]
Misc. fruit and vegetables	—	1.89–2.56	Valley of Mexico	[79]
Grass meadows	Sandy loam	0.13–0.24	Wroclaw, Poland	[80, 81]
Citrus[a]	Sandy loam	2.0	Sicily	[82]
Sugar beet	Calcareous	0.38–0.42	Tel-Adashim, Israel	[83]
Rhodes grass	Calcareous	0.22–0.40	Zora, Israel	[83]
Pasture, trees, sugarcane	—	0.1–0.3	Pretoria, Africa	[84]

[a] Boron toxicity symptoms noted

plants[1]. Boron retention in soil depends on boron concentration in the soil solution, soil pH, texture, organic matter, cation exchange capacity, type of clay and mineral coating on the clay. Furthermore, the degree of boron fixation will be influenced by moisture content, wetting and drying cycles and temperature.

The precipitation: evaporation ratio of an area is a key factor in determining the degree to which boron can concentrate in soils and reach toxic levels. In climates where the yearly precipitation: evaporation ratio is less than unity, the

---

[1] The total boron content of a soil is of little value for diagnosing boron status and experimental work by Gupta [90] suggests that less than 5% of the soil boron is available for plant uptake

Table 12. Irrigation water quality guidelines for boron

Criterion	Irrigation water boron limit (mgB l^{-1})	Source
Nature of crop		
Sensitive	0.3–1	[58]
Semi-tolerant	1–2	
Tolerant	2–4	
Sensitive	0.5–1	[85]
Semi-tolerant	1–2	
Tolerant	2–10	
Sensitive	1	[40]
Intermediate	1.5	
Tolerant	2.0	
Sensitive	0.75	[86]
Nature of soil/time		
All soils/long-term	1.0	[87]
Fine textured soils for 20 years	2.0	
Degree of problem		
No problem	<0.5	[88]
Increasing problem	0.5–2	
Severe problem	2–10	

probability of boron toxicity is far higher. Apart from any other factor, it is the dryness of such climates which makes irrigation necessary, and which will tend to cause water to be evaporated from the soil leaving salts behind.

Boron Toxicity to Grazing Animals

Boron toxicity data available for grazing animals are summarised in Table 14. Sheep have been found to develop enteritis when exposed to naturally occurring boron enriched soils (30–300 mgB kg^{-1}) and associated water levels of 1–20 mgB l^{-1} [91]. It is thought that the ingestion of high concentrations of boron causes decreases in intestinal proteolytic enzyme activity and blood nitrogen.

In 1973 the U.S. EPA proposed that 5 mgB l^{-1} should be the maximum safe level of boron in livestock drinking water. However, this has been challenged by Green and Weeth [92] and Weeth et al. [93], who propose that the level be raised to 40 mgB l^{-1}. They have found no overt signs of toxicosis in heifers when exposed to 120 mgB l^{-1} and that 300 mgB l^{-1} is not acutely toxic to this species when consumed via drinking water. They estimate the safe tolerance level to lie within the range 40–150 mgB l^{-1}.

Table 13. Safe boron concentrations in irrigation water according to seasonal water need [89]

Tolerance group	Crops	Seasonal irrigation need (mm)		
		50	100	200
		Safe boron concentration (mg l^{-1})		
Sensitive	Plums Pears Apples Cherries Blackcurrants Strawberries Raspberries	4.0	2.0	1.0
Intermediate	Barley Wheat Maize Oats Potatoes Peas Radish Tomatoes	6.0	3.0	1.5
Tolerant	Asparagus Beet Mangolds Lucerne Broad beans Turnips Onions Cabbage Lettuce Carrots	8.0	4.0	2.0

Table 14. Summary of boron toxicity data for grazing animals

Species	Boron compound tested	Type of test	Soil/feed/water characteristics	Test response reported	Boron concn.	Ref.
Sheep	—	Evidence in the field	—	Young sheep developed enteritis	Soil 30–300 mgB kg^{-1}	[91]
Sheep	Boric acid H$_3$BO$_3$	Daily feed	—	Sheep developed enteritis	40 mgB kg^{-1}	[91]
Cattle	Boric acid H$_3$BO$_3$	Cattle exposed to high concentrations of boron in their drinking water	—	Swelling and irritation of legs, lethargy, diarrhoea with 30 day exposure	150–300 mgB l^{-1}	[92]
Cattle	Boric acid H$_3$BO$_3$	Cattle exposed to high concentration of boron in their drinking water	—	No overt signs of toxicosis with 10 day exposure	120 mgB l^{-1}	[93]

Safety of Boron in Aquatic and Terrestrial Environments

Generally, environmental concentrations of boron found in surface water are below levels identified as toxic to aquatic organisms; a summary of boron concentrations in European surface waters is given in Table 15. All recent data (post-1978) show boron levels to be lower than 1 mgBl^{-1}. Concentrations high enough to produce toxic effects in laboratory tests are found in areas where weathering of boron-rich formations and deposits occurs, such as in southwest U.S.A.; where levels of 5–15 mgBl^{-1} have been found. However, reproducing populations of the most sensitive species, rainbow trout, have been observed in surface waters in these regions, indicating no cause for concern.

The prime concern for effects on terrestrial plants centres on the use of irrigation water with elevated levels of boron. At the present time, there is no evidence of widespread damage to crops resulting from this practice. In some areas, wastewater is used for irrigation and crops grown under these conditions are generally confined to those relatively insensitive to boron toxicity. Good irrigation practices will be necessary, however, in arid regions with high evapotranspiration rates and care will be needed when using wastewater, particularly in areas with naturally high boron levels.

It is not anticipated that there will be any significant increase in the discharge of boron associated with the use of perborate to the environment in the foreseeable future. The use of boron in detergents in Europe is expected to decrease due to the introduction of bleach activators and liquid detergents. There is a possibility that perborate may be used in detergents in the U.S.A.; however, the estimated increase in boron concentrations is likely to be environmentally insignificant.

Clinical Toxicology

The main human exposure route to boron from the use of perborate is through ingestion of drinking water and foodstuffs. Workers may also be exposed to borax dust during manufacture of perborate but this has not been considered here. Evidence of industrial toxication by borax and boric acid are first considered, as these give an insight into the level of harm to the general public expected from exposure through normal diet.

Evidence of Industrial Intoxication

Despite the widespread applications, no cases of industrial intoxication on exposure to inorganic borates have been reported [101]. The European Communities' Directive on the Classification, Packaging and Labelling of Dangerous

Table 15. Concentrations of boron in European surface waters

Country	Year	Number of samples	Boron mgB l^{-1}	Ref.
Germany	1974	300	0.100–2.000	[94]
	1982	15	0.078–0.272	[95]
Italy	1975	6	0.400–1.000	[96]
Netherlands	1972	17	0.100–0.330	[97]
	1981	3	0.040–0.090	[98]
Sweden	1972	147	0.001–1.046	[99]
U.K.	1978	20	0.060–0.650	[100]
	1980	63	0.046–0.822	[40]

Substances, in its Sixth Amendment [102] has adopted the acute oral toxicity in the rat as an approximate guide to classification. Substances with a median lethal dose (LD_{50}) greater than 2000 mg/kg bodyweight are not classified as dangerous. In this context, the acute oral toxicity of boric acid in the rat is LD_{50} 3000–4000 mg kg^{-1} bodyweight, and that of borax is LD_{50} 4500–6000 mg kg^{-1} [103]; they are therefore not classified as dangerous. For simple comparison, sodium chloride (common salt) has an LD_{50} 3750 mg kg^{-1} bodyweight, a similar order of magnitude to borax and boric acid.

A recent kinetic study on human volunteers [104] has shown that when a single intravenous dose of an aqueous solution containing 500 mg of boric acid is injected, the boric acid is excreted rapidly in the urine with a half-life of 21 h, and completely within 96 h. Symptoms of intoxication are therefore not encountered up to this level of intake, but occur only if such a massive quantity is absorbed internally that the kidney function is temporarily overloaded. These symptoms of intoxication (not encountered in industrial applications) consist of vomiting and diarrhoea, followed by an erythmatous rash, shock, coma and, in extreme cases, death [105].

Chronic Studies

The chronic effects of boron are difficult to recognise because of other social and environmental factors. Possible concerns about the long-term chronic effects, and in particular the mutagenicity/carcinogenicity of inorganic borates have been allayed by the results of a recent study in the U.S.A. [106]. The results show no evidence of carcinogenicity from a full 2-year bioassay on boric acid in mice at feed doses of 2500 and 5000 ppm in the diet. A variety of mammalian cell culture tests were also negative. Furthermore neither boric acid nor borax is a mutagen in the Ames test. There is report [107] of elevated blood sugar levels in areas of Russia where the population is exposed to 6 mgB l^{-1} in drinking water. In a study of the effects of 4.6 mgB l^{-1} [108], it was found that the incidence of gastric hypoacidity was approximately twice as high in adults, and faecal enterokinase activity was 2–8 times lower in children than in a comparable group exposed to

1.6 mgB l^{-1}. It was concluded that a high concentration of boron in the drinking water decreases the functional activity of the gastrointestinal tract. However, no clear relationship emerged between faecal enzyme activities and boron concentration [109] in patients with enteritis who were living in a boron rich area.

Safety of Boron in the Diet

Boron is ingested in the normal diet at a rate of about 10–20 mg d^{-1} [110]; the chief source being fruit and vegetables. The boron content in a typical serving of various foods is shown in Table 16. The amount that an individual consumes is however strongly dependent on dietary characteristics and the level of boron in the drinking water consumed, and there is a wide variation in boron intake across the population. The relative contribution of food stuffs and drinking water to boron intake is also expected to vary significantly from region to region. At present, no firm conclusions can be drawn about the total boron concentra-

Table 16. Boron content in a serving of various foods[a] [111]

Apple sauce, bottled	0.279
Grape juice, bottled	0.202
Apple juice, bottled	0.188
Peaches, canned	0.187
Broccoli, frozen	
flowers	0.185
stalks	0.089
Cherries, frozen	0.147
Pears, canned	0.122
Carrots, canned	0.075
Green beans, frozen	0.046
Orange juice, frozen	0.041
Lettuce, iceberg	0.039
Noodles	
egg	0.037
spaghetti	0.006
Cornflakes, fortified	0.031
Bread, white enriched	0.020
Ice cream, vanilla	0.019
Potatoes, canned	0.017
Chicken breast, ground	0.005
Coffee, freeze dried	0.005
Rice, minute	0.003
Milk, 2%	0.002
Beef, ground round	ND[b]
Cheese, cheddar or cream	ND
Eggs, frozen	ND
Sugar, granulated	ND

[a] Milligrams boron per 100 millilitres (3$\frac{1}{3}$ fluid ounces or 0.42 cup) or 100 grams (3$\frac{2}{3}$ ounces, dry weight).
[b] ND—non-detectable.

tion to which different populations are exposed. The contribution of boron from use of perborate is also difficult to assess. While use of perborate represents a major use of boron and a use which leads to direct discharge of boron to the aqueous environment (unlike in the case of fibre glass where the boron is effectively tied up), boron also occurs naturally and boron in the diet from natural sources will vary by region.

Boron present in food and water is rapidly and completely absorbed by the human system but there is no accumulation of boron within the body, and the normal dietary intake is excreted at the same rate as it is ingested [112]. 10–20 mgB d^{-1} corresponds to about 60–120 mg of boric acid, and this is well within the capacity of the kidneys for excretion.

No daily acceptable intake values have so far been established for boron. Water quality limits for boron in water intended for human consumption have however been defined. In the EC Drinking Water Directive surface freshwater used for drinking water must have a boron concentration less than or equal to 1 mgB l^{-1} [39]. From the environmental concentration data already presented (Table 15) it is seen that this standard is met in most cases.

With regard to overall exposure, chronic effects have yet to be fully established (particularly in view of the presence of many other compounding factors such as effects of other pollutants, diet, etc.); however, the fact that there is no evidence of any problems associated with occupational exposure would indicate that there is no major cause for concern. As already noted, it is not anticipated that there will be any significant increases in the discharges of boron into the environment through the use of perborate in the near future.

References

1. Interox Chemicals Ltd (1982) Product data sheets for sodium perborate, tetrahydrate and monohydrate. Available from Interox Chemicals, P.O. Box 7, Warrington, Cheshire WA4 6HB, U.K.
2. Thorpes dictionary of applied chemistry (1950) 4th edn, vol X: Sodium. Longmans, p 874
3. Mellor's comprehensive treatise on inorganic and theoretical chemistry (1980) vol V: Boron, part A: Boron-oxygen compounds
4. SRI International: Chemical economics handbook marketing research report (1986) Boron minerals and chemicals. Stanford Research Institute, Menlo Park, California 94025 U.S.A.
5. Waggot A (1969) Water Res 3:749
6. Hem JD (1970) Study and interpretation of the chemical characteristics of natural water. 2nd edn, Geol Surv Supply. Paper 1473. U.S. Geol Survey
7. Broekaert JAC, Leis F (1979) Analytica Chimica Acta 109:73
8. HMSO (1980) Boron in waters, effluents, sewage and some solids, 1980: Methods for the examination of waters and associated materials. Her Majesty's Stationery Office, London, U.K.
9. Turnbull H et al. (1954) Ind Eng Chem 46:324
10. Thompson JAJ et al. (1976) Water Res 10:869
11. Procter and Gamble (1979) (unpublished data)
12. NAS (1973) National Academy of Sciences National Academy for Engineering. Water quality criteria 1972. EPA Ecol Res Ser EPA-R3-73-033. U.S. Environmental Protection Agency, Washington, DC

13. McKee JE, Wolf HW (1963) Water quality criteria. The Resource Agency of California (2nd edn) State Water Quality Control Board, Publ No 3A
14. Wallen IE et al. (1957) Sew Ind Waste 29:695
15. Wurtz A (1945) Annales de la Station Centrale de Hydrobiologie Appliquee 1:179
16. Birge WJ, Black JA (1977) Sensitivity of vertebrate embryos to boron compounds. Report No EPA-560/1-76-008. Environmental Protection Agency, Office of Toxic Substances, Washington DC
17. Birge WJ, Black JA (1981) Toxicity of boron to embryonic and larval stages of largemouth bass (*Micropterus salmoides*) and rainbow trout (*Salmo gairdneri*). Completion Report prepared for Procter and Gamble Company
18. Procter and Gamble Company (1987) Aquatic safety assessment for boron. Report for the U.S. EPA, Washington DC
19. Taylor D et al. (1985) Aquatic Toxicology 7:134
20. Birge WJ et al. (1984) Toxicity of boron to embryonic and larval stages for rainbow trout (*Salmo gairdneri*). Completion report prepared for Procter and Gamble Company
21. Bingham FT (1982) The boron concentration of wild trout streams in California. Report to Procter and Gamble Company, p 9
22. Avetisyan LT (1983) Biol Zh Arm 36:22
23. Sen PR, Chatterjee DK (1979) Enhancing production of Indian major carp fry and fingerlings by use of growth promoting substance. Adv Aquacult FAO Tech Conf Aquaculture 1976 (Published 1979), p 134
24. Gersich FM (1984) Environ Toxicol Chem 3:89
25. Lewis MA, Valentine LC (1981) Environm Contam Toxicol 27:309
26. Bringmann G (1978) Zeitschrift für Wasser and Abwasser Forschung 11:210
27. Fay RW (1959) J Econ Entomol 52:1027
28. Kobayashi N (1971) Fertilised sea urchin eggs as an indicatory material for marine pollution bioassay, preliminary experiment. Publication of the Seto Marine Biological Laboratory 3, No 18, p 379
29. Takeuchi T (1958) Sci Rep Tohoku Univ Series 4 biol 24:33
30. Subba Rao DV (1981) Can J Fish and Aquat Sci 38:52
31. Antia NJ, Cheng JY (1975) J Fish Res Board Can 32:2487
32. Stanley RA (1974) Arch Environ Cont and Toxicol 2:331
33. Bowen JE, Gauch HG (1966) Plant Physiol 41:319
34. Gerike P et al. (1976) Tenside Detergents 13:149
35. Speece RE (1987) Boron toxicity levels in anaerobic digestion. Unpublished report to Procter and Gamble
36. Forstner U, Wittman GTW (1979) Metal pollution in the aquatic environment. Springer, Berlin Heidelberg New York
37. Rogers HR, Van Den Berg CMG (1988) Talanta 35:271
38. Rogers HR (personal communication)
39. Council Directive on Pollution Caused by Certain Dangerous Substances Discharged to the Aquatic Environment of the Community. Official Journal, L129/24 (1976)
40. Mance G et al. (1988) Proposed environmental quality standards for List II substances in water: Boron. Environmental Strategy, Standards and Legislation Unit. Report No. TR256, Water Research Centre, U.K.
41. U.S. EPA (personal communication)
42. Sommer AL (1927) Science 66:482
43. Gupta UC et al. (1985) Can J Soil Sci 65:381
44. Wilcox LV (1960) U.S.D.A. Information Bull 211:3
45. Maas EV (1984) Salt tolerance of plants. In: Christie BR (ed) Handbook of plant sciences in agriculture, CRC Press, Ohio
46. Keren R, Bingham FT (1985) Boron in water, soils and plants. In: Advances in soil science, 1, Springer, Berlin Heidelberg New York
47. Eaton FM (1944) J Agric Res 69:237
48. Haas ARC (1929) Bot Gaz 88:113
49. Woodbridge CG (1955) Con J Agr Sci 35:282
50. Singh RN, Singh JR (1974) Indian J Hort 31:255
51. Bingham FT (unpublished data)
52. Khudairi AK (1961) Arid Zone Res 1961:175
53. Yosif YH et al. (1972) Soil Sci Soc Am Proc 36:923

54. El-Sheikh AM (1971) J Amer Soc Hort Sci 96:536
55. Ryan J et al. (1977) Plant Soil 47:253
56. Francios (unpublished data)
57. Vlamis J, Ulrich A (1973) J Am Soc Sugarbeet Technol 17:280
58. Eaton FM (1935) Boron in soils and irrigation waters and its effect on plants, with particular reference to the San Joaquin Valley of California, U.S. Dep Agric Tech Bull No 448, Washington DC
59. Stone R (1980) Long term effects of land application of domestic wastewater. Mess, AR, slow rate irrigation site. EPA 600-280-079
60. Bouwer H et al. (1980) J Water Pollut Control Fed 52:2457
61. Bouwer H et al. (1981) The Flushing Meadows Project—Wastewater renovation by high rate infiltration for groundwater recharge. In: D'Itri FM, Aquirre Martinez J, Athie Lambarrie M (eds) Municipal Wastewater in Agriculture. Academic, New York, p 195
62. Stone R, Rowlands J (1980) Long term effects of land application of domestic wastewater. Camarillo, California Irrigation site. EPA 600-2-80-080
63. Stenquist RS et al. (1979) Three California water reclamation case histories. In: Proc Water Reuse Symp 3, 1693–1736. Am Water Works Assoc Res Found Denver CO, U.S.A.
64. Pound CE et al. (1978) Long term effects of land application of domestic wastewater. Hollister, California, Rapid infiltration site. EPA 600-2-78-084
65. Eccles LA (1979) Groundwater quality in the Upper anta Ana River Basin, So, CA. U.S. Geological Survey, Menlo Park, CA, Water Resources Div, USGS/WRD/WRI-80/012, USGS/WRI-79-113, PB80-161888
66. Younger VB (1974) Ecological and physiological implications of Greenbelt irrigation with reclaimed wastewater. In: Sopper WE (ed) Conf Recycling Treated Municipal Wastewater through Forest and Cropland. p 375. EPA 660/2-74-003
67. Carriker N, Brezonik PL (1978) J Environ Qual 7:516
68. King LD, Morris HD (1972) J Environ Qual 1:425
69. Cuadra Moreno J (1981) Agricultural land irrigation with wastewater in the Mezquital Valley. In: D'Itri FM, Aquirre Martinez J, Athie Lambarrie M (eds) Municipal Wastewater in Agriculture. Academic, New York, p 217
70. White DP et al. (1975) Changes in vegetation and surface soil properties following irrigation of woodlands with municipal wastewater. NTIS, PB-244, 798
71. Ball RC (1977) Pollut Eng Technol 3:205
72. Olson JV, Fuog RM (1981) Year-round land application in cold climate combines reuse, reclamation and disposal. In: Proc Water Reuse Symp II, 2, p 1007, Am Water Works Assoc Res Found, Denver, CO
73. Koerner EL, Hans DE (1979) Long term effects of land application of domestic wastewater. Roswell, New Mexico, slow rate irrigation site. EPA 600-2-79-047
74. Kardos LT (1974) Renovation of municipal wastewater through land disposal by spray irrigation. In Conf Recycling Treated Municipal Wastewater through Forest and Cropland. EPA 660/2-74-003
75. Brockway DA (1982) An overview of the current status on the selection and management of vegetation for slow rate systems to treat municipal wastewater in the North Central United States. In: D'Intri FM (ed) Land treatment of municipal wastewater: vegetation selection and management. Ann Arbor Science, p 5
76. Giordana PM et al. (1975) J Environ Qual 4:394
77. Mendoza H (1981) An evaluation of the use of Mexico City wastewater on the irrigation of crops. In Proc. Water Reuse Symp II, 2, p 952. Am Water Works Assoc Res Found, Denver, CO
78. Burton JM (1982) Oldfield management studies of the water quality management facility at Michigan State University. In: D'Intri FM (ed) Land Treatment of Municipal Wastewater: vegetation selection and management, p 107
79. Mendoza Gamez G, Flores Herrera F (1981) Mexico City's master plan for reuse. In Proc Water Reuse Symp II, 1, Am Water Works Assoc Res Found, Denver CO, p 308
80. Cebula J (1980) Environ Protect Eng 6:145
81. Hossner LR et al. (1978) Sewage disposal on agricultural soils: chemical, microbiological implications, vol 1—Chemical implications PB-285-857
82. Indelicato S (1981) A case of raw wastewater irrigation in Sicily. In Proc. Water Reuse Symp. II, vol 1, Water Reuse in the Future, Aug 23–28, 1981. Washington DC. Am Water Works Ass Res Found, Denver CO
83. Feigin A (1979) Prog Water Technol 11:151

84. Odendaal PE, Van Vuuren LR (1974) Reuse of wastewater in South Africa—research and application. In: Proc of Water Reuse Symp II. Am Water Works Assoc Res Found, Denver, CO, p 896
85. FAO (1976) Soils Bull 31:268
86. EPA (1973) Water Quality Criteria for Irrigation Water
87. NAS and NAE (1973) Water Quality Criteria 1972, EPA-R3-73-033, U.S.A., p 592
88. Ayers RS, Westcot DW (1976) Water quality for irrigation. Irrigation and Drainage Paper 29. Food and Agriculture Organisation of the United Nations, Rome, Italy
89. MAFF (1981) Water quality for crop irrigation: Guidelines on chemical criteria. Leaflet 776. Ministry of Agriculture, Fisheries and Food, U.K.
90. Gupta UC (1986) Soil Sci Soc Am Proc 32:45
91. Koval'skii VV et al. (1965) Agrokhimiya, 153–169 (Chem Abs 64:10148)
92. Green GH, Weeth HJ (1977) J Anim Sci 46:812
93. Weeth HJ et al. (1981) Am J Vet Res 42:474
94. Graffman G (1974) Chem Zeit 98:499
95. Ruhrverband (1982) Ruhrawassergute im Wasserwirtschaftsjahr 1982
96. Manfredi F et al. (1975) Igiene Moderna 68:400
97. Brinkman FJ, Dekker K (1972) H_2O 5:525
98. Rijkswaterstaat (1981) Kwaliteitsonderzoek in de rijkswateren
99. Ahl T, Jonsson E (1972) Ambio 1:66
100. HMSO (1980) Twentieth and Final Report on the Standing Technical Commitee on Synthetic Detergents. Her Majesty's Stationery Office, London
101. Cassarett and Doull's Toxicology (1980) 2nd edn, MacMillan, New York, p 440
102. Directive 79/831/EEC, 18 September 1979
103. Weir RJ, Fisher RS (1972) Toxicology and Applied Pharmacology 23:351
104. Jansen JA et al. (1984) Arch Toxicol 55:64
105. Kliegel W (1980) Bor in Biologie, Medizin und Pharmazie, Springer, Berlin Heidelberg New York
106. US Department of Health and Human Services (1987) Toxicology and Carcinogenesis Studies of Boric Acid in $B6C3F_1$, Mice. NTP Technical Report No 324 (Publication No 88-2580). National Toxicology Program PO Box 12233 Research Triangle Park NC27709, U.S.A.
107. Verbitskaya GV (1975) Gig Sanit 17:49 (Chem Abs 83:158653)
108. Boreina AI et al. (1972) Gig Sanit 5:19 (Biol Abs 55:40660)
109. Koval'skii VV et al. (1962) Dokl Akad Nau SSSR 146:967 (Chem Abs 58:2690)
110. Ploquin J (1967) Bul de la Societe Scientific d'Hygiene Alimentaire 55:70
111. Hunt CD (1987) USDA Agricultural Research: Nov/Dec: 13
112. Browning E (1961) Toxicology of industrial metals. Butterworths, London, p 76

TAED—Tetraacetylethylenediamine

P. A. Gilbert

Unilever Research, Quarry Road East, Bebington, Wirral, Merseyside, L63 3JW

Introduction 319
Analysis for DAED and TAED 321
Environmental Fate of DAED 321
    Physical Chemical Properties 321
    Biodegradability 322
    Behaviour During Sewage Treatment 322
Levels in the Environment 324
Possible Effects on the Environment 325
    Toxicity to Aquatic Organisms 325
    Effects on Sewage Treatment 328
Mammalian Toxicology 328
References 328

Summary

TAED is incorporated into fabric washing powders as a bleach activator. In the wash it undergoes almost quantitative perhydrolysis to diacetylethylenediamine (DAED) and to peracetate, which provides efficient bleaching and hygiene benefits at low wash temperatures. DAED, TAED and Triacetylethylenediamine (TriAED) are readily and completely biodegradable and are substantially removed during sewage treatment. In consequence, levels in the aquatic environment will be very low. It is estimated, for example, that concentrations of DAED in rivers immediately below a treated sewage effluent outfall should be in the range 3–6 $\mu g\,L^{-1}$, further downstream these levels will fall rapidly. Their toxicity to aquatic organisms and to mammals is also very low, providing large safety margins. The use of TAED in detergents will not therefore result in any adverse effects on the environment or on man.

Introduction

TAED is a white crystalline amide (mp 149–154 °C) having the structural formula:

$$\begin{array}{c} CH_3CO \\\\ \diagdown \\\\ NCH_2CH_2N \\\\ \diagup \\\\ CH_3CO \end{array} \begin{array}{c} COCH_3 \\\\ \diagup \\\\ \\\\ \diagdown \\\\ COCH_3 \end{array}$$

Molecular weight 228 Empirical formula $C_{10}H_{16}N_2O_4$
CAS Registry Number 10543-57-4
EINECS Number 2341238
Name (-IUPAC nomenclature) N,N'-1,2-ethanediylbis(N-acetyl-)acetamide
Other Chemical synonyms are:

NN' ethylenebis-diacetamide
N,N,N',N' Tetraacetyl-1,2-diaminoethane
N,N,N',N'-tetraacetylethylendiamin
N,N,N',N'-tetraacetylethylenediamine

TAED is produced industrially by the acylation of ethylene diamine with acetic anhydride and total production in Western Europe was estimated in 1989 [1] to be approximately 30 000 tonnes per year. Its main use, which will account for most of this production, is as a bleach activator in fabric washing powders. A much smaller quantity is used in a similar role in wash performance boosters.

Sources of hydrogen peroxide, such as sodium perborate or percarbonate, have long been used in fabric washing powders to provide a gentle oxygen bleaching action and to assist in the removal of oxidisable stains. Such bleaches are, however, only really effective at wash temperatures above 60 °C. In the modern world, with the need to conserve energy and the use of man-made fibres which cannot be washed at high temperatures, the boil wash has almost disappeared and average wash temperatures have declined steadily. A means of boosting the bleach performance at lower temperatures is therefore required. TAED meets this need. In the wash liquor, at alkaline pH and in the presence of a source of peroxide such as sodium perborate, TAED undergoes rapid perhydrolysis (Fig. 1) to DAED-diacetylethylenediamine and the peracetate anion. Peracetate is an efficient oxidising agent at lower temperatures (30–40 °C) and the incorporation of TAED therefore allows the delivery of effective bleaching of stains under modern low temperature wash conditions. Since peracids also possess antimicrobial properties the use of TAED can also deliver significant hygiene benefits. A number of studies [2, 3] have confirmed that products containing TAED can reduce microbial contamination at wash temperatures of 30 °C to levels which could otherwise only be achieved in a boil-wash.

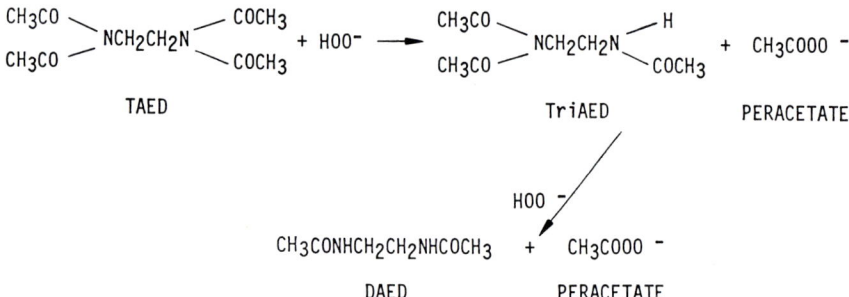

Fig. 1. Perhydrolysis of TAED

The conversion of TAED to DAED during the wash is almost complete. In consequence it is the latter which will predominantly be present in the spent wash liquor which will be discharged to the sewer and whose environmental fate and effects should particularly be addressed. The peracetate formed will be largely reduced to acetate during the wash and this process will rapidly be completed in the sewer when any remaining peracid comes into contact with raw sewage. Since acetate is a natural product of low toxicity and ready biodegradability its environmental impact is of no concern and will not be discussed further.

Analysis for DAED and TAED

A range of methods, both wet chemical and instrumental, is available for the determination of TAED in the raw material and formulated products. The preferred technique [2] is gas chromatography using cyanopropyl-phase packed columns and dual flame-ionisation detectors. Using this method the levels of DAED and any TriAED—triacetylethylenediamine, can also be determined.

No sufficiently sensitive and reliable method is yet available for the determination of DAED or TAED in environmental samples. Here the challenge is to quantitatively concentrate and separate very low concentrations of a highly water soluble material from a complex and uncharacterised matrix before presentation to the gas chromatograph.

To date, reverse-phase solid phase extraction, and cation exchange of protonated DAED have been investigated and have shown some promise but difficulties remain. Development of the required analytical technique is, however, continuing and it is hoped that a sensitive and reliable method will be available soon.

Environmental fate of DAED

In order to predict the fate of DAED and any traces of TAED or TriAED in the environment and to estimate the concentrations likely to arise at key points in the environment it is necessary to take account of a number of factors.

Physical Chemical Properties

Both DAED and TAED are highly polar, water soluble compounds of low volatility. Their high affinity for the water phase is confirmed by their very low octanol: water partition coefficients. Low values of 4.7×10^{-2} for DAED and 1.9×10^{-1} for TAED have been reported [2].

In water, under environmentally relevant conditions (pH 7, 15 °C) DAED is resistant to hydrolysis. Under these conditions TAED undergoes hydrolysis to TriAED and then to DAED. These processes are, however, relatively slow with a half-lives of approximately 40 and 80 days respectively [2].

On the basis of this information it can be assumed that DAED and any TAED or TriAED will remain in the aqueous phase with minimal adsorption onto sewage solids or sediments and will not bioconcentrate in the biota. Hydrolysis is unlikely in practice to have a significant influence on their environmental concentrations in surface waters.

Biodegradability

Very high levels of mineralisation have been reported for both DAED and TAED and TriAED in simple tests of inherent and ready ultimate biodegradability (Table 1). River die-away tests with DAED and TAED labelled with ^{14}C in the ethylene moiety also showed [2] that in both cases well over 80% of the label was evolved as $^{14}CO_2$ within 3 weeks.

These studies indicate that DAED, TAED and TriAED can be classed as readily and ultimately biodegradable. In consequence they should rapidly and completely biodegrade under the various aerobic conditions present in the environment, leaving no residues.

Behaviour During Sewage Treatment

The key processes operated in a modern sewage treatment plant are indicated in Fig. 2. From a consideration of the physical chemical properties discussed above it is clear that DAED, TAED and TriAED will not be associated with the sewage solids and should remain in the aqueous phase. This was confirmed [2] by distribution studies in raw sewage. They will therefore pass almost quantitatively with the settled sewage to the activated sludge plant and it is the extent to which

Table 1. Biodegradability test results for DAED and TAED

	Inherent biodegradability tests % Ultimate Biodegradation	Test	Ref.	Ready biodegradability tests % Ultimate Biodegradation	Test	Ref.
DAED	95	Zahn-Wellens (DOC)	[2]	100	STURM (CO_2)	[4]
				91	STURM (CO_2)	[2]
				76–96	Closed bottle (O_2)	[4]
TAED	100	SCAS (DOC)	[2]	52–64	Closed bottle (O_2)	[4]
				76–89	MOST (DOC)	[4]
				79	STURM (CO_2)	[2]
TriAED	99	Zahn-Wellens (DOC)	[2]			

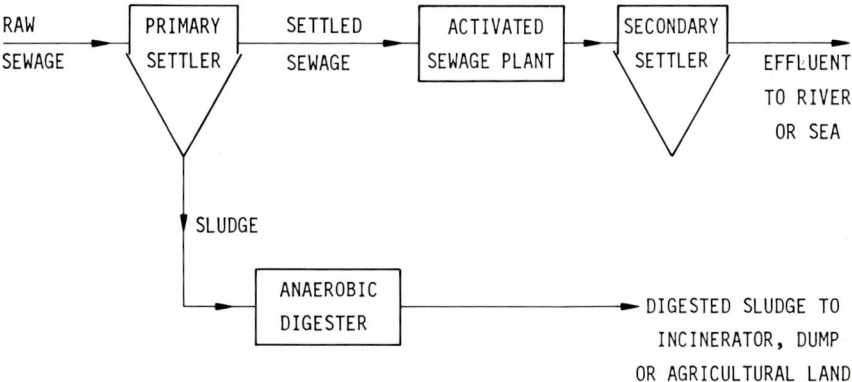

Fig. 2. Conventional Sewage Treatment Process

they undergo biodegradation here that will determine their concentrations in treated effluents and receiving waters.

Since a sufficiently sensitive analytical technique is not yet available for the determination of DAED or TAED in sewage liquors a detailed study of their treatability was made using radiocarbon labelled material [2]. Settled domestic sewage containing 3 mg L^{-1} DAED or TAED was continuously dosed into a laboratory scale model activated sludge plant [5] adapted to allow collection of all the carbon dioxide evolved.

Separate experiments were made using DAED and TAED labelled with ^{14}C in the acetate or the ethylene moieties. The quantities of ^{14}C remaining in the treated effluent, evolved as $^{14}CO_2$ and associated with the activated sludge were then determined over two separate 24 h periods. The results of these studies showed (Table 2) that between 29 and 59% of the ^{14}C label was evolved as $^{14}CO_2$ and a further 33 to 70% was associated with the activated sludge. There were no consistent differences between DAED and TAED or due to the position of the radiolabel. Under these conditions the bacterial biomass is growing rapidly and it is to be expected that a large proportion of the carbon from a readily biodegradable chemical will be used for growth and incorporated into natural bacterial cell material. In similar experiments with [U-^{14}C] labelled glucose, for example, approximately 40% of the label was evolved as $^{14}CO_2$ with more than 50% being incorporated into the activated sludge. Very high levels of removal (92–99%) of radiolabel from solution were also observed. Even so it is likely that not all of the low levels of ^{14}C present in the effluent were due to the presence of DAED or TAED. A significant proportion of this ^{14}C was probably present as natural cell metabolites produced by bacteria which had used the test compounds as a source of carbon for synthesis of new cell material. Again, in the case of the similar experiment on [U-^{14}C] glucose, between 5 and 10% of the ^{14}C was present in the effluent. These results are therefore consistent with the almost complete ultimate biodegradation of DAED and TAED during treatment of sewage by the activated sludge process. Confirmation of this conclusion was obtained from similar studies [2] in which DAED and TAED were continuously

Table 2. Fate of ^{14}C-labelled DAED and TAED in activated sludge study

	Test compound	Distribution (%) of ^{14}C found in			Overall removal of ^{14}C during treatment (%)
		Effluent	CO_2	Activated sludge	
EXPERIMENT 1	DAED [Ethylene-^{14}C]	2	44	54	98
	DAED [Acetyl-^{14}C]	2	52	45	98
	TAED [Ethylene-^{14}C]	4	30	65	96
	TAED [Acetyl-^{14}C]	1	29	70	99
EXPERIMENT 2	DAED [Ethylene-^{14}C]	2	46	51	98
	DAED [Acetyl-^{14}C]	8	59	33	92
	TAED [Ethylene-^{14}C]	5	42	53	95
	TAED [Acetyl-^{14}C]	4	51	45	96

dosed at a concentration of 20 mg L^{-1}, into model activated sludge plants operating on settled domestic sewage. The dissolved organic carbon (DOC) concentrations present in dosed and control plants were then monitored over a period of several weeks. After a period of acclimatisation no significant difference was observed between the DOC levels present in the control plant and those dosed with DAED or TAED.

Levels in the Environment

When TAED is used in fabric washing powders the level of incorporation typically ranges from 1 to 3% and a reasonable estimate for the average level of TAED in such products is 2%. Of course, not all fabric washing powders contain a bleach activator and in one report [1] it was estimated that approximately half of the Western European fabric washing powders contained TAED. Consumption of TAED in France, Germany, Italy and the U.K. in 1987 was estimated [2] to range from 4000 to 6000 tonnes per annum for each country. Per capita consumption in other Western European countries is expected in most cases to be similar and not to exceed this range. Using these estimates for TAED and estimates for the population and per capita water use it is possible to calculate the levels of DAED in raw sewage which are likely to result from its use in detergent products. Assuming, conservatively, that 95% biodegradation of DAED occurs during secondary sewage treatment and a dilution factor of 1:10 for sewage effluents entering receiving waters it is also possible to calculate the levels of DAED in treated effluents and rivers (Table 3). It can be seen that levels of DAED in raw sewage are expected to be below 1 mg L^{-1} in general, rising to 1.15 mg L^{-1} in the U.K. Levels in treated effluents are likely to be in the range 30 to 60 µg L^{-1} and in receiving waters, immediately below a sewage outfall, 3 to 6 µg L^{-1}. Further downstream continuing biodegradation and further dilution should rapidly and significantly reduce these concentrations. In situations where

Table 3. Estimated levels of DAED in the environment (1987)

Country	France	Germany	Italy	U.K.
Population ($\times 10^6$)	55	61	57	58
Water consumption (L/inhabitant/day)	220	225	225	200
Consumption of TAED (tonnes/year)	4000	4000	5000	6000
Level of DAED (mg L^{-1}) in:				
Raw sewage	0.74	0.65	0.87	1.15
Treated sewage effluent (assuming 95% removal)	0.037	0.032	0.044	0.058
Rivers receiving treated effluents (1:10 dilution)	0.004	0.003	0.004	0.006

sewage is discharged directly without secondary treatment, the concentrations of DAED in receiving waters immediately below the point of discharge may of course be an order of magnitude higher than these estimates. DAED will, however, biodegrade at least as rapidly and completely as the other components of the sewage and as self-purification proceeds as the river flows downstream, DAED concentrations will fall rapidly.

Levels in waster-waters of any TAED and its hydrolysis intermediate TriAED are likely to be one or two orders of magnitude lower than those of DAED and in consequence levels in receiving waters will be well below the ug L^{-1} level. Levels of DAED, TAED or TriAED is soils amended with sewage sludge are also expected to be insignificant.

Possible Effects on the Environment

In order to determine the possible significance of the levels of DAED, TAED and TriAED expected to be present in the aquatic environment, information is required on the concentrations at which they may cause adverse effects.

Toxicity to Aquatic Organisms

Short term, acute toxicity tests have been carried out on DAED, TAED and TriAED using a range of aquatic organisms representing the three trophic levels in the aquatic environment. The results, expressed in terms of EC$_{50}$ values, are summarised in Table 4. The EC$_{50}$ is the concentration of the test compound producing an effect, usually death or immobilisation, in 50% of the test organisms exposed.

Table 4. Acute toxicity of DAED, TAED and TriAED to aquatic organisms

Species	Test duration	EC_{50} (mg L^{-1}) for DAED	TAED	TriAED	Comments	Ref.
Algae						
Chlorella vulgaris	14 day	>500	>500	>500	No effects on growth rate at highest concentrations tested	[2, 4]
Invertebrates						
Daphnia magna	48 h	>800	>800	>400–800	Insufficient mortalities at highest concentration of DAED and TAED to allow calculation of EC_{50}	[2, 4]
Gammarus pulex	72 h	>800	>800	>800	No deaths at highest concentration tested	[2, 4]
Fish						
Carrasius auratus	96 h	40 000–75 000	>2500	>21200	No deaths at highest concentration of TAED and TriAED tested	[4, 6]
Idus leusiscus	48 h	—	>200	—	No mortalities at highest test concentration	[2]
Brachydanio rerio	96 h	—	>1500	—	No mortalities at highest test concentration	[2]

These studies clearly demonstrate the remarkably low toxicity of DAED, TAED and TriAED to aquatic organisms since in the majority of tests no effects were observed even at the highest concentrations tested. The EC_{50} of DAED for the goldfish (*Carrasius auratus*) was found to lie between 40 and 75 *grammes* per

Table 5. Summary of mammalian toxicity data on TAED and DAED

Studies	TAED	DAED
Acute oral toxicity		
LD_{50} (Mice)	5.9 g/kg body weight	Not tested
LD_{50} (Rat)	10 g/kg body weight	Not tested
Classification	"Practically non-toxic"	Expected to be less toxic than TAED (See sub-acute oral toxicity)
Skin irritant		
Intradermal injection (Rabbit)	Virtually non-irritant	Not tested
3 h occlusive patch test (Rabbit)	Slightly irritant	Not tested
Eye irritation		
Dry powder (Rabbit)	Marginally irritant	Not tested
Sensitisation		
Magnusson-Kligman tests (Guinea pig)	Non-sensitiser	Non-sensitiser
Sub-acute oral toxicity		
13 week feeding study (Rat)	No effect level 25 mg/kg body weight/day	No effect level 5.7 g/kg body weight/day
Mutagenicity		
Ames test (*Salmonella typhimurium* strains TA98, TA100, TA1537, TA1535, TA1538) with and without metabolic activation (rat liver enzymes, S9 mix)	Non-mutagenic	Non-mutagenic
Sister Chromatid Exchange Test (Chinese Hamster V79 cells) with and without metabolic activation (rat liver enzymes, S9 mix)	Non-mutagenic	Not tested
Metabolism		
Radiotracer study following oral intubation (Rats)	Rapidly absorbed from intestinal tract, metabolised by hydroxylation and deacetylation to N,N' Diacetyl N glycolyl ethylenediamine, TriAED, N-acetyl N' glycolylethylene diamine and DAED which are excreted via the urine	Rapidly absorbed from intestinal tract, and rapidly excreted unchanged via the urine.

litre. The concentrations at which no effects were observed in these studies are at least 5 orders of magnitude higher than the concentrations of these substances likely to be present in the aquatic environment. It may therefore be concluded that adverse effects will not occur and that further aquatic toxicity studies are unnecessary.

Effects on Sewage Treatment

Given the low toxicity of DAED, TAED and TriAED no adverse effects would be expected on the microorganisms present in sewage treatment plants. This is confirmed by the observation that laboratory scale activated sludge plants continued to mineralise and nitrify efficiently when dosed with DAED or TAED at 20 mg L^{-1} [2]. It has also been shown that levels of DAED as high at 8% of dry digester solids had no effect on anaerobic gas production [2].

Mammalian Toxicology

Data [6] on the toxicology of TAED and DAED are briefly summarised in Table 5. These data demonstrate the relatively low toxicity of TAED and DAED which are also shown to be non-mutagenic, non-sensitising and to have a low irritancy potential. The estimated environmental concentrations reported above suggest that exposure of birds and mammals to DAED in the environment will be extremely low. Similarly, levels in drinking water derived from rivers receiving sewage effluents will be negligible. In consequence the safety factors will be many orders of magnitude.

References

1. Stanford Research Institute—Chemical Economics Handbook 1989
2. Unilever Research Laboratory, Port Sunlight, U.K. (unpublished results)
3. Graf E, Rasche P (1979) EMPA, St Gallens, Switzerland (unpublished results)
4. Schoberl P, Huber L (1988) Tenside Surfactants Detergents 25:99
5. Painter HA, King EF (1978), WRC, Porous pot method for assessing biodegradability. Technical Report TR70: Water Research Centre, U.K.
6. Unilever Environmental Safety Laboratory, Sharnbrook, U.K. (unpublished results)

Carboxymethylcellulose (CMC)

J. G. Batelaan, C. G. van Ginkel, and F. Balk

Akzo International Research, Department of Organic, and Polymeric Chemistry, P.O. Box 9300, NL-6800 SB Arnhem

Introduction 329
   Water Soluble Polymers. 329
   Carboxymethylcellulose 330
   Laundry 330
Biodegradation of CMC 331
   Enzymatic Hydrolysis of CMC. 331
   Aerobic Biodegradation 333
   Anaerobic Biodegradation. 334
   CMC Removal in Biological Waste Water Treatment Systems 335
Ecotoxicity of CMC. 335
References 335

Summary

Carboxymethylcellulose (CMC), a water soluble polymer, is produced by etherification of cellulose. One sixth of the world production (350 000 t in 1988) is used in laundry formulations. CMC from laundry wash ends up in waste water treatment systems and/or the environment.

   CMC is biodegraded in nature and biological waste water treatment systems by aerobic and anaerobic microorganisms. After an initial hydrolysis by extracellular enzymes the resulting products are metabolized by cellulotic and non-cellulotic microorganisms. The biodegradation rate of CMC varies inversely proportional with the degree of substitution. The few ecotoxicity data available clearly show that CMC is a harmless substance.

Introduction

Water Soluble Polymers

Water-soluble polymers can be distinguished into three main types: Natural (e.g. Starch), synthetic (e.g. polyacrylic acid) and semisynthetic like carboxymethylcellulose (CMC). The cellulose ethers are volumewise the most important class of semisynthetic polymers, with CMC as the most important representative. Thanks to their easy availability, versatility, non-toxicity and price, these materials are used in numerous applications, in food as well as in more technical uses such as oil-drilling. In water-based end-use formulations many functions have to be performed: control of rheology (viscosity), metal ion sequestration, suspension stabilization or flocculation, etc. In most instances these water-soluble polymers or gums have proved to be useful, even when used in quite small amounts.

Carboxymethylcellulose

Cellulose, the most abundant organic material in nature, is completely insoluble in water. For increasing its solubility the carboxymethyl group is well suited: easily introduced into the cellulose molecule, it eliminates the hydrogen bonds that render the parent molecule insoluble. Because of its bulk and charge it hinders close alignment of CMC molecules. Cellulose with more than about one carboxymethyl group per two glucose units is soluble in neutral water. The degree of substitution (DS), that is the number of carboxymethyl groups per unit, and the average degree of polymerization (DP) are the main parameters to characterize a CMC. The uniformity of distribution is expected to be a key factor determining performance, but cannot be measured easily. CMC production requires pure cellulose, available from cotton linters up to DP 6000 or hard and soft wood after chemical pulping.

The cellulose (shredded sheet or powder) is treated with concentrated sodium hydroxide in water or an aqueous alcohol. During this alkalization step the crystalline cellulose swells and becomes accessible to the etherification agent. Chloroacetic acid as such or as sodium salt is then added. Depending on the amount of alcohol present, mostly ethanol or isopropanol, one distinguishes "dry" and "slurry" processes. Heating to about 60 °C starts the exothermic etherification reaction. As nucleophiles not only the cellulose anion is present, yielding CMC after attack on chloroacetate anion, but also hydroxide and alkoxide, yielding glycolate and glycolate ether. The crude reaction mixture obtained after drying (60% CMC-Na salt) is suitable for laundry use.

For some applications such as food, the reaction mixture is washed free of low molecular salts by aqueous alcohol. After removal of the alcohol by evaporation the effluent, free of significant levels of residual chloroacetic acid, is fed to the sewage treatment plant, without significant problems [1, 2, 3].

Laundry

Laundry formulations are complex mixtures of many functional agents. CMC is included in amounts of 0.5–1% with the aim of inhibiting the redeposition of soil. Especially with cotton, CMC was found to be effective but the modern fabric

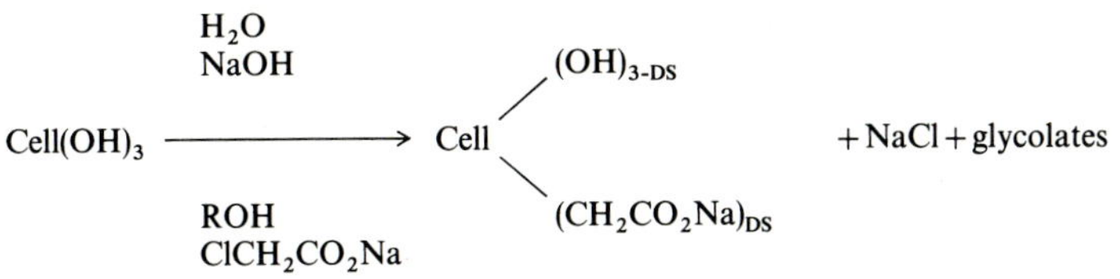

$$\text{Cell(OH)}_3 \xrightarrow[\substack{\text{ROH} \\ \text{ClCH}_2\text{CO}_2\text{Na}}]{\substack{\text{H}_2\text{O} \\ \text{NaOH}}} \text{Cell} \begin{matrix} (\text{OH})_{3\text{-DS}} \\ \\ (\text{CH}_2\text{CO}_2\text{Na})_{\text{DS}} \end{matrix} + \text{NaCl} + \text{glycolates}$$

DS: Degree of substitution

Fig. 1. Scheme of the reaction of cellulose with chloroacetic acid yielding CMC

blends (polyester/cotton) also gain from the addition of CMC to the laundry powder. Stabilization of colloidal suspensions, in this case of soil, is a well known property of CMC. Molecules of CMC absorbed onto the soil particles might keep the soil particles from agglomeration and deposition onto the fabric. An other explanation of the beneficial action of CMC comprises the adsorption of CMC onto the cotton fibers. So, the net surface charge of the fabric became more negative. The soil particles, negatively charged at the prevailing alkalinity of laundry liquid (pH 10), are repelled from the vicinity of the fabric.

Detergent grade CMC has a degree of substitution of 0.5 to 0.75. The estimated world production amounts to 350 000 t (1988) of which about one sixth finds its way into detergents [4, 5].

Biodegradation of CMC

In nature, the biological degradation of cellulose involves the enzymatic hydrolysis of the glycosidic linkages by extracellular enzymes. The resulting oligomers and monomers in turn are metabolized by microorganisms. The biodegradation of CMC is also initiated by an enzymatic attack followed by the mineralization of saccharides by either aerobic or anaerobic microorganisms. In addition, the removal of CMC by these extracellular enzymes and microorganisms in waste water treatment systems will be discussed.

Enzymatic Hydrolysis of CMC

The group of hydrolytic enzymes which degrade cellulose and cellulose derivatives are known as cellulases. These cellulases are mixtures of extracellular enzymes produced by a large variety of fungal and bacterial species growing on cellulosic substrates like cellulose powder, cotton Avicel and CMC. The cellulases of these microorganisms are characterized by three main enzymes: the

1,4-β-D glucan glucano hydrolase or endoglacanase (EC 3.2.1.2), the 1,4-β-D-glucan cellobiohydrolase or exoglucanase (EC 32.1.9.1) and the β-D-glucoside glucohydrolase or β-glucosidase (EC 32.1.2.1). The mode of action of these enzymes on cellulose and cellulose derivatives has not yet been completely clarified. Endoglucanases are known to effectively hydrolyze substituted celluloses but not crystalline cellulose so the role of endoglucanase in the degradation of cellulose remains to be established. A number of reviews dealing with the production and properties of cellulase and their action of cellulose hydrolysis has been published [6, 7, 8].

Endoglucanases from sources like *Trichoderma, Aspergillus, Penicillium, Streptomyces, Cellulomonas, Clostridium* [8, 9] are capable by hydrolysing CMC. The enzymatic degradation of CMC by cellulase is determined by the degree of substitution [10, 11, 12, 13, 14] while the degree of polymerization plays a minor role. Figure 2 shows the hydrolysis of several CMC samples of varying substitution by the increase in reducing glucose equivalents.

Susceptibility of substituted cellulose derivatives to enzymatic hydrolysis increases as they become more water soluble and less crystalline upto the complete solubility point (DS 0.5). After this point the polymer stability improves with increasing DS, while at high substitution levels CMC appears to be stable to the enzymatic attack. The rate of CMC hydrolysis levels off at increased degrees of conversion and a limit is approached [11, 13]. This limit and the decreased activity with highly substituted CMC is caused by the inability of the endo-glucanase to split substituted glucose units. Only linkages between non-

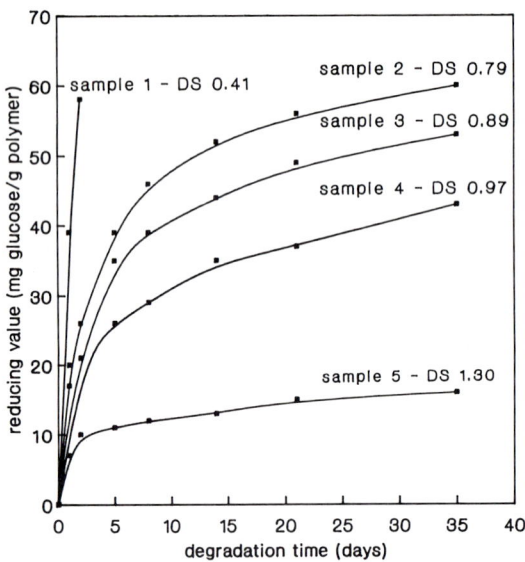

Fig. 2. Progress of enzymatic hydrolysis of CMC samples of varying substitution as indicated by reducing equivalents measurements [11]

substituted glucose units and one of the bond linking an unsubstituted to two neighboring substituted ones can be hydrolyzed by the action of cellulases [12]. The enzymatic hydrolysis products of CMC are a mixture of glucose, disaccharides, trisaccharides and substituted saccharides [12]. Using high resolution exclusion chromatography (SEC), Hamacher and Sahm [14] analyzed hydrolysis products of CMC. Glucose and substituted lower oligomers were detected. The proportion of low oligomers essentially decreases on increasing the degree of substitution as large fragments survive [14].

No enzymes have been found which are capable of splitting the carboxymethyl groups from the cellulose derivatives. Nevertheless some microorganisms biodegrade carboxymethylated cellobiose [15]. It is not known why the action of these enzymes responsible for the biodegradation cellobiose derivatives is only limited to short chains.

Aerobic Biodegradation

Biochemical oxygen demand (BOD) tests are widely used to measure the biodegradation by aerobic mixed cultures. These tests indicate a very slow biodegradation rate of CMC (Fig. 3) and confirmed the resistance to biodegradation of CMC with increasing substitution.

CMC (DS 0.4–0.5) is partly oxidized in a BOD test and the oxidation is accompanied by a little cellular synthesis whereas CMC (DS 1.2–1.4) is recalcitrant [16, 17, 18, 19].

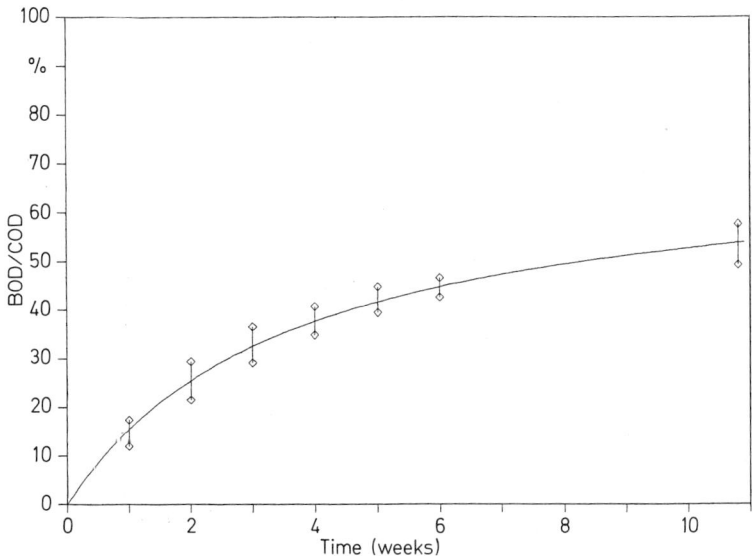

Fig. 3. BOD curve for CMC (DS = 0.6) [19]

The duration of these BOD tests is often limited to four weeks. However, after an adaptation period of 16 weeks also highly substituted CMC is biodegraded and no stable intermediates are produced [16, 18].

Freeman et al. isolated a Gram negative bacterium which is capable of liquefying CMC [20]. Woodward isolated two *Aspergillus* spp [10] and Siu et al. [21] demonstrated that CMC is metabolized by *Myrothecium verrucaria*. These isolates show that the capacity to utilize CMC by aerobic microorganisms is widely distributed.

Anaerobic Biodegradation

Anaerobic digestion of substituted cellulose and cellulose is a complex multistep process. The anaerobic biodegradation is initiated by the hydrolysis of the polymers catalyzed by cellulases. The resulting saccharides are metabolized fermentatively by cellulolytic and non-cellulolytic bacteria. Finally, fermentation products such as acetic acid, formic acid, hydrogen gas and carbon dioxide are utilized by methanogenic bacteria to produce methane.

Anaerobic biodegradation of CMC by an anaerobic cellulose degrader, a carbohydrate fermenter and a methanogenic bacteria is shown in Fig. 4 [22, 23]. Methane formation from CMC by a mixed culture is demonstrated also by Ferrara et al. [24].

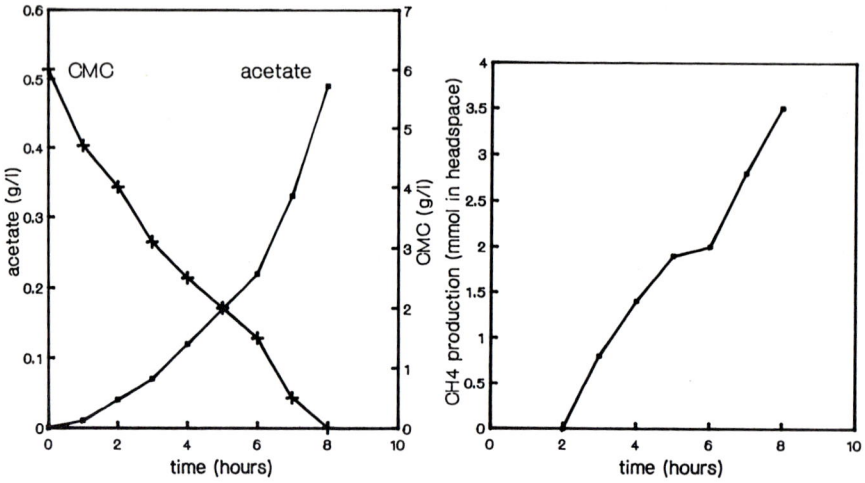

Fig. 4. CMC degradation and acetate and methane production by a coculture [23]

CMC Removal in Biological Waste Water Treatment Systems

Activated sludge systems are widely used as biological waste water treatment systems. Aerobic CMC utilizing microorganisms which are capable of maintaining themselves in an activated sludge systems have been described [18]. In these laboratory scale activated sludge systems, the removal by adsorption on activated sludge is not substantial. CMC is biodegraded in an activated sludge system while owing to the acclimatization of the microorganisms the CMC removal rate increased. The operation of activated sludge plants of paper mills and textile plants apparently hardly causes any problems. Therefore, problems in a sewage treatment plant with laundry wash water are not expected because the CMC concentration in the wash water will be in the order of $\mu g/dm^3$.

Although no data are available of CMC removal in bioreactors with anaerobic sludge, the formation of methane by anaerobic sludge shows the possiblities of anaerobic treatment systems to remove CMC.

Ecotoxicity of CMC

The few ecotoxicity data available for CMC have been obtained in studies on the environmental effect of drilling fluids for oil exploration. The CMC tested is not characterized. In a 96 h acute test with rainbow trout (*Salmo gardneri*), no LC_{50} could be determined (> 10 g/dm^3) [25]. Toxicity data to invertebrates were not found. In relation to the disposal of used drilling fluids on land, the effects on plants were studied. The sodium salt of CMC did not influence the growth of green beans (*Phaeseoulus vulgaris*) and sweet corn (*Zea mays*) for eight weeks. CMC was tested at concentrations of 1.1 and 3.4 g/kg soil (silt loam) [26]. Predicted environmental concentrations of CMC originating from washing powders are far lower than the concentration of CMC studied in relation to drilling fluids, and will probably never exceed a few $\mu g/dm^3$. Therefore, toxic effects of CMC are not expected. Food grade CMC is approved for human consumption, implying that the compound is considered to be non-toxic. Therefore, CMC is used as an additive in many commercial food preparations, including fish food.

References

1. Brandt L (1986) In: Ulmann's Encyclopedia of Industrial Chemistry, vol A5, VCH Verlag, Weinheim, p 477

2. Greminger GK (1979) In: Kirk-Ottmer Encyclopedia of Chemical Technology, vol 5, John Wiley, New York, p 145
3. Stelzer GI, Klug ED (1979) Carboxymethylcellulose In: Davidson RL (ed) Handbook of water soluble gums and resins 4.1–4.24 McGraw Hill, NY
4. Worlds production and consumption figures, Chemical marketing reporters 25:54 (1988)
5. Cutler WG (1972) In: Davis RC (ed) Detergency, theory and test methods. Marcel Dekker, New York, p 287
6. Marsden WL, Gray PP (1986) Crit Rev Biotechnol 3:235
7. Lee YH, Fan LT (1980) Adv Biochem Eng 17:101
8. Finch P, Roberts JC (1985) Enzymatic degradation of cellulose In: Nevell TP, Zeronian SH (eds) Cellulose chemistry and its application, Ellis Horwood, Chichester, p 372
9. Thomas KG, Zeikus JG (1981) Appl Environ Microbiol 231:42
10. Reese ET, Siu RGH, Levinson HS (1950) J Bact 59:485
11. Wirick MG (1968) J Polymer Sci 6:1965
12. Bhattacharjee SS, Perlin AS (1971) J Polymer Sci 36:509
13. Kasulka U, Dautzenberg H, Polter E, Philipp B (1983) Cellulose Chem Technol 17:423
14. Hamacher H, Sahm K (1985) Carbohydrate Polymers 5:319
15. Reese ET (1957) Industrial Eng Chem 49:89
16. Rosall B (1974) Int Biodetn Bull 10:95
17. Wirick MG (1974) Journal WPCF 46:512
18. Bryan CE, Harrison PS (1973) Text Asia 5:56
19. Blok J, Hnatiuk-De Rooij E: Akzo Research Report D86/27/0389B
20. Freeman GC, Baillie AJ, MacInnes CA (1948) Chem and Ind 279
21. Siu RGH, Darby RT, Burkhobler PR, Baarghoorn ES (1949) Tex Research 19:484
22. Marty DG (1985) Biotechnol Lett 7:895
23. Marty DG (1985) Biotechnol Lett 7:990
24. Ferrara R, Barbaris R, Jodice R, Vicenzino E (1984) Agric Wastes 11:79
25. Spraque JB Wogan WJ (1979) Environ Pollut 19:269
26. Miller RW, Honavar S, Hunsaker B (1980) J Environ Qual 9:547

Polymeric Materials Polycarboxylates

Hans-Joachim Opgenorth
BASF AG, D-6700 Ludwigshafen

Introduction 338
Physical and Chemical Properties of the Polycarboxylates 339
Environmental Behavior of the Polycarboxylates 341
    Sewage Treatment 342
    Surface Water 344
    Soil 345
    Drinking Water Preparation 346
    Concentrations in the Environment 346
Toxicology of the Polycarboxylates 348
References 349

Summary

Polycarboxylates used in detergents are watersoluble polymers which have a particularly high density of anionic carboxylate groups along a polymer chain. In low-phosphate and phosphate-free detergents they are used for avoiding incrustation and soil redeposition. They are not complexing agents but dispersants, which are adsorbed on calcium carbonate and calcium phosphate and thus inhibit crystals in the washwater and therefore prevent deposition in the fabric.

    Although polycarboxylates are not readily biodegraded in sewage treatment plants, surface water and the soil, they generally have extremely low toxicity.

    In particular, higher molecular weight polycarboxylates are readily immobilized by adsorption on solids or precipitation as sparingly soluble calcium polymer salts. It is therefore certain that the copolymer P(AA-MA) 70 000 is virtually completely eliminated from sewage in clarification tanks and treatment plants. Hence, very little enters the surface water via the treated sewage. The polycarboxylates have no effect on the operation of treatment plants, in particular the removal of heavy metals from sewage.

    In agricultural use, the polycarboxylates adsorbed on sewage sludge enter the soil with the sewage sludge. However, they are fixed in the uppermost soil layer and therefore cannot be washed into the groundwater by rain.

    Where polycarboxylates do enter the untreated water used for drinking water preparation, they are reliably removed by treatment processes such as ground water infiltration, filtration over sand and active carbon and flocculation with iron and aluminum salts, so that polycarboxylates have not been detected analytically in drinking water to date.

    Polycarboxylates are toxic to warm blooded animals only in very high doses. They have virtually no irritant effect on skin and mucous membranes and no sensitizing properties. There has been no indication of possible irreversible genetic damage.

Introduction

In connection with detergents, the term "polycarboxylates" is widely used for water-soluble linear polymers, i.e. salts of polycarboxylic acids, whose common basic building block is acrylic acid. They are therefore sometimes also referred to generally as "polyacrylates". However, "polyacrylates" includes the esters of polyacrylic acid. Since this has frequently led to confusion, the term "polycarboxylates" is preferred in this article.

By copolymerization, variation of the degree of polymerization and crosslinking of the polymer chains, the properties of the polycarboxylates can be adapted to a large number of applications. Thus, polycarboxylates have been used for many years, for example, for dispersing, flocculation and thickening. Low molecular weight polycarboxylates are used, among other things, as dispersants in detergents and cleaning agents, for pigment pastes and pharmaceutical preparations and in water treatment, while high molecular weight polycarboxylates serve as thickeners, for example for cosmetics, and color pastes, for flooding in oil production and as flocculants in ore, coal and water treatment. Other important applications are in the paper and textile industries [1, 2].

It is only since the beginning of the 1980s that polycarboxylates have been used in detergents and cleaning agents. In low-phosphate and phosphate-free detergents, they are used as assistants especially for avoiding incrustation and soil redeposition. Their effect is based not on complexing with the hardness producers in the water but on the dispersion of calcium carbonate or of calcium phosphate and the soil detached during washing. Between 2 and 5% are used in phosphate-free detergents based on zeolite/sodium carbonate [3].

The polycarboxylates used in detergents and cleaning agents are virtually exclusively homopolymers of acrylic acid (AA) and copolymers of acrylic acid (AA) and maleic acid (MA). For the sake of simplicity, the abbreviations P(AA) and P(AA-MA) will be used below for the polymers (Fig. 1). In the past, the use of polyhydroxyacrylic acid and of copolymers of acrylic acid with allyl alcohol, vinyl ether and other olefins has also been considered. These materials are not used in detergents due to insufficient performance. The mean molecular weights Mw of these polymers are in general between 1000 and 100 000 and are thus very much lower than those of flocculants and thickeners, whose mean molecular weights are more than 1 million. For brief characterization of the polycarboxylates, the mean molecular weight Mw—where it is known and considered relevant—is added as a number to the abbreviation, e.g. P(AA-MA) 70 000.

Polycarboxylates are generally produced industrially by free radical polymerization of the monomer acrylic acid or the monomers acrylic acid and maleic anhydride. By adding sodium hyroxide solution, the sodium salts of the polycarboxylates are formed [1]. However, they can also be prepared by hydrolysis of the polyacrylates and polyacrylonitriles. But products of this type usually contain unhydrolyzed ester and nitrile groups as impurities [2].

```
Homopolymer (P(AA))    ---[-CH----CH₂-]----
                          |
                          COONa        m

Copolymer (P(AA-MA))   ---[-CH----CH₂-]--------[-CH-----CH---]---
                          |                    |       |
                          COONa        m       COONa   COONa  n
```

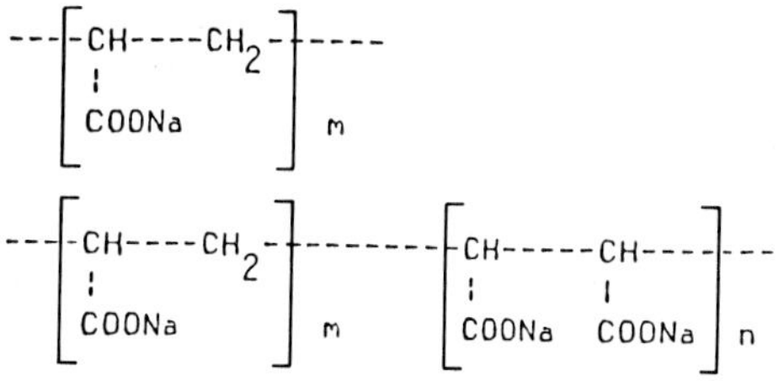

Fig. 1. Polycarboxylates based on acrylic acid (AA) or acrylic acid/maleic acid (AA-MA) and used in detergents and cleaning agents

Because of the wide range of applictions and the considerable variability of the products, information about total amounts of polycarboxylates consumed is very uncertain. Furthermore, such information rapidly becomes obsolete owing to the continuing trend toward phosphate-free detergents. For the Federal Republic of Germany, the amount of polycarboxylates consumed in detergents in 1987 was estimated at about 15000 metric tonnes (based on active substance) [4]. By the beginning of 1990, annual consumption had increased to about 19000 tonnes, with a market share for phosphate-free detergents of over 95% [5]. Complete replacement of the phosphates in detergents would thus result in consumption of about 20000 metric tonnes per year, or 0.32 kg per inhabitant per year. In most other Western European countries, too, the market share of phosphate-free detergents formulated using polycarboxylates is increasing, whereas the use of polycarboxylates is not yet widespread overseas.

Polycarboxylates can be advantageously used in cleaning agents too [6]. However, since consumption in detergents is substantially higher, this application is only of minor importance with respect to the total consumption of polycarboxylates.

Physical and Chemical Properties of the Polycarboxylates

Polycarboxylates are polyelectrolytes, i.e. polymers which have ion-forming, functional groups along a polymer chain [7]. In particular, the polycarboxylates used in detergents and cleaning agents have a particularly high density of anionic carboxylate groups. This strutural characteristic substantially determines the physical and chemical properties of this class of substances.

Because of the electrolyte character of the polymer molecules, the alkali metal salts of the polycarboxylates are very readily soluble in water but insoluble in organic solvents, such as alcohols or hydrocarbons [7]. At neutral and alkaline pH, polycarboxylates form sparingly soluble coagulates with excess calcium ions, these coagulates forming all the more rapidly the greater the calcium excess [8].

The behavior of the polycarboxylates toward calcium ions is very important both for the performance characteristics and for the behavior in the environment. Thus, the copolymer P(AA-MA) 70 000 is, with the exception of low molecular weight constituents, completely precipitated as a sparingly soluble calcium compound in water which contains 1.43 mmol of calcium, at initial polymer concentrations between 100 and 0.05 mg/l. At a polycarboxylate concentration of 1000 mg/l, however, this type of precipitation does not occur since calcium is no longer present in excess [9]. Since the undissolved calcium polycarboxylate has no dispersant effect, a builder must be present in the wash process to reduce the water hardness to such an extent that precipitation of the polycarboxylate is no longer possible. At the same time, however, this means that the conditions are again reversed when the washwater is diluted in the sewage, and the polycarboxylate is precipitated with calcium, loses its activity and is eliminated with other solids from the sewage. The tendency to form insoluble calcium polycarboxylates increases with the charge density and the molecular weight of the polymers. Low molecular weight polycarboxylates with a molecular weight of 1000 form virtually no precipitates with calcium ions.

Typical for all polyelectrolytes is their pronounced tendency to adsorb on solid surfaces. Electrostatic effects are primarily responsible for this. Thus, it is not surprising that the anionic polycarboxylates are readily adsorbed on substances with a positive zeta potential, such as calcium carbonate [10] and calcium sulfate [11]. This is associated with a reversal of the surface charge, which stabilizes the disperse state. The dispersant action of polycarboxylates is in this case independent of the molecular weight. A different situation is encountered in the case of adsorbents with a negative surface charge. Provided that the electrostatic repulsion, as is encountered with dirt particles in the washwater, is not too great, adsorption occurs as a result of molecular forces [12]. However, water hardness may influence this process. Whereas in the case of low molecular weight polycarboxylates the dispersant action is retained even in the presence of calcium ions, polycarboxylates having a higher molecular weight may induce the opposite effect, flocculation, under these conditions.

The incrustation-inhibiting action of the polycarboxylates, is based on adsorption on calcium carbonate which is still microcrystalline, and in phosphate-containing detergents also on calcium phosphate. This inhibits crystal growth. Since this process is observed even when substoichiometric amounts of polycarboxylates are used, the term threshold effect is used [13]. It should be taken into account here that the magnitude of the effect is dependent on various parameters of the test methods used. Even the determination of the "calcium binding", by the Hampshire test [14], gives as a result only the dispersing behavior under the special test conditions in the case of polycarboxylates. It does not indicate the amount of calcium which is capable of binding by polycarboxylates but the amount of calcium carbonate which is dispersed. The calcium carbonate dispersing capacity determined by Richter and Winkler [15] is more comparable with the results of practical tests.

Polycarboxylates bind calcium and divalent heavy metal ions poorly in comparison with typical complexing agents, such as NTA, EDTA, or phosphonates [16]. As described above, they are, for example, not able to prevent the formation of calcium carbonate at pH 10 but only to slow down crystal growth. Thermodynamic binding constants cannot be derived, owing to the nonuniform molecular structure. The literature values for conditional binding constants also vary as a function of the method of calculation, by several powers of ten [15].

The literature describes different analytical methods for the various purposes [17], although these methods are not generally suitable for environmentally relevant problems. For determining polycarboxylic acids in cooling waters and wastewaters of power stations, a method based on the formation of a sparingly soluble compound with a quarternary ammonium salt has been proposed. Quantitative determination is carried out photometrically by turbidity measurement [18]. Another method, which was developed especially for analysis of polycarboxylates with average molecular weights of between 800 and 10000, uses a colorimetric detection method after removal of water-soluble impurities by adsorption/desorption [19]. The two last-mentioned methods are presumably mainly suitable for concentration ranges above 1 mg/l. For the quantitative determination of polycarboxylates in concentrations below 1 mg/l, polyelectrolyte titration is suitable [20]. This method utilizes the high charge density of the polycarboxylates by titration with a polycation in the presence of an anionic, metachromatic indicator dye, so that the polyanion/polycation reaction takes place preferentially. By optimizing the process conditions, it was possible to reduce the detection limit to 0.01 mg/l [21]. Problems may arise in the presence of natural anionic polyelectrolytes, but some of these problems can be eliminated by oxidative pretreatment of the samples.

Environmental Behavior of the Polycarboxylates

Since polycarboxylates have been used to a significant extent in detergents only for the last few years, research into the ecological and toxicological properties does not extend back very far in time. Some publications contain brief overviews of environmentally relevant data [4, 22]. Other publications of recent years have come to the conclusion that detergent polycarboxylates have not been sufficiently investigated with regard to their environmental compatibility, since the scientific literature on this subject currently provides little information [23]. Such conclusions fail to take account of the fact that manufacturers and users have carried out extensive investigations into the environmental compatibility of the polycarboxylates, only some of which have so far been published [9]. In the very recently completed report by the main committee "Phosphates and water" of the Technical Group on Water Chemistry in the Gesellschaft Deutscher

Chemiker [5], it was also possible to include results of investigations about to be published [24, 25, 26].

The following discussion of the environmental behavior of the polycarboxylates follows the route of entry and the possible propagation in the environment. Since polycarboxylates, as detergent ingredients, inevitably enter the sewage and, through the sewage, the environment, the retention and the behavior in sewage treatment, in surface water, in the soil and in drinking water preparation will be considered in detail successively.

Sewage Treatment

In the highly industrialized, densely populated countries of the world, domestic wastewater, which inevitably also contains the dirt-bearing washwater, is preferably treated centrally in biological treatment plants. Where this is not the case, solids and, where possible, suspended substances are removed by settling in clarification tanks or mechanical treatment plants before the sewage is discharged into surface water or the subsoil. Polycarboxylates too can be deposited from sewage in this way, provided that they are precipitated as a result of water hardness. In the case of the copolymer P(AA-MA) 70 000, it was found that sedimentation of the sparingly soluble calcium polymer salt in the untreated sewage is substantially complete after a few days (>80% in 4 days) [26]. In clarification tanks, good elimination is therefore expected for this polycarboxylate. Whether the residence time in mechanical treatment plants is sufficient or not depends on the extent to which particles capable of settling out have already formed beforehand in the sewer system.

Conditions similar to those in mechanical treatment plants are also encountered in primary sedimentation for biological sewage treatment. In a model plant for simulating primary sedimentation, 70–84% of the suspended substances and 8–29% of the polycarboxylates P(AA) 4500, P(AA-MA) 12 000 and P(AA-MA) 70 000 were eliminated during a hydraulic residence time of 2 h in a settling tank [25]. Here too, the settling properties play a decisive role, in addition to the solubility.

In biological sewage treatment, elimination of the polycarboxylates takes place largely through adsorption on the organic biomass. Only a small amount undergoes biodegradation.

In respirometric tests, such as the determination of $BOD_5/ThOD$ [27] or BOD_{30}/COD [13, 24], the modified MITI test (I) [28] and the "closed bottle test" [29], various P(AA) showed only low degrees of degradation, generally <20%, which is presumably due to low molecular weight constituents or impurities. Even test methods which use bacteria concentrations higher than those employed in the respirometric tests lead to incomplete biodegradation in the case of P(AA) and P(AA-MA) (Table 1). However, degrees of degradation of up to 45% are found with low molecular weight materials (P(AA) 1000), so that it

Table 1. Biodegradation and elimination of polycarboxylates in tests with activated sludge [25]

Polycarboxylate	Biodegradation (CO_2 formation)* [%]	SCAS test	Elimination [%] Confirmatory test without $FeCl_3$	with $FeCl_3$
P(AA) 1000	45	45	n.d.	n.d.
P(AA) 2000	20	21	n.d.	n.d.
P(AA) 4500	9**	40	27	98
P(AA) 10 000	16	58	n.d.	n.d.
P(AA) 60 000	n.d.	93	n.d.	n.d.
P(AA-MA) 12 000	31**	83	70	96
P(AA-MA) 70 000	20**	95	82	n.d.

n.d. = not determined
* CO_2 formation after contact with activated sludge for 30–90 days
** Mean value of several measurements

may be assumed that, in the conventional test systems, only short-chain polycarboxylates with a molecular weight of < 1000 undergo substantial biodegradation.

Elimination of the polycaboxylates under treatment plant conditions has been investigated for many examples in various test systems. Thus, in static tests similar to the Zahn-Wellens test, more than 90% of P(AA-MA) 70 000 are bound to the activated sludge after contact for only two hours, if the water hardness is sufficient. The unadsorbed remainder readily undergoes biodegradation [24]. P(AA)s shows a range of behavior. There are products which undergo no elimination at all in this test, and others which are virtually completely eliminated with activated sludge [28, 30], since the tendency to undergo elimination increases continuously with increasing chain length of the polycarboxylates. Table 1 confirms this dependence with corresponding results in the SCAS test. The comparatively high value for P(AA) 1000 is due not to adsorption on the sludge but to biodegradation. In continuous treatment plant simulation tests (carried out similarly to the OECD confirmatory test), a similar picture is found. Here too, higher molecular weight polycarboxylates are eliminated more readily than low molecular weight ones. A complete balance of the retention in biological wastewater treatment was determined for P(AA-MA) 70 000 with the aid of a ^{14}C-labeled product [31], the polycarboxylate being added firstly as a shock load then continuously over several days (Table 2). Only 2–3% activity was found in the discharge, so that P(AA-MA) 70 000 may be assumed to have a total elimination of 97–98% in treatment plants. Similarly high degrees of elimination are also found for low molecular weight polycarboxylates if iron salts are added to the wastewater, as is usual for phosphate precipitation (Table 1).

Various investigations have shown that polycarboxylates in concentrations relevant in practice do not interfere with the operation of treatment plants. Thus, the activity of the bacteria in activated sludges is not adversely affected by the presence of polycarboxylates in the sewage [26]. Furthermore, laboratory tests

Table 2. Retention of polycarboxylate P(AA-MA) 70 000 in biological sewage treatment [31]

Retention	Addition as shock load	Continuous addition
Adsorption	about 96%	>90%
CO_2 formation	<2%	about 5%
Discharge	<3%	2%

with various P(AA) and P(AA-MA) have shown that there is no effect on the settling behavior of the sludges during primary and final sedimentation, the mobility of heavy metals, simultaneous phosphate precipitation and the dewatering of excess sludges, at least up to concentrations of 10 mg of polycarboxylate per liter of wastewater [25]. These laboratory results were confirmed with P(AA-MA) 70 000 in two pilot treatment plants in a field trial lasting several months [9, 24].

As under aerobic conditions, polycarboxylates undergo only slight biodegradation under anaerobic conditions too. Investigations with P(AA-MA) 70 000 have shown that polycarboxylate adsorbed on activated sludge remains undissolved in the sludge in the digestion tower, and the activity of the digestion tower is not adversely affected by the presence of the polycaboxylate [24].

Surface Water

Although the higher molecular weight polycarboxylates are preferentially precipitated in treatment plants and seepage pits, a certain proportion of individual polycarboxylates can enter the surface water. Further biodegradation is possible in the surface water, but laboratory tests show that the effect is generally small [25].

Remobilization of heavy metals from the sediments of rivers and lakes by polycarboxylates can be ruled out, since no significant effects have been found in shaking tests with various polycarboxylates in concentrations up to 10 mg/l [25]. This concentration of 10 mg/l is well above the concentrations expected in surface water, even under unfavourable circumstances (see the section on concentrations in the environment).

Information about the aquatic toxicity of polycarboxylates which can be used in detergents has been rarely published to date [9, 13]. Table 3 gives an overview of selected test results obtained with the polycarboxylates P(AA) 4500 and P(AA-MA) 70 000. Other investigations using other polycarboxylates have led to comparable results. The same applies to many toxicity tests with bacteria [9, 25]. In general, no effects on the test organisms were found even at the highest concentrations tested. The fact that the maximum concentrations tested differ so greatly in some cases is due to the fact that maximum concentrations are

Table 3. Test results for aquatic toxicity of polycarboxylates [5, 9, 25, 32]

Test system/test organism		P(AA) 4500	P(AA-MA) 70 000
Acute toxicity			
Daphnia magna	EC_{50} (48 h):	>200*	>200*
Fish: bluegill	LC_{50} (96 h):	>1000*	n.d.
zebra fish	LC_{50} (96 h):	>200*	n.d.
golden ide	LC_{50} (96 h):	n.d.	>200*
Subchronic and chronic toxicity			
Algae: *Scenedesmus subspicatus*	EC_{10} (96 h):	180	32–>200*
Daphnia magna	NOEC (21 d):	1000*	1.3–350
Fish: zebra fish	NOEC (14 d):	n.d.	40*
eggs and larvae of zebra fish	NOEC (28 d):	1000*	n.d.
and young zebra fish	NOEC (42 d):	n.d.	40*

n.d. = not determined
* Maximum concentration tested

recommended or prescribed in many test specifications. The wide spans in different investigations into the chronic toxicity of P(AA-MA) 70 000 with respect to algae and to daphnia are striking. The fact that P(AA-MA) 70 000 is precipitated as an insoluble calcium polymer salt in the presence of excess water hardness in the test system i.e. at low polycarboxylate concentrations, plays a role here. The environmental relevance of these results is limited by the fact that the polycarboxylate in reality is preferentially adsorbed on solids and is therefore not present in the tested form in surface water [5].

Soil

Since polycarboxylates in treatment plants are eliminated to a great extent by adsorption on the sludge and remain in the sewage sludge during sludge digestion, whether or not polycarboxylates enter the environment by this route is dependent on the disposal of the sewage sludges. The most frequently used disposal methods are: storage in landfills, incineration and spreading over land utilized for agriculture. Particularly in agricultural utilization of the sewage sludges there is a likelihood of the polycarboxylates entering the environment, i.e. the soil.

As in treatment plants and surface water, polycarboxylates also undergo only slight biodegradation by soil organisms [24, 25]. The high affinity of polycarboxylates for solids, for example soil materials, is at least partly responsible for the fact that they are much more capable of withstanding attack by microorganisms in the soil compared with dissolved substances. However, the high tendency to adsorption results in the polycarboxylates being substantially immobilized in the soil.

The adsorption properties of some polycarboxylates were determined by lysimeter tests for soils of different origins. For example, for P(AA) 4500, P(AA-MA) 12 000 and P(AA-MA) 70 000, it was found that 84–93% of the products used were retained in the first 15 mm of the 100 mm thick soil layer [25]. The mobile remainder, which was detected in the eluate of the lysimeter, is biodegradable and therefore does not enter the soil with the sewage sludge. A corresponding investigation with sewage sludge which contained adsorbed P(AA-MA) 70 000, no mobile constituent of the polycarboxylate was in fact observed. It remained completely in the layer in which it was applied [24]. It is therefore unlikely that polycarboxylates which are applied together with sewage sludge to areas used for agricultural purposes will be washed by rain or irrigation into lower strata or the groundwater.

In addition to the aforementioned tests on aquatic toxicity, some tests were also carried out with higher plants and soil organisms. In growth tests with corn, soybeans, wheat and grass, P(AA) 4500 had no adverse effect on germination, growth and yield, up to the maximum test concentration of 225 mg/kg of soil [25]. Furthermore in a similar test with P(AA-MA) 70 000, no effect on the growth and the appearance of oats was observed up to 400 mg/kg of soil [24]. The toxicity test with chironomids (P(AA) 4500) and earthworms (P(AA-MA) 70 000) also gave no indication of adverse effects on organisms in or on the soil, in the tested concentration range up to 10 000 or 1600 mg/kg of soil respectively [24, 25].

Drinking Water Preparation

Owing to the high tendency of polycarboxylates to be adsorbed on solids, it is likely that drinking water preparation methods such as bank filtration, groundwater infiltration and filtration over sand and active carbon are also capable of removing small residual amounts from the untreated water. Furthermore, flocculation with aluminum and iron salts likewise leads to elimination of more than 90% in the case of P(AA) 4500, P(AA-MA) 12 000 and P(AA-MA) 70 000 [25].

Concentrations in the Environment

The report by the main committee "Phosphates and water" [5] contains an estimate of the environmental concentrations of polycarboxylates from detergents. Table 4 gives an overview. The conditions in the Federal Republic of Germany at the beginning of 1990, with a specific consumption of 0.31 kg of the polycarboxylate P(AA-MA) 70 000 per inhabitant per year, were taken as a basis.

Table 4. Estimate of the environmental concentrations of polycarboxylates from detergents for the Federal Republic of Germany [5]

	Concentrations [mg/l]	[g/kg]
Untreated sewage	2.9	
digested sewage sludge		14.2
soil after application of sewage sludge*		0.013–0.027
treated sewage	0.26	
highly contaminated surface water (dilution 1:5)	0.05	
drinking water	0.005	

* After application of the maximum amount legally permitted in the Federal Republic of Germany, over a period of three years, on soils used for agricultural purposes

The mean concentrations in untreated sewage are calculated as 2.9 mg/l, assuming a production of 300 l per inhabitant per day in the Federal Republic of Germany. When smaller amounts of sewage are produced, the concentrations are correspondingly higher. Huber [22] calculated mean sewage concentration of about 3.5 mg/l under similar conditions.

Taking into account the technical facilities for sewage treatment in the Federal Republic of Germany, a mean degree of elimination of 91% was assumed in sewage treatment. This leads to a reduction of the polycarboxylate concentration in the treated sewage to 0.26 mg/l. Since P(AA-MA) 70 000 is eliminated with the activated sludge, except for a small biodegradable amount, the digested sewage sludge contains about 1.4% of polycarboxylate, based on the dry weight of the sewage sludge.

The polycarboxylate enters the environment in particular when sewage sludge is applied as a fertilizer to agricultural areas. Because of the heavy metals which accumulate in the sewage sludge, in many countries application is limited in terms of time and quantity. If the allowed heavy metal concentrations are not exceeded, the legally admissible maximum amount of sewage sludge in the Federal Republic of Germany may be applied to the same soil for only three years, after which sludge must not be applied to the soil for a period of several years. The calculation shows that the uppermost soil layer can reach a polycarboxylate concentration of from 0.013 to 0.027 g per kg of soil in three years, depending on the nature of the soil.

Because of the high degree of elimination in sewage treatment, the concentrations of polycarboxylate P(AA-MA) 70 000 in the surface water are correspondingly low. At a dilution of 1:5, the result obtained is 0.05 mg/l However, most rivers carry much more water so that the concentrations are in general very much lower. Under these circumstances it is necessary, in a more precise analysis, also to take into account possible introduction into surface water through soil erosion.

Because of their chemical structure, polycarboxylates are extremely hydrophilic substances, so that they are unlikely to accumulate in aquatic organisms. Furthermore, biological concentration is very improbable because it is assumed that only compounds with a molecular weight of less than 600 will be absorbed through biological membranes [33].

Estimation of possible polycarboxylate concentrations in drinking water is based on an untreated water with a concentration of 0.05 mg/l. Since high degrees of elimination are to be assumed in various drinking water preparation methods, such as bank filtration, groundwater infiltration, filtration over sand and active carbon and in flocculation with iron and aluminum salts, a reduction to 0.005 mg/l may generally be regarded as a very cautious estimate. In the drinking water samples tested to date, no synthetic polycarboxylates were detectable with the very sensitive method of polyelectrolyte titration [5, 25].

Toxicology of the Polycarboxylates

The salts of linear polyacrylic acids do not show particularly striking toxicological behavior [34]. The very extensive test results for higher molecular weight, slightly crosslinked polyacrylates which are used as thickeners in cosmetics also give no indiction of specific toxic properties [35]. An IARC working group gave no information on the toxicology of polyacrylic acid [36]. Table 5 summarizes results for the toxicology of two defined polycarboxylates, which are suitable for the production of detergents and cleaning agents, in warm blooded animals.

The acute oral toxicity of polycarboxylates in the rat is generally very low [37]. Even the repeated administration of vey high doses (1000 mg/kg/d) of P(AA-MA) 70000 is tolerated without symptoms [9].

Polycarboxylates do not irritate the skin [37]. Furthermore, dermal application for four weeks results in little or no reddening [9, 34, 37]. Systemic effects were not observed afterwards; this is not unexpected since polycarboxylates can scarcely penetrate the skin in sufficient amounts, owing to their high molecular weight. In humans too, P(AA) was found to cause no irritation in the patch test after 48 hours [34].

In the tests on the tolerance by mucous membranes, the polycarboxylates are generally found to be nonirritant [37]. However, dilute sodium polyacrylate solutions may cause slight reddening and lachrymation for a short time in the rabbit, these symptoms subsequently disappearing. 0.3% of P(AA) in water is reported as the maximum concentration for which there is no irritant effect in the eye, while 2–3% of P(AA) are designated as only slightly irritant and 4.5% of P(AA) are designated as slightly irritant after being washed out immediately with water [34].

Sensitization reactions with polycarboxylates have not been observed to date. In the Magnusson and Kligman maximization test, none of the animals

Table 5. Test results for toxicity of polycarboxylates to warm blooded animals [5, 9, 25]

Test system/test organism	Test substance/test result P(AA) 4500	P(AA-MA) 70 000
Toxicity (rat)		
acute (oral) LC_{50} [mg/kg]:	>5000*	>5000*
subchronic		
(90 d)** NOEL [mg/kg/d]:	n.d.	1000*
Skin tolerance (rabbit)		
acute	n. irrit.	n. irrit.
subacute (25% in water, 28 d)	n.d.	m. irrit.***
Tolerance by mucous membrane (rabbit)	n. irrit.	n. irrit.
Sensitization		
maximixation test (guinea pig)	n.d.	n. sens.
humans	n.d.	n. sens.
Teratogenicity/embryotoxicity	n.d.	n. terat.
Mutagenicity		
Ames test	neg.	neg.
mouse lymphoma test	neg.	neg.
UDS test (unscheduled DNA synthesis)	neg.	neg.
chromosome analysis in bone marrow	n.d.	neg.
point mutation	n.d.	neg.

n. = not; m. = mildly; n.d. = not determined
* Maximum concentration tested
** Administration in drinking water (max. 6400 mg/l)
*** No systemic symptoms

showed skin changes with P(AA-MA) 70 000 [9, 37]. Patch tests with various sodium polyacrylates also gave no reaction in humans [25, 34].

Teratogenicity and embryotoxicity tests with P(AA-MA) 70 000 were negative [25]. The many in vitro and in vivo mutagenicity tests also give no indication of any irreversible effects by polycarboxylates [9, 25, 34, 38]. In this connection, it is remarkable that many polyanions significantly slow down the growth of tumors when they are injected directly into the blood stream [39].

References

1. a. Jung D, Penzel E, Wenzel F (1980) In: Ullmanns Encyklopädie der technischen Chemie, 4th edn, Verlag Chemie, Weinheim, vol 19, p 1
   b. NN, Chemische Industrie, XXXVI/May 1984, 266–268
2. Teot AS (1982) In: Kirk-Othmer—Encyclopedia of Chemical Technology, 3rd edn, John Wiley, Chichester, vol 20, p 207
3. Krings P, Vogt GH (1988) Chimia 42:245
4. a. Berth P, Krings P (1989) In: Kompendium—Auswirkungen der Phosphathöchstmengenverordnung für Waschmittel auf Kläranlagen und Gewässer, Academia Verlag, Sankt Augustin, p 69
   b. Rohe D (1988) Chemische Industrie, Heft 4, p 28
5. Stellungnahme des Hauptausschusses "Phosphate und Wasser" der Fachgruppe Wasserchemie in der Gesellschaft Deutscher Chemiker zur Beurteilung der Umweltverträglichkeit von

Polycarboxylaten aus Waschmitteln, Schreiben der Fachgruppe Wasserchemie vom 02. 02. 1990 an das Bundesministerium für Umwelt, Naturschutz und Reaktorsicherheit
6. Perner J, Neumann H-W (1987) Tenside Surfactants Detergents 24:334
7. Armstrong RW, Strauss UP (1969) In: Encyclopedia of Polymer Science and Technology, John Wiley, Chichester, vol 10, p 781
8. Jekel M, Sontheimer H (1978) Vom Wasser 51:81
9. Opgenorth H-J (1987) Tenside Surfactants Detergents 24:366
10. BASF AG, unpublished results
11. Nestler CH (1968) J Colloid Interface Sci 26:10
12. Zini P (1987) Seifen-Öle-Fette-Wachse 113:45, 187
13. Jakobi G (1984) Angew Makromol Chem 123/124:119
14. Haschke H, Morlock G, Kuzel P (1972) Chemikerzeitung 96:199
15. Richter F, Winkler WE (1987) Tenside Surfactants Detergents 24:213
16. BASF AG, unpublished results
17. a. Wimberley JW, Jordan DE (1971) Anal Chim Acta 56:308
    b. Crisp S, Lewis BG, Wilson AD (1975) J Dent Res 54:1238
    c. Schaffer JF, Woodhams RT (1979) Tenside Detergents 16:240
18. Angenend FJ, Schulte-Wieschen U (1979) VGB Kraftwerkstechnik 59:995
19. Hach Company, Technical Information: Polyacrylic acid (PAA) Test, 1984
20. a. Horn D (1978) Progr Colloid and Polymer Sci 65:251
    b. Schempp W, Tran HT (1981) Wochenblatt für Papierfabrikation 19:726
    c. Horn D, Heuck C-C (1983) J Biolog Chem 258:1655
21. a. Schroeder U, Horn D, Wassmer KH (1991) Seifen-Öle-Title-Wachse 117:311
    b. Wassmer KH, Schroeder U, Horn D (1991) Makromol Chem 192:553
22. a. Huber L (1987) Seifen-Öle-Fette-Wachse 113:393
    b. Schöberl P, Huber L (1988) Tenside Detergents 25:99
23. a. Giger W, Conrad T (1985) Wasser Berlin '85, Wissenschaftsverlag Berlin, p 362
    b. Hunter M, da Motta Marques DML, Lester JN, Perry R (1988) Environ Technol Lett 9:1
    c. Chiaudani G, Premazzi G, Vismara R, Butelli P, Poltronieri P, Ingegneria Ambientale Inquinamento e Depurazione No. 8, Oct. 1988
24. Opgenorth H-J (1990) Münchener Beitr Abwasser-, Fisch-, Flußbiol, vol 43, p 338
25. Procter and Gamble, unpublished results
26. BASF AG, unpublished results
27. Abe V, Matsumura S, Yajima H, Suzuki R, Masago Y, (1984) Yakugaku 33:228
28. a. Schefer W, Romanin K (1988) Textilveredlung 23:340
    b. Schefer W, Romanin K (1989) Gas-Wasser-Abwasserf 69:131
29. a. Fischer WK, Gerike P, Schmid R (1974) Wasser-Abwasserf 7:99
    b. Fischer WK (1975) Tenside Detergents 12:53
30. a. Schefer W (1982) Textilveredlung, 17:541
    b. Schefer W (1983) Seifen-Öle-Fette-Wachse 109:423
31. Schuman H, Institut für Wasser-, Boden- und Lufthygiene, Berlin, unpublished results
32. Bundesgesundheitsamt Berlin, Institut für Wasser-, Boden- und Lufthygiene, unpublished results
33. Zitko V (1981) Uptake and excretion of chemicals by aquatic fauna. In: Stokes PM (ed) Ecotoxicology and the aquatic environment, Pergamon, New York, p 67
34. NN, Toxicity Profile, Polyacrylic acid and its sodium salt, BIBRA (British Industrial Biological Research Association), 1987, and references cited therein
35. NN, Cosmetic Ingredient Review 8, Final report on the safety assessment of Carbomers-934, -910, -934P, -940, -941, and -962, Journal of the American College of Toxicology, 1, 1982, 109–141
36. NN, Acrylic acid, methyl acrylate, ethyl acrylate and polyacrylic acid, In: IARC Monographs on the evaluation of the carcinogenic risk of chemicals to humans, vol 19, February 1979
37. BASF AG, unpublished results
38. Ishidate M Jr, Sofuni T, Yoshikawa K, Hayashi M, Nohmi T, Sawada M, Matsuoka A (1984) Fd Chem Toxic 22:623
39. Hodnett EM, Amirmoazzami J, Tai JTH (1978) J Med Chem 21:652, and references cited therein

Fluorescent Whitening Agents

J. B. Kramer

Chemiewinkel, University of Amsterdam, P.O. Box 20242, 1000 HE Amsterdam, The Netherlands

Introduction	351
Chemical and Physical Properties	352
Production Methods	352
Amounts and Uses	355
Analytical Methods	356
Sewage Treatment	357
Discharge to the Environment	358
Environmental Fate	359
Adsorption on Sludge	359
Degradation	360
Bioaccumulation	361
Toxicology	362
Ecotoxicology	362
Mammalian Toxicology	363
References	364

Summary

Fluorescent whitening agents (FWAs) are used in laundry detergents to make the fabrics whiter and colors bright to compensate the yellowish shade of washed fabrics. The amount in washing powders is about 0.15% on dry weight basis, of which 20–95% binds to the fabrics. FWAs are readily eliminated from household effluents in sewage treatment plants due to strong adsorption on activated sludge. They are only sporadically detected in natural water systems. The concentrations found are well below the concentrations toxic to aquatic organisms. Uptake in fish and plants is low and reversible. Photodegradation occurs and biodegradation is possible but only in activated sludge after an adaptation period.

Introduction

Most natural organic fabrics have an absorption band in the short UV-region extending partly into the blue visible region, causing a yellowish shade. How much is absorbed in the visible region depends mainly on the number and extension of conjugated systems in the polymer molecules. Especially impurities, dirt and degradation products include such extended conjugated systems. Most of these compounds can be removed by chemical bleaching.

Another method of eliminating this yellowish shade is by the use of blue dyestuffs. A more modern method is to compensate the yellow surplus by increasing the blue light emitted by means of fluorescent whitening agents (FWAs, also called optical brighteners). FWAs have the property of absorbing UV-light and emitting blue light; this phenomenon is known as fluorescence. Thus not only is the yellowish shade compensated for but also more visible light is emitted than absorbed, which gives the colors more brightness.

FWAs in their present form have been used in laundry detergents since the 1940s [1]. The amount ranges from 0.03%–0.3% (dry weight basis), the mean value is 0.15% [2]. Between 20–95% is bound to the fabrics during the washing process, depending on the washing conditions and types of fabric. The remainder is discharged with the washing water [3].

Most research on the human and environmental safety of FWAs was done in the early 1970s. The literature on this subject has been reviewed by Burg [4] and Coulston [5].

Chemical and Physical Properties

The structures and names of the seven most commonly used FWAs are given in Fig. 1 [6]. They are anionic derivatives of diamino stilbene or distyryl biphenyl; only FWA 6 is a triazolyl stilbene derivative. Due to their anionic nature they are more or less water soluble.

The FWAs have an UV absorption maximum between 340–360 nm in water and transform UV energy almost quantitatively into blue fluorescent light [6].

Under the influence of light, all seven FWAs can undergo E,Z-isomerization. Only the E-form has fluorescent properties, because of its rigid π-electron system. In the Z-form it is not possible to bring the benzene rings into one plane [7].

Production Methods

The syntheses of FWA 1–4 progresses by means of condensation reactions of 4,4'-diaminostilbene-2,2'-disulfonic acid (**9**) with cyanuric chloride (**8**), aniline (**10**) and primary or secondary amines (Fig. 2).

The condensation reactions can be effected 'in one pot' since a Cl atom in dichlorotriazine compounds is less reactive than a Cl atom in cyanuric chloride.

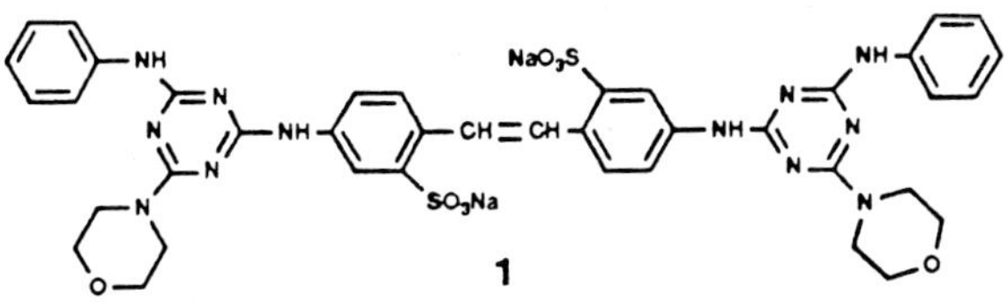

1
Disodium 4,4'-bis[(4-anilino-6-morpholino-1,3,5-triazin-2-yl)amino] stilbene-2,2'-disulfonate

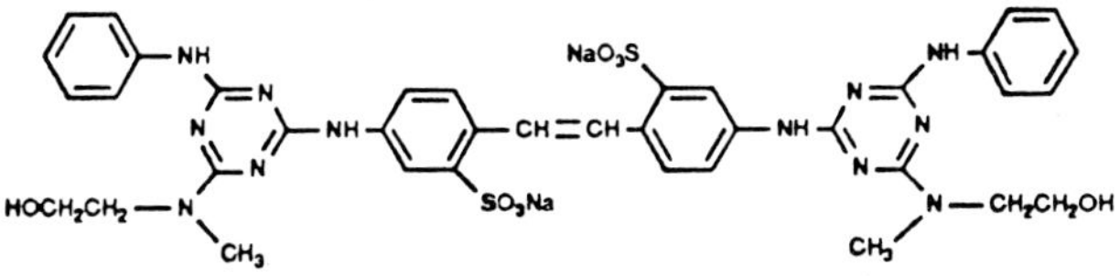

2
Disodium 4,4'-bis{[4-anilino-6-(N-methyl-N-2-hydroxyethyl)amino-1,3,5-triazin-2-yl] amino}stilbene-2,2'-disulfonate

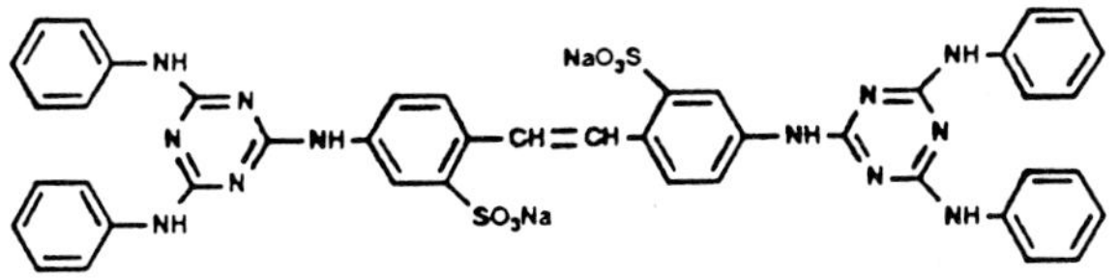

3
Disodium 4,4'-bis(4,6-di-anilino-1,3,5-triazin-2-yl)amino-stilbene-2,2'-disulfonate

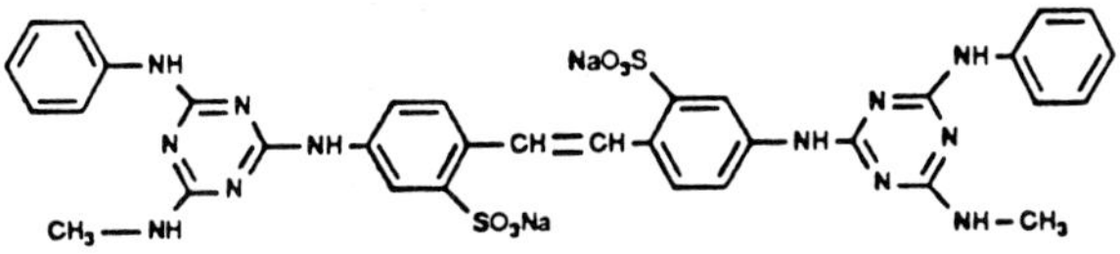

4
Disodium 4,4'-bis[(4-anilino-6-methylamino-1,3,5-triazin-2-yl)amino]stilbene-2,2'-disulfonate

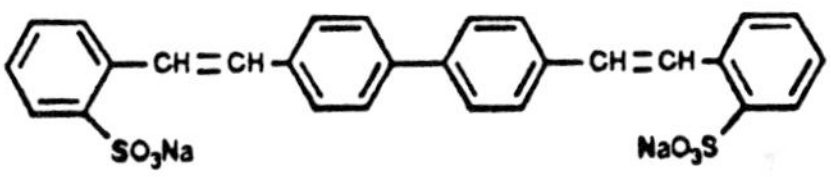

5
Disodium 4,4'-bis(2-sulfostryryl)biphenyl

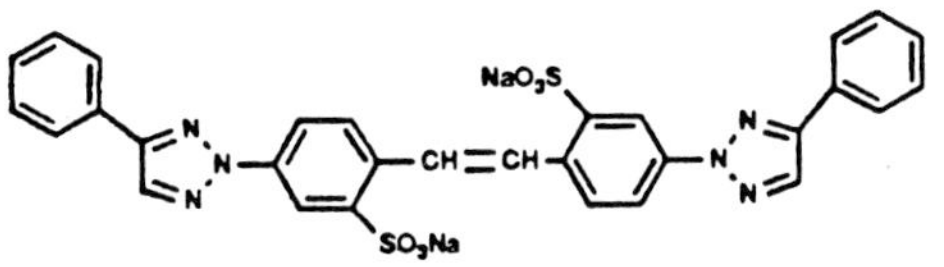

6
Disodium 4,4'-bis(4-phenyl-1,2,3-triazol-2-yl)stilbene-2,2'-disulfonate

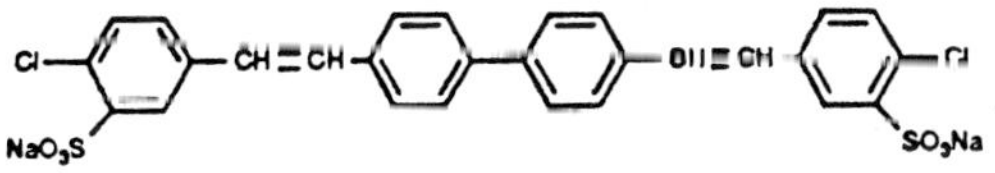

7
Disodium 4,4'-bis(3-sulfo-4-chlorostryryl)biphenyl

Fig. 1. The structures and names of the seven FWAs

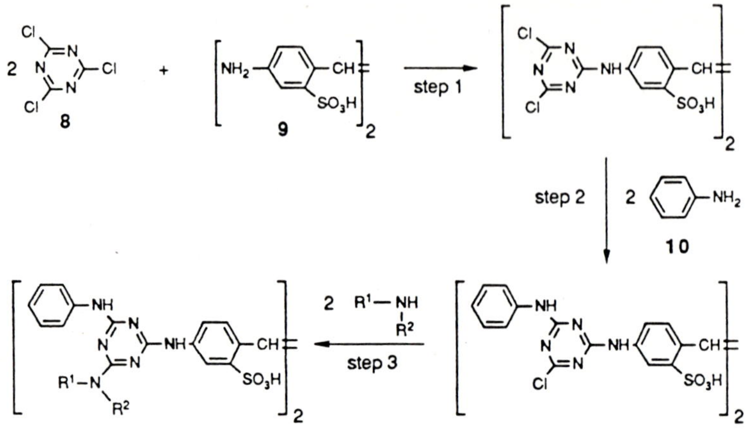

$R^1,R^2 = CH_2CH_2OCH_2CH_2$; FWA 1
$R^1 = CH_2CH_2OH$ $R^2 = CH_3$; FWA 2
$R^1 = C_6H_5$ $R^2 = H$; FWA 3
$R^1 = CH_3$ $R^2 = H$; FWA 4

Fig. 2. The synthesis of FWA 1, 2, 3 and 4

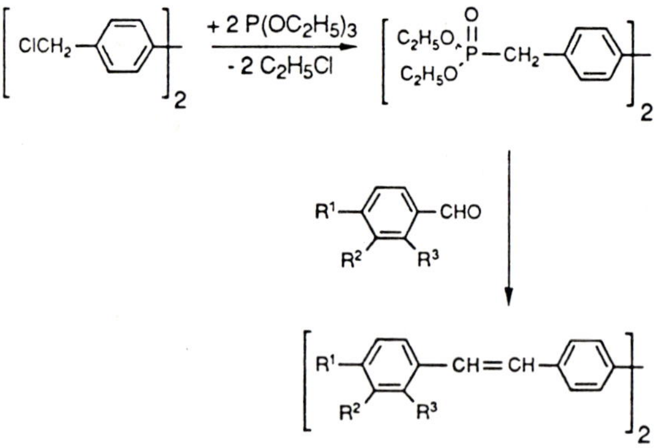

$R^1 = H$ $R^2 = H$ $R^3 = SO_3Na$; FWA 5
$R^1 = Cl$ $R^2 = SO_3Na$ $R^3 = H$; FWA 7

Fig. 3. The synthesis of FWA 5 and 7

The reaction steps 1, 2 and 3 of Fig. 2 can be achieved in temperature ranges of 0–5 °C (step 1), 40–50 °C (step 2) and at 80–100 °C (step 3). A drawback of this method is the high reactivity and multifunctional ability of the cyanuric chloride.

FWA 5 and 7 are produced by way of a Wittig reaction (Fig. 3). The synthesis is described by Stilz and Pommer [8] and proceeds without any complications.

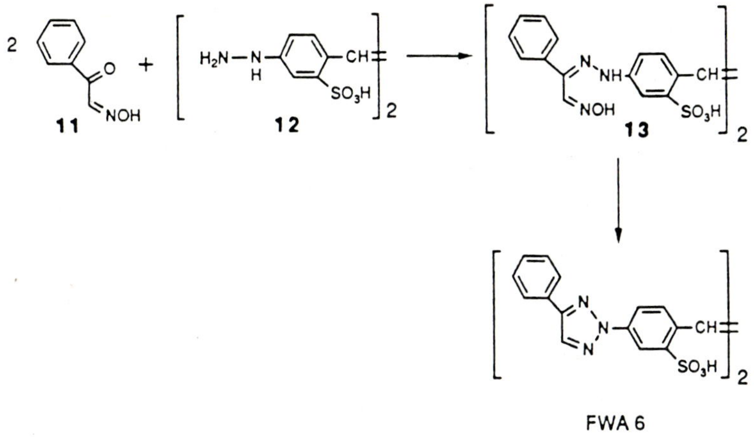

Fig. 4. The synthesis of FWA 6

FWA 6 is produced by transforming 4,4'-diaminostilbene-2,2'-disulfonic acid into 4,4'-dihydrazinostilbene 2,2'-disulfonic acid (**12**). The reaction of (**12**) and oximoacetophenone (**11**) gives the oximohydrazone (**13**) which leads to the product FWA 6 by a dehydrolizing cyclization with acetic acid anhydride or in the urea melt (Fig. 4).

All the syntheses are described in [9].

Amounts and Uses

As already described in the section on properties, mostly seven FWAs are used in laundry detergents. Sporadically others are used but their amounts are so small and their structures so similar to the seven mentioned that they will not be treated here. Exact amounts of FWA production are unknown, but estimates show a world wide production of 14 000 tonnes of FWA 1 and 3000 tonnes of others in 1989 [10].

The FWAs used have hardly changed over the last 10–15 years [5, 11], only FWA 7 is relatively new in this series. Due to changing demands, however, a shift has occurred in the share of the individual FWAs. FWA 1 is still the most used because of its relatively large range of application.

The washing temperature has an important influence on the performance of FWAs. Table 1 lists the FWAs commonly used in detergents in different parts of the world.

Below 40 °C the dissolution rate may be a problem; FWA 1 especially is difficult to dissolve at these temperatures. Often the dissolution rate is increased by grinding the crystals.

Table 1. World wide washing habits

Region	Detergent used	FWAs	Washing temp. (°C)
North America	Conventional powder	1, 2, 4	10–50
	heavy duty liquid	2, 3, 6	10–50
Europe	compact and convent. powder	1, 5, 7	30–95, preferably 60
	heavy duty liquid	3, 4, 5	
Far East	compact powder	1, 5	15–25
	convent. powder	1, 5	15–25
Australia	compact and convent. powder	1, 2, 5	15–25
	heavy duty liquid	5	15–25
Africa	convent. powder	1, 4, 5	15–25

Ref. [12, 13]

In liquid detergents the water solubility in the washing liquor is less important than in a concentrated surfactant solution. For example FWA 6 has a solubility of 1 g/l in water at 20 °C, but of 20 g/l in a solution of 90% nonionics in water at 20 °C [12].

The effect of FWAs also strongly depends on the formulation of the washing-powder. When hypochlorite is used as a bleaching agent, only FWA 5, 6 and 7 are stable in the washing liquor. On the fiber all FWAs have good stability to oxidizing bleaches. Photobleaches of the phthalocyanine type have a marked mutual effect with stilbene-type FWAs under the influence of light. In this case FWA 5, 6 and 7 are also more stable than the others [12]. Cationic substances can react with FWAs which results in a quenching of the fluorescence. The extent of this quenching is partly dependent on the solubility of the complex formed [12].

Which FWA will be used in a detergent therefore depends on several factors because all FWAs have, despite their great structural resemblance, their own specific properties.

Analytical Methods

Analyses of FWAs need to distinguish compounds with very high structural resemblance. A quantitative analysis is mostly in the trace range. Therefore, the demands on the efficiency of chromatographic systems are very high. A number of publications propose specific chromatographic methods to define a limited number of FWAs [14–22]. Qualitative analysis of FWAs is first done by determining retention (Rf) values in thin layer chromatography (TLC) or in high performance liquid chromatography (HPLC). Theidel [23] has developed a number of TLC-systems to separate FWAs. By combining the results of several of these systems individual FWAs can be detected. In the same article Theidel describes a method for the extraction of FWAs in textiles and detergents.

Important in the determination of FWAs with TLC and HPLC is that the Z-form of the stilbene type FWAs is non-fluorescent. Mostly these forms have Rf values different from the E-forms. The resulting disadvantage is that TLC analysis always has to be carried out in the dark to prevent uncontrolled transformations which can lead to tails on the chromatograms. An advantage is that the Z-form spots can be made visible by drying the chromatograms in daylight, hereby transforming a part into the E-form. Hence a FWA can have two specific Rf values in both TLC and HPLC which can be used to identify individual FWAs.

Further identification of FWAs can be done with absorption measurements in situ on the TLC plates [24], or by infrared spectroscopy [23]. In the latter method FWAs are scraped off the chromatograms and transferred to a KBr layer.

Fluorimetry is used for quantitative analysis of thin layer chromatograms [25]. Detection limits are about 1 to 0.5 ng. The advantage of direct determination on TLC plates is that the spots do not have to be removed, for then a comparatively large quantity of substance would be required [24]. Quantitative analyses are also possible with HPLC.

It is also possible to analyse FWAs directly in solution. Eisenbrand [26] and Klauk [27] described some examples of low fluorescence intensity for the trace analysis of FWAs. Anders [28] has found that a fast recorder is able to analyse 10 ppb of FWA with a statistical standard deviation of $\pm 15\%$ for a single measurement and a reproducibility of $\pm 50\%$ with a statistical certainty of 99%. The largest problem of such a low concentration of FWA in solution is the limited light stability of the compounds. Especially the E,Z-isomerization causes a rapid decrease in fluorescence. A test with a diamino stilbene disulfonic acid type of FWA showed after only 10 seconds a loss of fluorescence intensity of 25% due to this E,Z-isomerization [28]. Thus a sample may not be exposed to a light source before measurement starts and values have to be determined immediately after placing the sample in the light path.

Sewage Treatment

The most important route by which FWAs enter the environment is with domestic wash water effluents containing on average 0.01–0.1 ppm FWA [29]. In industrialized countries most of these household effluents are treated in a sewage plant before reaching the surface waters. Therefore, the quantities of FWAs entering the natural environment strongly depend on the removal rate in the sewage treatment plants.

Removal of FWA 2 in semi-continuous activated sludge (SCAS) tests was 90% [30]. Field monitoring of sewage treatment plants in the U.S.A. showed different rates of removal at different stages of the sewage treatment operation

Table 2. Removal of FWAs in sewage treatment plants

Treatment stage	%
primary	55
primary and secondary	83–95
primary, secondary and tertiary	98

Ref. [31]

[31]. Primary or mechanical treatment reduced the FWA concentration by more than 50%, the main removal mechanism in this phase is thought to be adsorption. Secondary (biological) and tertiary (chemical) treatment removed FWAs almost completely (Table 2).

FWAs strongly accumulate in sewage sludge. Tests showed values for total amounts of FWA of 13–74 ppm in wet sludge and 140–1080 ppm in sludge on a dry weight basis [32].

Discharge to the Environment

FWAs enter the surface waters with the sewage treatment plant effluents and with direct discharges from households not connected to a sewage treatment plant. The FWA concentration will decrease when the effluents enter the surface waters because of dilution. Effluents of production plants mostly contain no FWAs due to effective wastewater treatment.

Measurements of FWA concentrations in surface waters are limited. In 1975 Anders [33] set up an extensive project to investigate concentrations in rivers of seven European countries. Only in the Viskan river in Sweden were concentrations of up to 8.2 ppb detected. In the other rivers no detectable concentrations were found. In all cases no distinction was made between individual FWAs; FWAs as a whole were analysed.

Samples of 35 U.S.A. rivers in 1974 showed concentrations of FWAs up to 0.6 ppb upstream of a sewage treatment plant [34]. Downstream concentrations were never above 40 ppb for FWA 1 as well as FWA 2.

Because of the lack of recent measurements, present concentrations can only be estimated. For Switzerland, with a consumption of 80 tonnes per year [35], a worst case estimate has been made assuming no binding to fabrics [36]. Of the total household effluents 90% is treated in sewage treatment plants with mechanical and biological treatment before entering the surface waters [37]. This gives a load to the environment of 3.6–12.7 t/y. The 10% untreated gives a load of 8 tonnes per year maximum, assuming that all non-treated domestic sewage is discharged to the surface waters. Total amounts entering the natural environment are therefore 11.6–20.7 t/y. The total waste water discharge is estimated to be 1.9×10^{12} l [37]. This gives a calculated concentration of 6.1–10.9 ppb. Assuming a minimal dilution factor of 3, the predicted concentration in surface water is approximately 3 ppb.

In other parts of the world, especially in Third World countries, only very few household effluents are treated before entering the environment. Therefore, concentrations of FWA can locally be much higher. Estimates show a use of 3350 tonnes of FWA 1 and 800 tonnes of others in 1989 in Africa, Central- and South-America and Asia (except Japan) [38].

Because FWAs are adsorbed by solids, Zinkernagel measured the FWAs in four river sediments [39]. Results showed concentrations between 1.14–2.21 ppb.

Environmental Fate

Adsorption on Sludge

Sewage sludge may be used as fertilizer, soil conditioner or land fill. Because FWAs are adsorbed by sludge it is important to determine the extent to which they may be leached from the sludge amended soil into the groundwater, which may be used as source for drinking water. However, an experiment by Esser and co-workers [40] showed that water containing FWA 2 does not carry the FWA with it as it diffuses through the soil: the FWA is strongly adsorbed by the soil, especially by the two upper sections of the soil (each section 10 cm high). The amount that reaches the lowest part of the soil is only 1.1% (Table 3). This indicates that pollution of groundwater by FWAs is unlikely.

A similar experiment led to the same conclusion. In this experiment [41] clean water was percolated through sludge containing FWA. After 68 days the total amount leached out from the original 216 mg FWA 1 and 240 mg FWA 2 in 20 kg sludge was only 0.02% for both FWAs. The sludge had been exposed to natural weather and non-agricultural conditions. No trend in the appearance of

Table 3. Distribution of FWA 2 in soil and water

Extracted	%
top section	52.6
2nd section	26.3
3rd section	9.2
4th section	7.2
bottom section	2.0
total retained	97.3
total eluted	1.1
total recovered	98.8

Ref. [40]

FWAs in the leachate versus time was observed which would have indicated a total elution of the FWAs.

Degradation

Because of their function FWAs used in detergents must not degrade too quickly. This raises questions about their degradation in the natural environment.

A test with FWA 5 showed that photodegradation starts with the cleavage of the ethylene bond of the stilbene section, resulting in o-sulfobenzoic acid and biphenyl-4,4'-dicarboxylic acid [42]. These products showed no further photodegradation. Photodegradation is rather fast; tests with 10 ppm solutions of FWA 2 and 5 showed, after 10 and 15 days respectively, only small amounts of the initial products [42].

Laboratory experiments showed that FWAs are not readily biodegradable. Tests which used the oxygen consumption over 5 days (BOD_5) as a measure of the ability of FWAs to serve as a substrate for bacterial growth, gave no evidence of biodegradation; all seven FWAs have a BOD_5 of zero [43]. This is due to their intended use on the fiber where they have to be stable to human sweat and resist microbial degradation.

In a river water die-away test at which the CO_2 production was measured in solutions of max. 1000 ppb over a period of 35 days, no biodegradability was found [44].

Guglielmetti [42] demonstrated that with the aid of activated sludge it is possible to degrade FWAs biologically. The experiments were carried out in the dark to prevent photodegradation.

Table 4 shows FWA 2 and FWA 5 needed an adaptation period to be finally degraded, the degradation was however incomplete after 30 days. Due to this relatively long time the elimination process of FWAs in a sewage treatment

Table 4. Biological degradation of FWA 2 and 5 and the photodegradation products of FWA 5 by activated sludge

	Percent degradation* after			
	5 days	10 days	15 days	30 days
FWA 2	0	0	38	78
	0	0	14	94
FWA 5	0	0	13	63
	0	0	15	62
biphenyl-4,4'-dicarboxylic acid	0	0	92	98
	0	13	96	99
o-sulfobenzoic acid	91	ns	ns	ns
	93	ns	ns	ns

* Parallel experiments; ns: not sampled
Ref. [42]

plant will be mostly adsorption. The adaptation period probably explains the BOD_5 values of zero. The photochemical products of FWA 5 were quickly and completely degraded.

Further reports on the degradability of FWAs are limited. In Japan, several experiments have been done which showed the possibility of degrading FWAs with activated sludge [45–48].

Bioaccumulation

Several experiments have been carried out to investigate the bioaccumulation of FWAs. A Bluegill test [49] showed that after an accumulation period of 70 days for FWA 1 and 35 days for FWA 2 only traces could be detected in the fish. The same study showed for FWA 5 detectable concentrations (<0.05 ppm) which, however, remained below the concentration in the water (0.1 ppm). When the fish were transferred to clean water the FWA was rapidly eliminated. An experiment with radiolabelled FWA 5, where Golden Ides were exposed to 0.1 mg/l FWA 5, indicated that the highest concentration appeared in the intestines. A concentration of 6 ppm was found which corresponds to a concentration factor of 60 [50]. An experiment with radiolabelled FWA 6 and Golden Ides showed that within one week an equilibrium between uptake and elimination of the FWA develops [51]. With water concentrations of 10 ppb and 100 ppb concentration factors between 7 and 14 were reported. Within two days concentration factors of 1 were found when the fish were transferred to clean water. Accumulation studies of FWA 1 and 2 with maximum concentrations of 125 ppb for a period of 90 and 105 days showed concentrations of FWA in Bluegill and Channel Catfish below the detection limit (concentrations less than 10 ppb) [52].

Field trials in the U.S.A. in 1974 showed concentration factors of 1.4, 1.3 and 21 for resp. FWA 1, 2 and 5. The fish were caught in the Passaie River [53]. Fish caught 1.5 and 8 km downstream from a sewage treatment plant showed no detectable amount of FWAs. In water and sediment of the river, no FWAs were detected either [54]. The effluent of the plant contained 2.3 ppb FWA 1, 3.7 ppb FWA 2 and traces FWA 5.

Indirect bioaccumulation of FWAs was studied through a model food chain comprised: algae→snails→fish [50]. The algae were grown in a nutrient medium containing 100 ppb ^{14}C-radiolabelled FWA 5 and fed to the snails. The algae and snails were fed to the fish. The results showed that accumulation through the food chain is very slight and not higher than found for the direct uptake. For the edible parts of the fish the combined exposure of both the direct and indirect route showed a concentration factor of 1.1.

FWAs are strongly adsorbed by sludge under non-agricultural conditions. Hence, it is necessary to determine the potential for bioaccumulation and translocation of FWAs in plants, grown in sludge amended soil, i.e. the

Table 5. Distribution of ^{14}C-labelled FWA 5 in bean plants grown for 40 days

	Nutrient solution 11.5 ppm		Soil 17.5 ppm	
	ppm	%	ppm	%
bean pods	0.05	1.1	0.11	0.1
leaves	0.08	1.4	2.75	2.7
stem	0.94	5.7	3.35	1.1
roots	18.61	91.8	121.6	96.1

Ref. [55]

biological availability of FWAs to plants. A study by Muecke et al. [55] determined the uptake of radiolabelled FWA 5 in bean plants. Plants grown in soil containing 17.5 ppm adsorbed only 2% after 40 days. In a nutrient solution of 11.5 ppm FWA 55% was adsorbed in the same period. In both experiments over 90% of the uptake was found in the roots (Table 5).

Similar findings were reported in uptake experiments with corn, soya and radish [56, 57]

All of the above studies indicate that accumulation by the roots due to ad- or absorption does occur, but concentrations due to translocation appear to be insignificant. FWAs present in the soil will very likely not be translocated to the aerial portions of the plant.

Toxicology

Ecotoxicology

Concentrations of FWAs present in the environment need to be compared with concentrations toxic to water organisms in order to assess the environmental safety of FWAs. Acute toxicity figures for fish are given in Table 6 as concentrations at which half of the population dies (LC_{50}-values) within 96 h.

Toxicity values (LC_{50}) for algae (*Selenastrum capricornutum*) and invertebrates (*Daphnia magna*) are given in Table 7. LC_0 values are given for FWA 1 and FWA 5 in Table 8.

These results show that concentrations in surface waters do not cause acute toxicity to water organisms. The concentrations which produce mortality exceed environmental concentrations by at least 1000 times. Concentrations in effluents of sewage treatment plants also produce no acute danger. Chronic toxicity tests on aquatic organisms are lacking.

Table 6. LC_{50} values after 96 h exposure (ppm)

FWA	1	2	3	4	5	6	7
Catfish	1060	86			126		
Trout	750	108		53	130	221	>1000*[2]
Bluegill	32	26	>100[1]	153	241	689	
Zebrafish	25[2]						

*(48 h)
Ref. [4] except: [1] [58], [2] [59]

Table 7. LC_{50} value of two FWAs for algae and daphnia

	Algae Conc. ppm	Time h	Daphnia Conc. ppm	Time h	Ref.
FWA 3	11	120	>1000	48	[60]
FWA 5	10	72	>1000*	24	[61]
FWA 5	8	96			[61]

*FWA 5 87% active substance

Table 8. LC_0 values for FWA 1 and 5

FWA 1	Conc. mg/l	Time h	Ref.	FWA 5	Conc. mg/l	Time h	Ref.
Trout	30	48	[62]	Daphnia	500	24	[61]
Zebrafish	10	48	[63]	Algae	3.1	24	[61]
	10	96	[63]				
Catfish	24	96	[64]				

Mammalian Toxicology

The section on bioaccumulation shows that oral intake of FWAs from fish is negligible. Studies of drinking water in seven European countries showed no detectable concentrations at a detection limit of 0.01 ppb [33]. Studies in the U.S.A. showed that average concentrations were never above 0.1 ppb [34]; these concentrations showed no effects in oral tests with rats [4].

Another route of exposure to FWAs is through skin exposed to detergent solution or to fabrics washed in detergents. In a hand washing test amounts of FWAs left on both hands ranged from 0.06 0.17 mg [65]. Several studies showed no irritation to skin [66–70], no skin sensitization potential [64, 67, 69, 71–73], no phototoxicity [74–76] and no photoallergy [67, 71].

Animal studies with FWAs showed no skin-, carcinogenic or photocarcinogenic effects. An overview of the toxicological studies is given in Table 9. Other toxicology tests of FWAs on aminals are reviewed in [4] and [5].

Table 9. Toxicological studies on the seven FWAs

	FWA 1	2	3	4	5	6	7
Acute oral	x	x	x	x	x	x	x
inhalation				x	x	x	x
intraperitoneal	x		x	x		x	
subcutaneous		x			x		x
percutaneous			x		x	x	x
Subacute oral					x		
percutaneous				x		x	
Subchronic oral		x	x	x	x	x	x
Chronic oral		x		x	x	x	x
epicutaneous	x	x	x			x	
Carcinogenicity							
oral		x		x	x		x
cutaneous	x	x					x
Mutagenicity	x	x	x	x	x	x	x
Teratogenicity			x	x	x	x	x
Skin irritation	x	x	x	x	x	x	x
Eye irritation	x	x	x	x	x	x	x
Sensitization	x	x	x	x	x	x	x
Photo-sensitization-				x			
carcinogenicity	x	x	x	x	x	x	x
Liver function				x			x

Ref. [77]

References

1. Anliker R (1975) In: [5] p 12
2. Procter and Gamble Company (1990) product information
3. Bode K-D (1975) Tenside Detergents 12:69 cited in [39]
4. Burg AW, Rohovsky MW, Kensler CJ (1977) CRC Crit Rev Env Contr 7:91
5. Coulston F, Korte F (eds) Anliker R, Müller G (guest eds) (1975) Fluorescent whitening agents; EQS environmental quality and safety, supp vol 4, Thieme, Stuttgart
6. Bayer AG and Ciba-Geigy AG (1990) product information
7. Theidel H (1964) Melliand Textilber. p 514 cited in [9]
8. Badische Anilin- and Soda-Fabrik AG (Stilz W, Pommer H, Koenig KH) Brit. 920988 (Ger Prior 12.8.1959) cited [9]
9. Gold H (1975) In: [5] p 25
10. Ciba-Geigy AG (1990) unpublished data
11. Siegrist AE (1978) J Amer Oil Chem Soc 55:114
12. Schuessler U (1986) Fluorescent whitening agents for detergents. World conference on detergents, Amer Oil Chem Soc. Champaign, Illinois
13. Ciba-Geigy AG (1990) unpublished data
14. Kiger J, Bon R (1960) Ann Pharm Franc 18:853 cited in [23]
15. Latinák J (1964) J Chromatogr 14:482 cited in [23]
16. Wandel M, Tengler H (1965) Deutchen Färberkalender 69:76 cited in [23]
17. Kurz J, Schuierer M (1967) Fette, Seifen, Anstrichm. 69:24 cited in [23]
18. Brown JC, J Soc Dyers Colour 80:185 cited in [23]
19. Lanter J (1966) J Soc Dyers Colour 82:125 cited in [23]
20. Schlegelmilch F, Abdelkader H, Eckelt M (1971) Text Ind 73:274 cited in [23]
21. Figge K (1968) Fette, Seifen Anstrichm 70:680 cited in [23]

22. Franc J, Kovar F, Mikes F (1966) Mikrochim. Acta p 133 cited in [23]
23. Theidel H (1975) In: [5] p 94
24. Theidel H (1975) In: [5] p 111
25. Anders G (1975) In: [5] p 104
26. Eisenbrand J (1962) Deut Lebensm Rundsch 8:230 cited in [28]
27. Eisenbrand J, Klauk A (1965) Deut Lebensm Rundsch 12:370 cited in [28]
28. Anders G (1975) In: [5] p 116
29. Anonymous; Waschen heute. Information zum Thema Waschen und Umweltschutz. Verband der Sweizerischen Seifen- und Waschmittelindustrie (SWI)
30. Procter and Gamble Company (1973) unpublished results
31. Ganz CR, Liebert C, Schulze J, Stensby PS (1975) J Water Pollut Control Fed 47:2834
32. Schulze J, Stensby P et al. personal communication cited in [39]
33. Anders G (1975) In: [5] p 143
34. Procter and Gamble Company (1974) Fluorescent whitening agents: occurrence in waste water, river waters and drinking waters, unpublished report cited in [4]
35. Ciba-Geigy AG (1990) unpublished data
36. Estimate made by Ciba-Geigy AG (1990)
37. Data from Bundesamt für Umwelt, Walt und Landschaft (Buwal)
38. Ciba-Geigy AG (1990) unpublished data
39. Zinkernagel R (1975) In: [5] p 129
40. Esser HO and collaborators: personal communication cited in [39]
41. Ciba-Geigy AG (1974) The fate of FWAs in sewage sludge under simulated environmental conditions—Final study unpublished report cited in [4]
42. Guglielmetti L (1975) In: [5] p 180
43. Bayer AG and Ciba-Geigy AG (1990) product information
44. Procter and Gamble Company (1988) unpublished report
45. Katayama M (1986) Kenkyu Kiyo-Tokyo Kasei Daigaku 26:11 (C.A. 106: 37889e)
46. Katayama M (1984) Nippon Nogei Kagaku Kaishi 58(5):449 (C.A. 101:97024a)
47. Katayama M, Kurata Y, Abe S (1982) Kaseigaku Zasshi 33(3):124 (C.A. 97:132835p)
48. Katayama M, Abe S (1979) Kaseigaku Zasshi 30(10):903 (C.A. 95 : 29812p)
49. Ganz CR, Schulze J; Stensby PS, Lyman FL, Macek K (1975) Env Sc and Tech 9:738
50. Feron J-P, Hitz HR (1975) In: [5] p 157
51. Hamburger B, Maul W, Patzschke K, Theidel H, Wegner L-A (1975) In: [5] p 165
52. Sturm RN, Williams KE (1975) Water Res 9:211
53. Procter and Gamble Company (1974) Fluorescent whitening agents: Relationship between aquatic levels and accumulation in fish, unpublished report cited in [4]
54. Ciba-Geigy AG (1974) Analysis of Westerly treatment plant (Cleveland, Ohio) and Lake Erie samples, unpublished report cited in [4]
55. Muecke, W, Dupuis G, Esser HO (1975) In: [5] p 174
56. Ciba-Geigy AG (1972) Corn and soyabean samples from Vero Beach pilot study, unpublished report cited [4]
57. Procter and Gamble Company unpublished data cited in [4]
58. Procter and Gamble Company (1987) unpublished report
59. Ciba-Geigy AG (1982) unpublished data
60. Procter and Gamble Company (1987) unpublished report
61. Ciba-Geigy AG (1989) unpublished data
62. Ciba-Geigy AG (1980) unpublished data
63. Ciba-Geigy AG (1982) unpublished data
64. Procter and Gamble Company unpublished data cited in [4]
65. Gloxhuber C, Bloching H, Kästner W (1975) In: [5] p 202
66. GAF Corporation unpublished data cited in [4]
67. Verona Corporation unpublished data cited in [4]
68. Lever Brothers Company (1966) Safety of compound 1 as shown by testing animals and in human subjects, unpublished report cited in [4]
69. Ciba-Geigy AG (1964) Repeated insult patch test with compound 2, Industrial Biology Laboratories, unpublished report cited in [4]
70. Ciba-Geigy AG (1971) Repeated insult patch test with compound 5, Food and Drug Research Laboratories, unpublished report cited in [4]
71. Griffith JF (1973) Arch Dermatol 107:728

72. Keplinger ML, Fancher OE, Lyman FL, Calandra JC (1974) Toxicol Appl Pharmacol 27:494 cited in [4]
73. Ciba-Geigy AG (1965) Repeated insult patch tests with Compound 1, unpublished report cited in [4]
74. Forbes PD, Urbach F (1975) In: [5] p 212
75. Ciba-Geigy AG (1971) Phototoxicity, unpublished report, Temple University cited in [4]
76. Ciba-Geigy AG, Acute study: Human skin, unpublished report by Urbach F cited in [4]
77. Bayer AG and Ciba-Geigy AG unpublished results

Anthropogenic Silicates

J. S. Falcone Jr.[1] and J. G. Blumberg[2]

[1] West Chester University, Department of Chemistry, West Chester, PA 19383
[2] PQ Corporation Research and Development Center, 280 Cedar Grove Road, Conshohocken, PA 19428

Introduction 367
Structure and Chemistry 368
    Structure of Silicates 368
    Soluble Silicate Chemistry 370
    Zeolite Chemistry 372
Silicates and the Environment 373
    Biological Interactions of Anthropogenic Silicates 374
    Plants 374
    Animals 376
Uses of Anthropogenic Silicates in Environmental Technology 378
    Hazardous Waste Mitigation 378
    Water Treatment 379
    Land Treatment 380

Summary

Anthropogenic silicates include slightly processed natural earths, glasses and metastable amorphous compounds, and highly ordered crystalline synthetic materials. Silicate building and glazing materials have been manufactured since ancient times. More recently, silicate products have been developed for separation, e.g. ion exchangers and adsorbant, and catalytic uses. The environmental chemistry of anthropogenic silicates is reviewed. Scientific studies of the environmental fate of anthropogenically generated silicate compounds are examined in the context of the natural biogeochemical silica cycle. Studies of the interaction of anthropogenic silicates with various elements of the biosphere, including human populations are discussed. Also reviewed, are the references documenting the emergence of new technologies for the use of silicates to avoid or mitigate adverse environmental aspects of nuclear, fossil fuel, and other technologies.

Introduction

The solid portions of the surface of the Earth consist predominantly of silicon and oxygen atoms in the form of silica and silicates. The use of Earth's silica and silicate resources by *H. Erectus* and later, by our own species, is the oldest technology on the planet [1]. For most of the first two million years, processing was limited to the hand shaping of natural silicates. With the domestication of fire came the first phase-modified anthropogenic silicates, ceramics a word whose etymologic root goes back through ancient Greek to the Sanskrit verb "to burn" [2]. The glazing of stones and pottery has been practiced for 14 000 years, and man-made glasses 9000 years old are known [3]. Vitreous enameling developed in Egypt 2800 years ago [4] and siliceous pozzolonic cement was developed in Roman times [5].

The preparation of silicates by chemical reaction began late in the industrial revolution with the preparation of portland cement in the 1830s [6]. It was shortly thereafter that soluble alkali silicates were produced on a commercial scale. By 1928, the adverse environmental effects of emissions from the sulfate process for soluble silicate production were in part responsible for its abandonment [7]. At about that time, processes were being developed for the use of soluble silicates in aqueous effluent treatment from other industries [8]. And shortly thereafter, the manufacture of synthetic amorphous silica and silicates by precipitation or gellation of aqueous alkali silicates began. The 1940s saw the development of aqueous silica sols and the beginnings of the production of crystalline silicates by hydrothermal synthesis. In the following decade, silicones, and organic silicates emerged as commercial products.

Beginning in the mid-1950s, a rapid growth occurred in the manufacture of synthetic zeolites, crystalline hydrated aluminosilicates, first as ion exchangers and adsorbants, then as shape selective or molecular sieves and petrochemical catalysts.

Structure and Chemistry

Structure of Silicates

Silicates have a complex structural chemistry. Liebau [9] has made a monumental contribution to the classification of the structures of crystalline silicates. Eight parameters are required to classify all conceivable silicate structures. Liebau's formalism for representing silicate structural information is demonstrated by the generic formula shown below for a typical silicate, $M_rSi_xO_y$:

$$M_r(B, M_\infty^D) (^PSi_xO_y)$$

B symbolizes the degree of branching in silicate anions, e.g. μB defines unbranched as in a chain of silicate tetrahedra and oB defines open branching where the anion contains pendent tetrahedra as in the structure

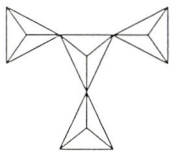

Fig. 1.

More complex branching exists, but is beyond the scope of this review.

M symbolizes the multiplicity or number of repeating single anion units forming a multiple anion with the same D value. D represents the dimensionality of the anion from 0 for a single polyhedra to 3 for a framework silicate anion. The ∞ symbol indicates an essentially infinite extension of the anion in the appropriate dimensions. Thus, when $D=0$ then t replaces $^{0}_{\infty}$ for multiple polyhedra and r for ring anions.

A P to the upper left of the Si symbol indicates the number of polyhedra in a repeating unit within an anion structure. This number is called by the German numeral followed by "er", e.g. a repeating pair of polyhedra in a chain would be called a zweier single chain. The coordination number (CN) of the silicon can be determined from the x and y values; however, in most cases it is 4. Another parameter not shown is linkedness (L) which for virtually all cases is either 0 for a discrete polyhedra or 1 for polyhedra sharing corners. Table 1 shows Liebau's summary of silicates observed to have the different possible combinations of L and CN values. We see that most examples have CN values of 4 and L values of 0 and 1.

Table 2 provides an estimate of the distribution of the 856 cases with $CN=4$ and $L<2$ according to their dimensionality, D, and multiplicity, M. It is in this chart that we observe the diversity of structures which make up the silicate system. Additionally, the chemical notation for these combinations of D and M values are presented in Table 2.

These charts show the natural abundance of 4 coordinate silicon in silicate anions. The predominant mode of condensation is via corner sharing of tetrahedra ($L=1$).

Table 1. The coordination number and linkedness of silicate anions

| Linkedness | Coordination number | | | Total |
	4	6	Mixed	
0	160	7	0	167
1	696	4	1	701
2	1	2	0	3
3	0	0	0	0
Mixed	0	0	0	0
Total	857	16	1	874

Table 2. The dimensionality and linkedness of anions

Anion type = D	M = 1 µB:br	M = 2 µB:br	M = 3 µB:br	M > 3 µB:br	Totals µB:br
Discrete = 0					
Oligosilicates (t)	160: monosilicates	90: disilicates	12:1 trisilicates	6:	268:1
Cyclosilicates (r)	73:4 monocyclosilicate	27:1 dicyclosilicates			100:5
Chains = 1					
Polysilicates	100:10 monopolysilicates	58:7 dipolysilicates	9: tripolysilicates	5:	172:17
Layers = 2					
Phyllosilicates	131:12 monophyllosilicates	12:9 diphyllosilicates			143:21
Framework = 3					
Tectosilicates	89:40 tectosiliates				89:40
Totals	553:66 619	187:17 204	21:1 22	11: 11	772:84 856

* Unbranched:branched

Soluble Silicate Chemistry

The soluble silicate glasses, powders, and liquids are among the largest volume synthetic chemicals, surpassed in volume only by commodity acids and bases. Worldwide volume for the sodium silicate form is estimated to be roughly 3×10^9 metric tonnes split evenly among North America, Western Europe, Japan, and the rest of the world. These products are manufactured primarily from sodium carbonate and relatively pure glass sand in open hearth regenerative or electric furnaces according to the reaction:

$$Na_2CO_3 + mSiO_2 \longrightarrow Na_2O \cdot (SiO_2)_m + CO_2$$

where m is called the ratio or modulus.

The glasses may be dissolved directly in water in rotary dissolvers to concentrations of roughly 25–30% based on SiO_2, or ground and sized for sale as powders or frits. Solutions of soluble silicates when spray dried form solid solutions containing roughly 20% water which dissolve in water orders of magnitude faster compared to equivalently sized powdered glass of the same modulus. Solutions of silicates, with values of m < 1.6 and solids ⩾ 20% are unstable with respect to the crystalline silicate, $Na_2H_2SiO_4 \cdot 8H_2O$. Two other silicates with m = 1 that are of commercial importance are the so-called anhydrous metasilicate, $Na_2O \cdot SiO_2$ and the sodium metasilicate pentahydrate, $Na_2H_2SiO_4 \cdot 4H_2O$. The latter is sodium dihydrogen monosilicate four hydrate and the former is an anhydrous sodium monopolysilicate.

The chemical properties of silicate anions, and the solution chemistry of these anions is at least as complex as their structures are diverse [10, 11]. Recent advances in our understanding of these complex solutions has come as a result of ^{29}Si NMR [12], trimethylsilylation (TMS) [13], and EMF [14] methods. These methods reveal a complex and likely changing distribution of silicate anions, sized from ionic monosilicates to large colloidal anions containing possibly as many as 50 Si atoms. In the presence of colloidal particles, the distribution of anions is seen as a metastable or quasiequilibrium where larger species are forming at the expense of less highly polymerized species. This view is consistent with the view that the reactivity or liability of silicate units decreases as the connectivity, n, of nearest neighbor silicons, expressed as Q^n, increases [15]. For example, the reaction of monosilicates to form a disilicate

$$2Q^0 \longrightarrow Q^1_2$$

is rapid compared to ring closure of linear trisilicate to form the cyclic trimer [16].

$$Q^1Q^2Q^1 \longrightarrow Q^2_3$$

Additionally, once formed, small rings are more stable than chains and species where silicons have ΔQ value greater than 1, e.g. a Q^4Q^1 structural component is very rare. Pendant groups are very reactive and are likely sources of building units for less highly polymerized anions. Finally, cages or cage-like structures are more stable than large rings [17]. This evidence results in a modification of the speciation diagram for silica in water originally presented by Stumm and Morgan to include species of increasing complexity. This diagram, shown in Fig. 2, is based on the application of NMR results to EMF data and represents

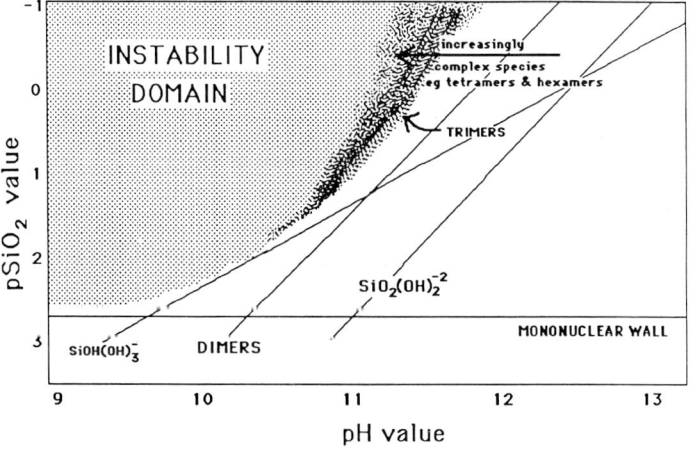

Fig. 2. Qualitative $pSiO_2$—pH Diagram

the solubility of silica in water at 25 °C versus pH value. At normal environmental pH values (pH < 9) dissolved silica has been thought to exist exclusively in monomeric form, since the silica concentration remains below the so-called "mononuclear wall" limit of 120 ppm for amorphous silica. Quartz has a solubility of only about 6 ppm, but its rate of crystallization is so slow at ordinary temperatures and pressures that the solubility of amorphous silica represents the upper limit of dissolved silica concentration in natural waters.

The depolymerization of polymeric dissolved silica species upon dilution occurs very rapidly to give smaller species, primarily monomer, that are indistinguishable from natural dissolved silica [18].

Recently, researchers have reported [19, 20] multimeric species below the mononuclear wall at a silica concentration of < 120 ppm ($pSiO_2 \simeq 2.74$) at pH values of 7.2 and 10.0, suggesting that the current model of the silica-water system can be further refined as we learn more about this system in the future.

Zeolite Chemistry

The zeolites are a group of aluminosilicates, which naturally occur and are also synthesized. They have a polymeric structure consisting of a framework made up of a three dimensional network of AlO_4 and SiO_4 tetrahedra linked by shared oxygen atoms. Cations, usually of group IA and IIA elements, such as sodium, potassium, and calcium, and water molecules reside in intracrystalline cavities, channels, or interconnected voids. This porosity may account for as much as 50% of the total volume of the crystals. Many of the cations are mobile and may undergo ion exchange. Water may be removed reversibly, usually by the application of heat.

In most zeolites, the primary structural units, the units AlO_4 and SiO_4 tetrahedra, usually assemble into secondary polyhedral building, e.g. single and double ring structures containing 4 to 12 polyhedra [21]. The ultimate structure of a zeolite consists of assemblages of these secondary units. The formulas used to designate zeolites are based on the crystal unit cell, the smallest repeating unit of these secondary assemblages, represented by:

$$M_{x/n}(AlO_2)_x(SiO_2)_y \cdot wH_2O$$

where n is the valence of the cation M.

The zeolite which has been studied most extensively for environmental interactions is a synthetic member of the zeolite group called zeolite Na-A (CAS 68989-22-0). A major use for this material is as a phosphate replacement in laundry detergents. It has the formula:

$$Na_{12}(AlO_2)_{12}(SiO_2)_{12} \cdot 27H_2O$$

It is a white powder which consists of cubic crystals with an average dimension of approximately 0.1 μm to 0.8 μm. The structure of the crystal unit cell of zeolite

Na-A is represented in the following stereogram. The apex of each angle is the location of a tetrahedral atom, either silicon or aluminum, the straight lines represent a shared oxygen atom. The sodium ions and water molecules in the intracrystalline voids are not represented.

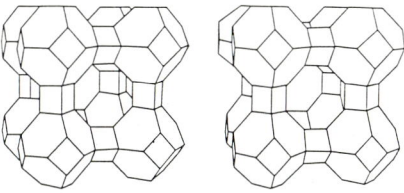

Fig. 3.

In contact with water, zeolite Na-A hydrolyzes and dissolves as sodium aluminate and sodium silicate [22]. The rate of hydrolysis depends on the composition of the surrounding medium, including factors such as cation types, concentration, presence of chelators and pH value [23]. Sodium aluminate and sodium silicate react to precipitate amorphous sodium aluminosilicate or, on a geological timescale, the minerals gibbsite, bayerite, halloysite, and kaolinite [24, 25]. The hydrolysis half-life of zeolite Na-A in aqueous media is about 55 days, based on kinetic studies [26]. What is observed on contact of zeolite Na-A with aqueous media, is the gradual conversion of the zeolite to amorphous sodium aluminosilicate.

Silicates and the Environment

What carbon is to biosphere, silicon is to the lithosphere. It is the fundamental structural element: compounds of silicon and oxygen comprise 60% of earth's crust [27]. These abundant mineral silicates dissolve to provide soluble silica to the hydrosphere. Ground water contains the highest concentrations: the median value in the U.S.A. is 17 ppm [28]. Of the surface waters, streams and rivers contain the most silica. For rivers the worldwide mean concentration is 13.1 ppm [29]. Lakes, seas, and oceans contain the least dissolved silica. The mean concentration for the world's oceans is about 6 ppm [30].

In addition to dissolution and other chemical processes, biochemical weathering processes decompose silicates in the natural environment. These processes, not yet fully understood, involve the transfer of energy from biological systems to silicates as well as ion substitution and chelate forming reactions which remove lattice cations [31]. The concentration of the resultant dissolved silica in natural waters is controlled by a buffering mechanism which is thought to involve the

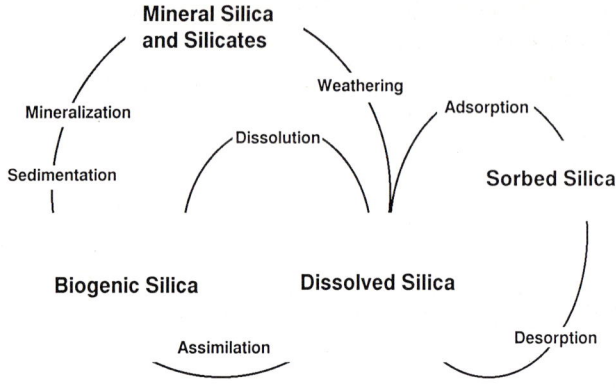

Fig. 4. The Natural Silica Cycle

sorption and desorption of dissolved silica on suspended particles [32]. An appreciation of the vast tonnage of silica that is mobilized by the natural silica cycle can be gained by contemplating the implications of the average silica weathering rate for watersheds of about 20 kg ha^{-1} y^{-1} [33].

Biological Interactions of Anthropogenic Silicates

It is surprising that the higher life forms that evolved on such a siliceous planet contain so little of the planet's natural abundance of silicates.

Silicon is most plentiful in lower plants and lower animals. Evidently, silica was first used by early life forms as a structural material before the cellulose-lignin and apatite-cartilage structural materials evolved [34].

Plants

The amount of silicon found in plants varies greatly with species and somewhat less so with soil factors. The grasses (*gramineae*) are rich in silica, with wet-land varieties containing the highest concentrations [35]. For example, the ash of the common horsetail (*Equisetum arvense*) contains 80% silica [36]. Diatoms, a class of unicellular plants (*Chrysophyta*), have a dry weight of as much as 50% silica [37]. At the other end of the continuum are yeasts, whose ash contains only about 1% silicon as silica [38]. The remainder of the plant kingdom ranges between these extremes. In general, legumes and dicotyledonous plants contain lower levels of silica than monocotyledonous species [39].

The great abundance of natural silicates limits the need for anthropogenic silicates as a fertilizer for crops. However, there appears to be some potential for supplementing the silica requirements of rice crops with industrial soluble silicates [40]. And manmade potassium silicates have been demonstrated to be useful as slow-release sources of potassium and silica [41]. These mineral supplements are particularly useful in geographical areas where limestone and other carbonate geology predominates. And there is considerable use of expanded perlite and vermiculite, two processed natural silicates, in potting soils to improve permeability.

Eutrophication

The potential impact of anthropogenic silicates on aquatic plants has been studied in great detail because of the discharge of these materials into wastewater systems arising from their use in laundry detergents, and the problem of premature eutrophication of inland freshwater lakes which is caused by excess nutrients, contributed, in part, by detergents. By eutrophication, we mean the premature ageing of lakes characterized by "blooms" of prokariotic, blue-green algae. This condition is very undesirable because blue-green algae produce unpleasant tastes, odors, colors, and floating clumps, and they are toxic to many organisms, including humans [42]. Whether detergent silicates contribute to eutrophication or cause other adverse ecological effects in inland waters is thus a question of considerable importance.

It has been demonstrated that at concentrations of less than 0.1 ppm, dissolved silica is a limiting nutrient for diatoms [43] and other algal species [44]. Thus, only in bodies of water that are an order of magnitude lower in silica concentration than average, could silica become a limiting factor to algal growth. The question of whether additional silica, added to waters that are not silica deficient, might nonetheless stimulate enhanced growth of diatom populations has also been investigated [45, 46], and it has been demonstrated that diatom population growth is independent of silica concentration, once the limiting concentration has been exceeded.

When a freshwater body becomes eutrophic due to large inputs of phosphorus a diatom "bloom", or rapid increase of diatom population, will occur, particularly in the Spring and Fall. Such blooms result in a decline in the dissolved silica content of the water, especially the water near the surface [47]. If this process is continued until the silica is depleted below the limiting concentration for diatoms, the diatom blooms are replaced by blue-green and green algal species which have much lower requirements for dissolved silica [48]. Thus, although it is most desirable to avoid nutrient overload of freshwater bodies, if such a condition does occur it is important to maintain the dissolved silica levels to encourage dominance of diatom species rather than blue-green and green algal species. Not only do diatom blooms lack toxicity and offensive visual and olfactory properties, a diatom bloom will deplete the available excess macronutrients, phosphorus, nitrogen, carbon [49]. This is because as the diatoms die off, they fall from the productive surface waters to the deeper waters where these nutrients will be unavailable for further algal growth until the fall turnover, whereas when blue-green and green algal species die, they lyse and immediately release their nutrients to the surface water where they are available for the next generation [50].

Animals

Nutritional Aspects

Relatively large amounts of silica occur in animals of the protozoan orders *Radiolaria* and *Heilozoa*, and in the siliceous sponges, *Hyalospongiae* [51]. However, the majority of animal species only contain silica in the parts per million range; with most of it being dissolved in the body fluids or interposed in the lungs. Most biogenous silica in animals is present as amorphous silica gel or dissolved silica [52], but some is bound to organic molecules whose precise structure has yet to be elucidated [53].

Although the essentiality of silicon has long been recognized in the primitive animals that use it as a structural material, only relatively recently, with the development of environmental isolators and synthetic diets, has it been realized that silicon is essential, in trace quantities, for most if not all animal species [54].

Toxicity

Soluble Silicates. The following table of LD_{50} values for sodium silicates was compiled from the results of a number of studies. It should be noted that these studies were conducted at different times, and vary somewhat as to the test conditions such as strain, number and sex distribution of animals, and length of observation period. A very closely controlled LD_{50} study would not yield data from which conclusions could be drawn with greater certainty unless a great number of animals were employed. Alkali silicates are not very toxic; in their lethal range, large doses are required and death results from nonspecific causes. The 95% confidence intervals are on the order of $0.5\,g\,kg^{-1}$, for the data reported below. Necropsy indicated acute gastroenteritis, vascular congestion and mottled livers [55], consistent with nonspecific causes of death, e.g. changes in pH value of body fluids, shock, chemical irritation or corrosion of the viscera.

Silicate modulus	Sodium silicate LD_{50} values Concentration, weight percent	LD_{50} $g\,kg^{-1}$ [55–58]
3.2	36	3.2
3	—	1.6–8.6
2	—	1.3–2.1
2.0	81	1.6
1.6	51	2.0–2.5
1.0	99	0.6
1.0	50	0.8
0.7	61	1.0
0.5	90	0.5

In man, the oral lethal dose of sodium silicates has been estimated as 0.5–5 g kg^{-1} [59]. Ingestion of 200 ml of sodium silicate egg preserving solution (these solutions typically contain 5–36% $Na_2O \cdot (SiO_2)_{3.2}$) caused severe vomiting, diarrhea and bleeding, elevated blood pressure, and renal damage, but was not fatal [60]. In the past, sodium silicate has been administered orally for medicinal purposes in doses of 1 to 3 g d^{-1} without reported adverse effects [61]; however, it is not presently known to be used as a drug.

Soluble silicates administered to dogs as 5% solutions had to be preneutralized or the dogs invariably vomited them [62]. Silica absorbed by the dogs from the neutralized solution was found to be quickly eliminated in the urine. The level of silica in the blood remained low, and it was suggested that these animals have a low renal threshold for dissolved silica. Feeding dogs and rats sodium silicate incorporated into a semisynthetic diet at SiO_2 levels equivalent to 0.8 g kg^{-1} d^{-1} produced polydipsia, polyuria, and soft stools. Renal lesions were observed in the dogs upon histopathological examination. Similar effects were not observed in the rats [63].

Long term drinking water studies [64, 65] with rats at dosage levels from 200 to 1800 ppm sodium silicate show small statistically unclear weight changes in one study and in the other an increase in serum alkaline phosphatase activity at a concentration of 1800 ppm in males, an increase of serum glutamic-pyruvic transaminase activity at 200 and 600 ppm sodium silicate in females, and a decrease in leukocyte count in both sexes at 600 ppm. No specific change in the rats due to the sodium silicate was observed in the latter study upon histopathological examination. One concluded that "soluble silica ... exerts biologically important effects on growth and reproductive performance" [66].

Studies of the urinary excretion of single oral dose of several silicates, including a 2.4 ratio sodium silicate, to rats conducted at doses of 40 and 1000 mg kg^{-1} indicated that the half-life of sodium silicate is 24 h [67]. Studies of the total silica eliminated (i.e., urinary and fecal SiO_2) by guinea pigs after oral single and multiple dose administration of alkaline silicates yielded roughly similar results [68].

Zeolite Na-A. The acute oral LD$_{50}$ of zeolite Na-A is reported as greater than 27 g kg^{-1} [69]. Rats tolerated a single oral dose of 10 g kg^{-1} "without any overt reaction" [70]. In dogs, 1 g kg^{-1} administered orally, produced slight hyperemia of the stomach, but no other adverse effects [71]. In a human stomach model, it was found that zeolite Na-A is hydrolyzed to silicates and aluminates as would be expected based on the previously cited chemical studies. Studies with rats fed 40, 200, and 1000 mg kg^{-1}, indicate the silicate is excreted by the urinary system and the aluminate in the feces [72]. In a 90 day subchronic feeding study, rats were administered 10 000 ppm zeolite Na-A in the diet. Diminished urine secretion, haematuria, ketone bodies in the urine, and siliceous urinary calculi were observed [73]. Where dietary concentrations exceed 0.125% (or thereabout), the urine becomes saturated with dissolved silica (the solubility of amorphous silica is about 120 ppm in the physiological pH range [74]), and precipitation occurs. The results of this study were confirmed in another study wherein rats were fed diets containing 0.1% to 2% zeolite Na-A for periods of time between 90 and 200 days [75]. The observed changes in the urinary system in studies of zeolite Na-A are consistent with observations in feeding studies of other sources of soluble silica [76].

The chronic oral toxicity of zeolite Na-A was tested in rats at 10 ppm, 100 ppm and 1000 ppm of the diet for 104 weeks [77]. Calculated doses were 0.62–0.65, 6.10–6.53 and 58.47–62.15 mg kg^{-1} d^{-1}, respectively. No significant incidence of a particular type of tumor or of spontaneous mortality was evident in any group. No treatment-related findings were seen in any of the organs examined histologically, and there was no indication of any treatment-related induction of neoplasms. In a second chronic feeding study wherein rats were fed at 0.001%, 0.01%, and 0.1% of the diet for 78–104 weeks, no compound-related effects on physical appearances, hematology, blood chemistry or urinanalysis were noted. Gross necropsy revealed no abnormal findings. The compound was not carcinogenic [78].

In a study which examined the toxicity of detergent ingredients [79], the acute toxicity of zeolite Na-A was tested on bacteria (*Pseudomonas fluorescens*), algae (*Chlorella vulgaris* and *Microcystis aeruginosa*), insects (*Aedes aegypti*), fishes (*Poecilia reticulata* and *Oryzias latipes*) and amphibians (*Xenopus laevis*). The L(E)C$_{50}$ values for the zeolite ranged from 180 to 320 mg/l.

Aquatic Interaction of Silicates and Aluminum. Research on the global environmental problem of acid rain has focused on the environmental interaction of acid, aluminum, and dissolved silica in fish toxicity. Acidification of natural waters results in an increased level of dissolved aluminum, and such dissolved aluminum is toxic to fish [80]. The solubilized aluminum binds to the fish's gill epithelia and causes loss of osmoregulatory function. However, when soluble silica is present in amounts such that there is a molar excess of silicon to aluminum, a hydroxyaluminosilicate species is believed to form which either limits the bioavailability of dissolved aluminum or blocks the toxic effects of aluminum on the fish gill epithelia [81].

Uses of Anthropogenic Silicates in Environmental Technology

Hazardous Waste Mitigation

Soluble silicates have been demonstrated on a laboratory scale to prevent the generation of acid mine water from waste coal refuse piles through neutralization, the development of a continuous gel seal on the surface of the refuse pile, and development of an internal silica/alumina gel which prevents percolation and retards erosion [82].

The use of silicates in grouts to seal hazardous waste sites to prevent the migration of contaminants to the groundwater has been extensively studied [83]. Sodium silicate and portland cement grouts are commonly used. Both materials are highly chemically tolerant, in that most waste materials will not inhibit their setting to form a seal. Portland cement grouts, being particulate materials are more suitable for coarse grained soils, while sodium silicate grouts are suitable for fine-grained or medium grained soils.

Another closely related application relating to environmental management is the use of silicate and or portland cement to stabilize or encapsulate hazardous waste in an insoluble low permeability silicate matrix [84]. Pojasek [85] has edited a multivolume treatise on this technology. The *EPA Guide to the Disposal of Chemically Stabilized and Solidified Waste* [86] also provides a valuable summary of silicate fixation and encapsulation technology.

Numerous highly specialized technologies exist for the silicate fixation of particular wastes. Heacock and Riches [87] discuss three processes for solidification of radioactive wastes. Other methods [88, 90] are used for fixing and transporting low-level nuclear wastes. In a review of twelve methods for treatment and disposal of oil well drilling wastes, silicate fixation is favorably evaluated [91]. The problem of dealing with high-level nuclear wastes is still of great concern and it currently appears that these materials will be placed in deep geological repositories sealed with silicate cements [92].

Water Treatment

Silica sol prepared from aged, partially neutralized sodium silicate solution when added to settling tanks, in conjunction with a primary coagulant such as aluminum sulfate or ferric salts, in water treatment greatly improve floc formation and settling [93]. This technology has developed and is used in many water plants throughout the world. Additionally, high modulus sodium silicates are added to municipal water to inhibit the development of red water problems associated with high levels of iron and manganese [94].

Several wastewater treatment technologies employing silicates have been patented. The use of soluble silicates to precipitate metallic ions as water-insoluble silicates is a common theme [95, 96]. Another broad area of technology is the use of silicates to adsorb or exchange metal ions from solution. Calcium silicates have been proposed for this application [97, 98]. Zeolites have also been demonstrated to be useful for exchanging lead and cadmium from wastewater [99]. This technology is developing for use in the treatment of the aqueous effluent from metal plating operations. An initial precipitation of metals from the water with a hydroxide is followed by "polishing" or "tertiary treatment" with zeolite to remove traces of metals prior to discharge [100].

Land Treatment

An expanded polysilicate and hydrogen peroxide solution has been developed to treat soil that has become contaminated with gasoline from leaking tanks [101]. The expanded polysilicate has frankel defects which are thought to act as catalysts to decompose the peroxide to singlet oxygen which then oxidizes the gasoline.

References

1. Lovejoy CO (1981) Science 211:341
2. Oldfather WA (1920) J Am Ceram Soc 3:526
3. Morey GW (1938) The properties of glass, ACS Monograph No. 77, Reinhold, New York, p 12
4. Andrews AI (1935) *Enamels*, Twin Cities Pub, Champaign, IL, p 2
5. Rochow EG (1987) Silicon and silicones, Springer, Berlin Heidelberg New York, p 26
6. Vail JG (1928) Soluble silicates, ACS Monograph No. 46, Chemical Catalog Co., NY, p 95
7. Vail JG (1928) Soluble silicates, ACS Monograph No. 46, Chemical Catalog Co., NY, p 410
8. Vail JG (1928) Soluble silicates, ACS Monograph No. 46, Chemical Catalog Co., NY, p 411
9. Liebau F (1985) Structural chemistry of silicates, Springer, Berlin Heidelberg New York
10. Hoebbel F, Ebert R (1988) Z Chem 28:41
11. Falcone JS (1989) Cements research progress, Am Ceram Soc, Westerville, OH
12. Harris RK et al. (1984) J Magn Res 57:115
13. Ray NH, Plaisted RJ (1983) J Chem Soc, DA 475
14. Svensson I et al. (1986) J Chem Soc F1 82:3635
15. Dent Glasser LS, Lackowski EE (1980) J Chem Soc DA 394, 399
16. Engelhardt G, Hoebbel D (1984) J Chem Soc Chem Comm 514
17. Harris RK (see Ref. [12])
18. O'Connor TL (1961) J Phys Chem 65:1
19. Cory LW et al. (1982) Geochim et Cosmochim ACTA 46:1317
20. Alvarez R, Sparks DL (1985) Nature 318:649
21. Lok BM et al. (1983) Zeolites 3:282
22. Cook TE et al. (1982) Environ Sci Technol 16:344
23. King JE et al. (1980) JWPCF 52:2785
24. Cook TE et al. (1982) Environ Sci Technol 16:349
25. Stumm W, Morgan J (1981) Aquatic chemistry, Wiley, NY, p 541
26. Llenado RA (1984) In: Olson D (ed) Procedings of the Sixth International Zeolite Conference, Butterworths, Guildford, UK, p 947
27. Iler RK (1955) The colloid chemistry of silica and silicates, Cornell Univ. Press, Ithaca, NY, p 3
28. Davis SN (1964) Am J Sci 262:870
29. Edwards AMC, Liss PS (1973) Nature 243:341
30. Kido K (1974) Marine Chemistry 2:277
31. Boyle JR, Voigt GK (1973) Plant Soil 38:191
32. Silverman MP (1979) In: Trudinger PA, Swaine DJ (eds) Biogeochemical cycling of mineral forming elements, Elsevier, p 445
33. Soukup MA (1974) Ph.D. Dissertation, Univ of MA. See also Soukup MA, The limnology of a eutrophic hardwater New England lake, with major emphasis on the biogeochemistry of dissolved silica, University Microfilm International, Ann Arbor MI, No. 75-27-527
34. Carlisle EM (1973) Trace Elem Metab Anim Proc 2nd p 407
35. D'Hoore J (1972) Soils of the humid tropics, Nat Acad Sci, Washington, DC, p 163
36. King EJ, Belt TH (1938) Physiol Rev 18:329
37. Soukup M (1974) U Mass Water Resources Res Ctr Pub 39:67

38. King EJ et al. (1933) Biochem J 27:1002
39. D'Hoore J (see Ref. [35])
40. Yoshida S (1975) Tech Bull ASPAC Food Fert Technol Cent 25
41. Miwa E et al. (1978) Soil Sci Plant Nutr 24:103
42. Schwimmer D, Schwimmer M (1964) In: Jackson D (ed) Algae and man, Plenum, New York, p 368
43. Kilham SS (1975) J Phycol 11:396
44. Klaveness D, Guillard RRL (1975) J Phycol 11:349
45. Jorgensen E (1957) Dansk Botanisk Arkiv 18:1
46. Schwartz AM, Davis AE (1974) The development of phosphate-free heavy duty detergents, Project 06080-FWE, U.S. Environmental Protection Agency, Washington, DC
47. Schleske CL, Stoermer EF (1971) Symposium on nutrients and eutrophication, Am Soc Limnology and Oceanography
48. Kilham P (1971) Limnology and Oceanography 16:10
49. Keating KI (1978) Science 199:971
50. Keating KI (1978) Science 199:
51. LeVier RR (1975) Bioinorganic Chemistry 4:109
52. Lewin J, Reimann BEF (1969) Annual Review of Plant Growth 20:289
53. Schwarz K (1974) Trace Elem Metab Anim, Proc Int Symp 2nd, p 355
54. Carlisle EM (1972) Science 178:154
55. Gaskins JR (1966) Analytical Report No. 016-2583, U.S. FDA, Div. Toxicological Evaluation, Bureau of Science, Washington, DC, 2
56. Calandra JC, Fancher OE (1972) Cleaning products and their accidental ingestion, The Soap and Detergent Association Scientific and Technical Report No. 5R, p 24
57. PQ Corporation (1961) Biological Study No. LH57085-1-4, 4
58. Hehir RM (1967) Research data on silicates—memorandum, EW Ligon/RM Hehir, U.S. FDA, Div Toxicological Evaluation, Bureau of Science, Washington, DC, 1
59. Joint FAO/WHO Expert Committee on Food Additives, Toxicological evaluation of some additives including anticaking agents, antimicrobials, antioxidants, emulsifiers, and thickening agents, WHO Food Additive Ser No 5, p 21
60. Eichhortst H (1920) Schweiz Med Wochschr 50:1081
61. Scheffler L et al. (1920) Comptes Rendus 171:416
62. King EJ et al. (1933) Biochem J 27:1002
63. Newberne PM, Wilson RB (1970) Proceedings National Academy of Science U.S. 65:872
64. Ito R et al. Toho Igakkai Zasshi 22:223
65. Smith GS et al. (1973) J Anim Sci 36:271
66. Smith GS et al. (1973) J Anim Sci 36:876
67. Benke GM, Osborne TW (1979) Fd Cosmet Toxicol 17:123
68. Sauer F et al. (1959) Can J Biochem and Physio 37:183
69. Beck LW et al. (1981) Regulatory Toxicology and Pharmacology 1:47
70. Gloxhuber C et al. (1983) Fd Chem Toxicol 21:210
71. International Joint Commission Task Force on the Health Effects of Non-NTA Detergent Builders, Health Implications of Non-NTA Detergent Builders, Windsor, Ontario, October 1980, revised March 1981, p 59
72. Benke GM, Osborne TW (1979) Fd Cosmet Toxicol 17:124
73. Gloxhuber C et al. (1983) Fd Chem Toxicol 21:213
74. Alexander GB et al. (1954) J Phys Chem 58:453
75. International Joint Commission Task Force on the Health Effects of Non-NTA Detergent Builders, Op Cit 62
76. Newberne PM, Wilson RB (1970) Proceedings National Academy of Science U.S. 6:872
77. Gloxhuber C et al. (1983) Op cit Fd Chem Toxicol 21:213
78. International Joint Commission Task Force on the Health Effects of Non-NTA Detergent Builders, Op cit 62
79. Canton JH, Sloof W (1982) Chemosphere 11:891
80. Witters HE (1986) Aquatic Tox 8:197
81. Birchall JD et al. (1989) Nature 338:146
82. Tyco Laboratories (1971) Silicate treatment for acid mine drainage prevention, Water Pollution Control Series No. 14010 DLI 02/71, U.S. EPA, Washington, DC
83. Malone PG (1984) Development of methods for insitu hazardous waste stabilization by injection grouting, EPA Land Disposal of Hazardous Waste 10th Research Symposium, p 33

84. Falcone JS Jr et al. (1984) Chemical interactions of soluble silicates in the management of hazardous wastes, In: Jackson et al. (eds) Hazardous and industrial waste management and testing: Third Symposium, ASTM Pub. STP851, ASTM, Baltimore, MD
85. Pojasek PB (1979) Toxic and hazardous waste disposal, Ann Arbor, MI
86. Malone PG et al. (1980) Guide to the disposal of chemically stabilized and solidified waste, Report SW-872, U.S. Army Eng Waterways Exper. Station, Vicksburg, MS
87. Heacock HW, Riches JW (1974) Waste solidification cement or urea formaldehyde, ASME Symp, New York
88. Granlund RW, Hayes JF (1979) Solidification of low-level radioactive liquid waste using a cement silicate process, Health Physics Soc Symp Williamsburg, VA
89. Phillips JW (1979) Applying techniques for solidification and transportation of radioactive wastes to hazardous waste sites EPA Conference on Management of Uncontrolled Hazardous Waste Sites, Washington, DC
90. Maugh TH (1979) Science 204:1295
91. Nesbitt LE, Sanders JA (1981) J Pet Tech 33:2377
92. Wakeley LD et al. (1984) Materials Res Soc Symp, Boston, MA, p 951
93. Baylis JR, Mrva AE (1963) J AWWA 55:1536
94. Dart FJ, Foley PD (1970) J AWWA 62:663
95. Kendall K (1967) U.S. Patent 3, 310, 370
96. Goto F (1975) Japanese Patent 75, 127, 448
97. Gushima Y (1975) Japanese Patent Ap JA 071570
98. Tohyama I et al. (1973) Isaka Kogyo 24:294
99. Hertzenberg EP, Sherry HS (1980) Lead and cadmium exchange of zeolite NaA, American Chemical Soc Symp, San Francisco, CA
100. Hertzenberg EP (1984) In: Olsen D, Biso A (eds) Proc of 6th International Zeolite Conf, Butterworths, London Great Britain, p 975
101. Smith DL, Sabherwal IH (1987) On-site remediation of gasoline-contaminated soil, Int Cong Env Profs, Chattanooga, TN, 241

Silicones

Jerry L. Hamelink

Dow Corning Corporation, Mail # CO3101, 2200 W. Salzburg Road, Midland, MI 48686-00994, USA

Introduction 383
Production 385
Manufacturing Chemistry 386
Analytical Methods 387
Physical Properties 388
Uses 389
Environmental Fate 390
Ecotoxicology 391
Toxicology 392
References 393

Summary

The chemical and physical properties of silicones are briefly summarized to provide a basis for assessing their fate and effects in the environment. Roughly one-half of the 400 000 tonnes of silicone produced during 1988 worldwide were believed to have been the potentially environmentally mobile fluids. About one-half of this fluid could conceivably gain entry to aqueous environments, via sludge disposed from Waste Water Treatment Plants, storm water run-off and other sources. Following introduction, the low molecular weight species are either lost to the atmosphere or adsorbed onto sedimenting particles. All of the higher molecular weight PDMS fluids are believed to be adsorbed out of the liquid phase, so none of them remain in the water. Sedimented silicones are believed to become permanent, innocuous, inert, minuscule components of the sediments. Since the use of these innocuous silicone fluids is expected to continue to increase, and, although they clearly pose no direct toxic hazards, their fate and effects in or on the environment will continue to be studied.

Introduction

Silicon resides directly below carbon in Group IV of the Periodic Table. Unlike carbon it has three shells of electrons ($1s^2$, $2s^2 2p^6$, $3s^2 3p^2$), rather than two ($1s^2$, $2s^2 2p^2$) [1]. It also has an atomic weight 2.3 ×, an atomic radius 1.6 ×, a covalent radius 1.5 × and a First Ionization Potential 72% that of carbon [1]. Its outermost shell could theoretically hold 10 more electrons (18 instead of 8, although 12 is the maximum known for Si) and the four valence electrons reside farther away from the nucleus than they do in carbon. However, loose electrons are not necessarily prone to form π bonds and silicon is not easily able to form unsaturated bonds. Rather, as Frederick Stanley Kipping, "a professor at University College, Nottingham, England" who pioneered the chemistry of the organosilicon compounds for 36 years [2], proved repeatedly, silicon will not form an isolable (or stable) double bond easily. Hence, although sterically hindered disilenes (Si=Si) and silenes (Si=C) have been recently prepared, true "silicones", i.e. silaketones (Si=O), have only been reported as transient species to date [1]. Yet, that is where their common name comes from, because if $(CH_3)_2CCl_2$ is hydrolyzed, the product is the ketone $(CH_3)_2C$=O (acetone), but whenever Kipping hydrolyzed a chlorosilane he made the "uninviting glues and oils" that he called silicones. This was/is the case because, rather than forming "silaketone" double bonds, the intermediate silanol hydrolysis products undergo intermolecular condensation to form the ether linkage, SiOSi, structures [1].

Although there are no naturally occurring organosilicon compounds, silicon containing sand and silicate minerals account for > 50% of the earth's crust. Thus, unlike carbon, which constitutes only $\approx 0.03\%$ of the earth's crust [1], the basic starting material/building block is extremely common. However, silicon cannot be pumped out of the ground (like oil) and its molar heat of oxidation is very high (i.e. 198 kcal, more than twice that of carbon) [3], so a considerable amount of energy is needed to reduce silica to silicon before it can be engineered into a silicone. Hence, the silicones generally have a higher unit cost than similar carbonaceous structures. Fortunately, for the industry, their extraordinary efficacy and reliability usually serves to offset their higher unit cost.

The bonds between most other atoms and silicon are stronger than those same bonds with carbon [4]. Silicon can be bound to both carbon and oxygen to form a silicone polymer. The Si–O–Si linkages (siloxanes) that constitute the backbone of a silicone is nothing like an organic ether. Rather, the bond with oxygen has a very large angle (i.e. about 150° or almost straight, versus 113° for dimethyl ether) [5], plus it is shorter and stronger than that with carbon, either because the unshared electrons from oxygen enter to some degree into silicons vacant d-orbitals [6] or because some kind of secondary bonding interaction between silicon and first row atoms exists [5]. These bonds also appear to be extraordinarily flexible [4], presumably because the substituents on the silicon, like the dimethyl groups on linear molecules, do not suffer any steric hindrances.

Silicones are engineered materials that possess several unique physical (and chemical) properties. These properties permit them to be used by many people to aid in the manufacture of or as part of the composition of a very large variety of different products. Polydimethylsiloxane (PDMS) fluids are the largest volume silicone polymers in commerce. Many of their applications are due to their unique surface properties. The diversity of interfaces and the contradictory nature of many of the surface property applications for which silicones are used are well discussed by Owen [4, 6]. PDMS is a dimethylsiloxane polymer, as shown below:

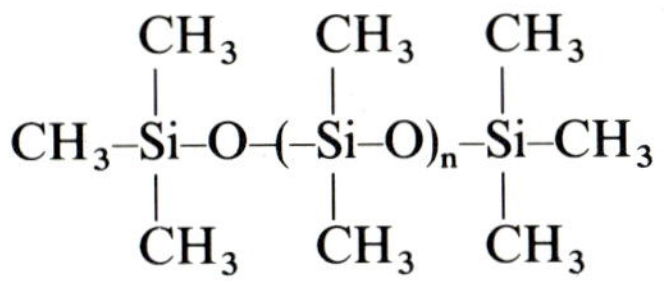

$$\text{CH}_3\text{-Si(CH}_3\text{)}_2\text{-O-(-Si(CH}_3\text{)}_2\text{-O)}_n\text{-Si(CH}_3\text{)}_2\text{-CH}_3$$

Siloxanes can be made with nearly any organic group (as represented by R), but because each oxygen links two silicons, the general formula for the siloxanes is $R_n SiO_{(4-n)/2}$ rather than $(R_2SiO)_n$, with n = 1 to 3. Attachment of only one organic group to the silicon makes it is trifunctional and cross-linked, networks can be formed. Attachment of two organic groups makes the silicon difunctional and either a relatively small cyclic or a linear structure can be constructed. The linear structures are usually "end-blocked" with monofunctional silicon having three organic groups attached. These silicone structures are commonly represented by the symbols T, D and M, respectively. Thus, the PDMS polymer shown above could also be depicted as MD_nM. A silicone polymer is most stable when end-blocked with a non-reactive group like $OSiMe_3$. Conversely, polymers used as silicone intermediates may be end-blocked with reactive groups like Cl, OH, OR, O2CR, etc. [7].

Production

World sales of non-semiconductor, metallurgical grade silicon were 122 593 short tonnes through September of 1988 [8]. Industrially, ground silicon is reacted with methyl chloride in a fluid bed reactor [7] to form primarily dimethyldichlorosilane. If the 122 593 short tonnes are assumed to be 75% of Annual World Production (AWP) for 1988, if 90% of this is assumed to be used by the silicone industry, if its weight is assumed to increase 4.6× upon conversion to Me_2SiCl_2 and if hydrolysis (which removes the chlorines and adds the oxygen) is assumed to reduce the weight by 43% (i.e. ×0.57), then 385 726 tonnes of dimethylhydrolyzate or "silicone" may have been synthesized in the world during 1988. Silicone production capacity in the United States, Western Europe and Japan, which are believed to possess 90% of AWP, was estimated to have reached approximately 435 000 tonnes in 1987 [9]. Consequently, AWP of silicones was believed to have been about 400 000 tonnes during 1988.

The AWP is believed to be approximately 55% fluids, 40% elastomers and 5% resins [9]. Since the elastomers and resins are not environmentally mobile, only the environmental fate and effects of the silicone fluids will be addressed. The most commonly used fluids are: 1) the Polyethermethylsiloxanes (PEMS), which may also be called copolyols or copolymers, and 2) the non-volatile linear and the volatile, cyclic PDMS's. The linear PDMS fluids are routinely distinguished from each other by their different viscosities. The low viscosity fluids range from 0.65 centistokes (cSt), for the smallest moiety, hexamethyldisiloxane, to 50 cSt fluids. They offer the maximum compressibility, the lowest temperature service range and the best heat transfer properties of the silicone fluids. They may be volatile and they are also soluble in more organic solvents, including the petroleum oils, than the higher viscosity fluids. The intermediate viscosity fluids (50–5000 cSt) are nonvolatile fluids that are suitable for a multitude of applications, including hydraulic and/or brake fluids, dielectric fluids, mold and paper release agents, antifoams, auto polishes and hand creams. Fluids in the 5000–2 500 000 cSt range are considered to be high-viscosity fluids. They are employed as internal lubricants and process aids for thermoplastics, as liquid springs in shock absorbers, as stationary phases for GLC columns and as gums to make elastomers [1, 9]. Only a small fraction (i.e. 2–3%) of the fluids entering the market consist of the small, cyclic, volatile oligomers. Less than 10% of the volatile tetramer octamethylcyclotetrasiloxane (OMCTS or D4) synthesized by hydrolysis of Me_2SiCl_2 enters commerce because most of the cyclics are stripped from the hydrolyzates and recovered for conversion to elastomers [7]. The PEMS and all of the other kinds of fluids, combined, currently comprise only a few percent of the total market.

Manufacturing Chemistry

The methyl silicones are manufactured with silicon, methanol and hydrochloric acid by a series of energy intensive processes involving various metal cataylsts [2]. PDMS's are produced by hydrolysis of chlorosilanes created by the Direct Process [1]. Platinum [7], copper, silver and iron catalyze syntheses while lead and bismuth often act as catalyst poisons. A fairly large amount of catalytic copper or silver (i.e. 5–10% by weight of silicon) may be required for the Direct Process [1]. Although only silicon and an organic halide should have to be added in a continuous operation [3], some of the metals may escape and hydrochloric acid is produced when the chlorosilanes are hydrolyzed. Thus, even though most of the acid is recycled and used in different production processes, some of it must still be neutralized (usually with limestone) and metal catalyst losses minimized before the waste waters may be discharged. The processing waste waters may be combined with the much larger volumes (approx. 20 ×) of the spent cooling waters which are needed for the energy intensive and/or exothermic manufacturing processes used to create the silicones. Hence, although some volatile organic halides may escape to the atmosphere, clear, nontoxic waste waters are generally discharged from silicone manufacturing operations.

The electropositive nature of silicon may cause vigorous reactions which are unknown in analogous carbon compounds [3]. The Direct Process consists of reacting powdered silicon with methyl chloride in a fluid bed reactor, as discussed above, to form primarily dimethyldichlorosilane and small amounts of the other chlorosilanes [1, 7]. The volatile silanes are separated and purified by distillation prior to being hydrolyzed to form the silicones. Chlorosilanes react vigorously with water and suitable measures are required to prevent them from escaping, since they are very corrosive due to the HCl vapors formed when they react with atmospheric humidity.

Once the silanes and siloxanes are formed, the other products are created by a variety of redistribution, addition or nucleophilic displacement reactions [1]. Two of the most commonly used processes are silylation and hydrosilylation. Hydrosilylation is the addition of a silicon-hydrogen bond to a multiple carbon-carbon bond that is usually catalyzed by very small amounts of chloroplatinic acid (CPA) which generally places the silicon on a terminal carbon [1]. Hence, hydrosilylation reactions do not proceed well with internal olefins and terminal double bonds react preferentially over the internal double bonds of dienes [1]. For example, if the silane $HSiCl_3$ were to be added drop-wise to the liquid olefin $CH_2{=}CHSiCl_3$ containing 10^{-5} to 10^{-8} M of CPA, a considerable amount of heat would be released and $Cl_3SiCH_2CH_2SiCl_3$ would be formed.

Silicon is more electropositive than hydrogen, carbon and oxygen [5], their respective Pauling electronegativities being 1.8, 2.1, 2.5 and 3.5 [3]. Silylation reactions capitalize on this electropositive nature. That is, silicon bonded to any good leaving groups, such as Cl, will silylate almost anything containing an acidic hydrogen such as alcohol, amines, etc. Thus, if the silylating agent trimethylchlorosilane (Me_3SiCl) is reacted with methanol (CH_3OH), trimethyl-methoxysilane (Me_3SiOMe) and hydrochloric acid are formed. As a general rule of thumb, Si bonded to almost any electronegative element or group will serve as a silylating agent. However, fluorine will not work under normal conditions because it is too electronegative. Reactions between the chlorosilanes and ammonia form the silazanes (Si–N–Si).

Analytical Methods

Many problems exist with analysis for trace amounts of silicones in any matrix because many silicones are contained in many personal care and chemical laboratory products. Hence, samples may easily be accidentally contaminated [10], making organosilicon analysis unusually challenging. IR spectroscopy is the best way to characterize reasonable amounts (i.e. >1 mg) of a silicone polymer. NMR and MS are also useful for structural determinations. GC/MS certainly provides the most powerful combination of selectivity, specificity and sensitivity [11], but is restricted to volatile compounds. A novel GLC analytical method, involving hexamethyldisiloxane (HMDS) as a selective, in situ, silylation and extraction agent, for low (i.e. ppm) concentrations of water-borne organosilicon substances was recently reported [12]. The GLC techniques which may be used to determine the kinds of alkyl substituents of organosilicon mixtures and/or for identification/quantification of the methyl-substituted linear and cyclic siloxanes, are well documented [11].

Volatile organosilicon substances, such as OMCTS $(Me_2SiO)_4$, may be easily assayed by GLC and ppb concentrations in water or air can be specifically detected using a purge and trap system [13]. However, most silicone products are large polymers which are not volatile at ordinary temperatures, so they cannot be directly assayed by GLC. Nonetheless, since PDMS is 37.8% silicon, its detection in an environmental compartment, like aquatic sediments, is a relatively trivial task [14]. This is the case because any silicon extracted from an environmental media with an organic solvent, like methylene chloride, that is detected by some means, such as atomic absorption, must be an organosilicon. Unfortunately, elucidation of the exact structure of the silicones extracted can be extremely difficult, if not completely impossible, to determine. The larger polymer mixtures can be assayed by Gel Permeation Chromatography (GPC), wherein a relatively short liquid chromatography column is eluted with a solvent, such as toluene, and the refractive index of the eluate recorded continuously. A molecular size distribution profile is obtained which can be correlated with the molecular weight profiles obtained through use of the appropriate standards. However, this technology requires a fairly substantial amount of silicone to work effectively. Consequently, although measuring the environmental fate and chemistry of an individual organosilicon species is extremely difficult, the environmental fate of the silicone fluids is generally well understood.

Physical Properties

The flexible Si–O–Si linkages of silicones permits the $Me_2SiO_{2/2}$ groups to rotate freely about the polar Si–O bonds of linear molecules. Hence, the very surface-active, non-polar methyl groups are routinely and effectively oriented along any non-polar interfaces and the oxygen molecule is oriented along any polar interface that may exist between two phases. The backbone may also flex freely, such that two molecules cannot reside close to each other. Thus, the intermolecular attractions between silicone polymers and other polymers is low. Their surface energy, like most other biocompatible polymers, is also very low [4]. Consequently, PDMS silicone fluids have low surface tension, low density, large molar volume and high affinity for the interfacial surfaces between two mediums.

The silicones are noted for being extremely hydrophobic. The most common linear siloxanes are completely insoluble in water. Their insolubility presumably arises because the random motions of the Si–O backbone and of the methyl groups prevent the alien water molecules from coming close and the Si bonded oxygen's lone pair of electrons is not available for hydrogen bonding [4, 5]. Thus, the simple silicones are insoluble in water and form very hydrophobic films when applied to any surface. The critical surface tension of wetting for PDMS (24 mNm^{-1}) is also higher than its liquid surface tension (20.4 mNm^{-1}). Hence, it will not form a running "droplet" but it will spread out over its own adsorbed film (i.e. "creep") whenever it is placed on a surface.

The flexibility of the Si–O bond, that was previously described, which leads to the alignment of the inorganic silicate along one side of the molecule and the organic substituents along the other, gives silicones their unique surface/interfacial properties [4, 6]. In other words, silicone fluids essentially prefer to reside at or on the interface between two polar and apolar mediums. The polar medium is usually water while air or a solid like laboratory glassware, sewage sludge, seston or aqueous sediments serve as the apolar medium. Consequently, as will become more evident, substantial amounts of the silicone fluids, regardless of how they are used by man, ultimately become sequestered in aqueous sediments.

Uses

Silicone fluids are very surface active agents and because of this characteristic, plus their biocompatibility, small amounts are effective as antifoam agents in Waste Water Treatment Plants (WWTP's), laundry detergents, in both food and drug synthesis fermentation processes and in a wide variety of personal care products [7]. For example, PDMS is known as Simethicone in antiflatulent tablets [2] and as Dimethicone to cosmetic chemists [15]. These physical properties also make the silicone liquids good light duty lubricants, mold-release agents, and useful components of auto polishes, numerous household products, metal protectants and moisture proofing agents [7]. The proportion of silicone fluids believed to be used as processing aids (which includes the antifoams, release coatings and surfactants) is about 30%. About 20% of the fluids are used for the formulation of cosmetics, toiletries, and a variety of pharmaceutical, medicinal or dental products. About 17% is used directly for polishes and coatings, about 10% is used to make the elastomeric/resinous paper coatings, textiles may be directly treated with about 8% and all other uses comprise about 15% of the annual world market [9].

PEMS consists of a PDMS "backbone" wherein some of the silicon substituents are polar side chains that are compatible with water. Type A have their polyether (polyalkyleneoxide) side chains attached to the polydimethyl silicone backbone by means of a silicon carbon bond. Type B have their side chains attached by way of silicon-oxygen-carbon bonds which may be hydrolyzed under the proper conditions to release the polyalkylene-oxide alcohols and silanol functional fluids [16]. They characteristically tend to reside at aqueous interfaces where the hydrophobic dimethylsilicone portion is pushed out of solution and the hydrophilic polyalkyleneoxide portion is pulled into the water

[15]. As a result of this alignment, the surface tension of an aqueous PEMS solution approaches that of a bulk dimethylsilicone, yet they are water soluble. Consequently, given this dual nature, they have proven to be useful for a wide range of applications such as a polyurethane foam control agents, home laundry fabric softeners and a variety of personal care hair products [15].

A substantial amount of PDMS is now being used as a dielectric coolant in power transformers. Soon after the undersirable ecotoxicological properties of PCB's were first recognized in the 1970s, several alternative transformer fluids and transformer designs appeared. Although PDMS fluids were also introduced as nontoxic, ecologically benign, alternative transformer fluids at that time, they are more expensive than the other alternatives and they did not command a substantial share of the market until recently, presumably, because it took time for the economic advantages offered by their unmatched combination of stability and reliability to be realized by the marketplace.

Environmental Fate

Much of the silicone contained in personal care products for hair or skin may be washed off during bathing and enter sewage systems as emulsions. Although most of the PDMS used in detergents, fabric softeners and in other textile treatments will stay on the fabrics, some of it may eventually enter sewage water. Low concentrations of PDMS may even be used directly in WWTP's to suppress foaming of the sewage without increasing the BOD. Thus, measureable amounts of PDMS may enter WWTP's, but it is removed from the water column by the flocculation-precipitation processes that are routinely used to treat sewage. Consequently, little, if any, of the water borne silicones introduced to WWTP's are contained in treated waste waters [14]. However, sludge disposal and other municipal waste water treatment operations may still introduce measurable amounts of silicones to aquatic environments [14]. PDMS may also reach terrestrial environments as the result of spills, when it is contained in an auto polish or when sewage sludge is used to amend arable land.

Silicone fluids are not believed to be persistent in/on arable lands because PDMS (which is renowned for its resistance to oxidation) may undergo extensive rearrangement on various clay soils [17]. Ironically, even though the high heat of formation (108 kcal/mole) of the Si–O bond makes it very resistant to homolytic cleavage, the Si–O bond rather than the Si–C bond is attacked [17]. Both volatile cyclics, which are the predominant kaolinite attack products, and non-volatile, linear, water soluble siloxanols, which are the predominant montmorillonite rearrangement products, are formed [17]. The cyclics are presumably lost rapidly to the atmosphere, while some of the siloxanols could be transferred to aquatic environments and undergo photolytic oxidative degradation [18]. Hence, PDMS is probably unstable in terrestrial environments. However, the catalytic activity of the clays is inversely related to the degree of hydration [17], so PDMS in aqueous sediments should not be as susceptible to these clay-catalyzed rearrangement reactions as in terrestrial soils. Thus, given their stability in wet environments and their affinity for solid-water interfaces (like those of sedimenting particles), any PDMS placed in an aqueous environment is believed to become a relatively stable resident of the sediments. Given the wide range of uses which currently exist for PDMS, some of it is expected to eventually become an innocuous, inert, low concentration (i.e. < 1–10 ppm) component of anthropogenically impacted aquatic sediments [17, 18, 19].

The volatile, cyclic oligomers are substantially different from the nonvolatile linear siloxanes. OMCTS or D4 is a fairly common, small (molecular size 1.08 × 1.03 nm), cyclic oligomer. The colorless, odorless, oily, 296 MW, liquid is sparingly soluble in water (approx. 0.05 mg L^{-1}) and volatile at room temperatures (calculated VP for 20 °C = 0.681 mm Hg). Hence, given its low aqueous solubility and modest volatility, it could be expected to have a high Henry's constant [20] and to reside primarily in the atmosphere at equilibrium [18], despite its molecular weight. Consequently, unlike the large, linear siloxanes, the small cyclic oligomers are expected to "behave" like low molecular weight organic solvents in water and they are rarely found in natural aquatic environments [21].

Ecotoxicology

The environmental fate and effects of silicones were recently (1986) reviewed by the multinational joint Group of Experts on the Scientific Aspects of Marine Pollution [22]. They evaluated the silanes and silanols; PDMS, polyphenylsiloxane, silicone polyether and other copolymer fluids (PEMS), and the silicone elastomers and resins. They and a more recent reviewer [19] both concluded that PDMS fluids are expected to become permanent residents of aqueous sediments but they should not have any adverse effects on the environment. This is true eventhough in laboratory studies conducted with PDMS and *Daphnia magna* using very "clean" laboratory water, the small invertebrates have been observed to float out and become physically "trapped" on the surface of water containing minute amounts of a non-volatile silicone fluid, eventhough no surface film is evident [23, 24]. This physical effect, a test method artifact, can be prevented either by introducing a fine screen below the surface of the water, as is commonly done to prevent floaters [25] or by adding a surfactant like Tween 80, which reduces the amount of silicone deposited on the exterior of the small invertebrates [26]. Thus, although small invertebrates, like daphnia, may appear to be exceptionally sensitive to a wide variety of silicones when studied under normal laboratory conditions, similar effects have not been observed [19] and are not expected to arise in physically complex microcosms or in natural environments because a large variety of other surfaces, natural "surfactants" (like

DOM) and interfaces exist that cause the silicones to be deposited and/or dissipated before they can have any adverse effects on planktonic organisms.

The molecular weight and size of D4 are small enough to warrant fish bioaccumulation tests [27] and in the laboratory, with a constant renewal, flow-through aqueous test system, it can be bioconcentrated 1000–5000 × by fish [28, 29]. This observation is consistent with its low aqueous solubility and log P of 4.45 [30] to 5.1 [31]. However, the low aqueous solubility in conjunction with its volatility should give it an exceptionally high Henry's constant and cause it to rapidly volatilize from water, so it is not expected to be bioaccumulated by fish in any natural environments. Thus, it is not believed to pose any ecological hazards and any D4 transferred into the atmosphere should rapidly undergo photo-oxidative decomposition [32] in the troposphere.

Fish are not able to accumulate comparable amounts of the analogous linear oligomers under laboratory conditions. This may arise because their lower aqueous solubility reduces their total "availability" or due to rapid depuration, as reported by Opperhuizen [29] (although that seems unlikely since no siloxanols nor any other water soluble organosilicon metabolites were found following depuration), or it may simply be due to the geometry and the unusual surface properties of the linear siloxanes. That is, MD_2M would presumably have a molecular size of about 2×0.5 nm, so its diameter should be small enough to pass through a cellular membrane. However, given their unusual affinity for residing on the interface between two phases, they may simply "prefer" to lie flat on a membrane, rather than pass through them directly, like a needle, such that fish can accumulate little directly from the water and that which does adhere to their exterior surfaces is rapidly depurated.

It has also been demonstrated that linear oligomers larger than MD_9M are not accumulated by any marine organisms when they have been intentionally added to their food and/or water [33]. Hence, silicones are highly unusual engineered materials and these unusual physical/chemical properties appear to make many of the extrapolations which are commonly used for organic chemicals inappropriate. For example, the apparent log P's, measured by C18-RP-HPLC with methanol/water (90:10), for the linear PDMS oligomers from MM to $MD_{14}M$ ranged from 4.2 up to 12.5, increasing approx. 0.6 units with each dimethyl-silicon fragment [30], so one might expect them to be fairly toxic to and highly accumulated by fish. However, they are biologically innocous to and poorly accumulated by fish when contained in their food and/or water [29, 30]. Consequently, the silicones are not believed to pose any ecotoxicological hazards.

Toxicology

Although PDMS and the other polymeric silicones are recognized to be biologically inert, this is not true for all organosilicons [7, 23, 33]. The volatile molecule, 2,6-*cis*-diphenylhexamethylcyclotetrasiloxane, which is not intentionally manufactured or sold, is known to be an androgen depressant, presumably because its aromatic carbons are in positions equivalent to those in estradiol and diethylstilbestrol [34]. Conversely, the dimethicones are generally unaffected by and have little effect on mammals [35]. Over 90% of the ^{14}C-labeled dimethicone injected intraperitoneally into rats was still present in the animals after 25 days and none of the labelled material was found in the faeces, urine or expired air [35]. All of the non-volatile PDMS's that have been investigated "are devoid of adverse reproductive, teratologic, and mutagenic effects" in rats, rabbits and mice [36]. Oral ingestion-fecal excretion balance data for rabbits, rats and man demonstrates that the high molecular weight silicone fluids do not undergo gastrointestinal absorption, even though silicones of six monomer units or less are absorbed [37]. Silicones are even less readily absorbed dermally than orally, and 20–100 cSt PDMS fluids were not fatiguing, irritating or sensitizing to the skin in repeated insult patch tests with humans [37]. Given these characteristics, plus their unique surface properties, many cosmetics contain silicones [35]. PDMS's may also be used for augmentation of soft tissues, as lubricants for disposable syringes and as a vitreous humor (jelly) for retinal detachment therapy [37]. Thus, although some of the chemical intermediates and all of the chlorosilanes pose potential occupational hazards, the silicone polymers themselves do not appear to pose any human health or environmental hazards. Nevertheless, efforts to quantify their fate and effects in the environment will continue.

References

1. Larson GL (1987) Petrarch systems silanes and silicones Cat. S-7, Petrarch Systems, Bartran Rd, Bristol, PA, 19007, p 9
2. Noll W (1968) Chemistry and technology of silicones, Academic, New York, p 702
3. Rochow EG (1951) Chemistry of the silicones, 2nd edn. John Wiley, New York
4. Owen MJ (1981) Chemtech 11:288
5. Abraham RJ, Grant GH (1988) J Comp Chem 9:244
6. Owen MJ (1986) Interfacial activity of polydimethylsiloxane. In: Mittal KL, Bothorel P (eds), p 1557 (Surfactants in solution, vol 6)
7. Stark FO, Falender JR, Wright AP (1982) Silicones, In: Wilkinson G, Stone FGA, Able EW (eds). Pergamon, Oxford, p 305 (Comprehensive Organometallic Chemistry)
8. U.S. Bureau of Mines, Mineral Industry Surveys, U.S. Dept of Interior, September 1988
9. Smart M, Lutz J, Sasano T (1988) CEH Product Review, Silicone Fluids 583.0100 A, March 1988 by Chemical Economics Handbook—SRI International
10. Frye CL (1987) Env Tox and Chem 6:329
11. Smith AL (1974) Analysis of silicones, John Wiley, New York (Chemical Analysis, vol 41)
12. Mahone LG et al. (1983) Env Tox and Chem 2:307
13. Simon PB (1988) D4 testing for Henry's Law Constant and "Volatilization from Water" Pers Comm Dec 15:1988
14. Pellenbarg RE (1979) Env Sci and Tech 13:565
15. Starch MS (1984) Drug and Cosmetic Industries 134(6):38
16. Vick SC (1984) Soap/Cosmetics/Chemical Specialties, May 1984:36
17. Buch RR, Ingebrigtson DN (1979) Env Sci and Tech 13:676
18. Buch RR, Lane TH, Annelin RB, Frye CL (1984) Env Tox and Chem 3:215

19. Frye CL (1988) The Sci of the Total Env 73:17
20. Mackay D, Leinonen (1975) Env Sci and Tech 9:1178
21. Simon PB (1985) Organosiloxanes in fresh water and salt water sediments, Final Report to Silicones Health Council, Washington DC
22. GESAMP (Group of Experts on the Scientific Aspects of Marine Pollution) (1986) Pub. WHO, p 26
23. Jarvie AWP (1986) In: Craig PJ (ed) Environmental aspects of organosilicon chemistry and use. Wiley, New York, p 229 (Organometallic Compounds in the Environment)
24. Hobbs EJ, Keplinger ML, Calandra JC (1975) Env Res 10:397
25. Dean JH, DeGrave GM (1986) Env Tox and Chem 5:1055
26. Annelin RB (1989) Silicone deposition on Daphnia, Pers Comm July 1989
27. Anliker R, Moser P, Poppinger D (1988) Chemosphere 17:1631
28. Annelin RB, Frye CL (1989) The Science of the Total Env 83:11
29. Opperhuizen A, Damen HWJ, Asyee GM, Van Der Steen JMD, Hutzinger O (1987) Tox and Env Chem 13:265
30. Bruggeman WA, Weber-fung D, Opperhuizen A, Van Der Steen J, Wijbenga A, Hutzinger O (1984) Tox and Env Chem 7:287
31. Chu HK (1987) Dow Corning Corp Tox Dept Internal Report, File No: 1362-12, Series No. I-0005-1666. Sept. 23, 1987
32. Abe Y, Butler GB, Hogen-esch TE (1981) J Macromol Sci-Chem A16(2):461
33. Aubert M, Aubert J, Augier H, Guillemaut C (1985) Chemosphere 14:127
34. Grigoras S, Lane TH, LeVier RR (1987) Main Group Met Chem 10(3):199
35. Elder RL (1982) J Am College of Tox 1:33
36. Kennedy GL Jr, Keplinger ML, Calandra JC, Hobbs EJ (1976) J Tox and Env Health 1:909
37. Howard PH, Durkin PR, Hanchett A (1974) Technical Report EPA-560/2-75-004

Volume 1 series: The Natural Environment and the Biogeochemical Cycles

Volume 1A: The Atmosphere (M. Schidlowski)
The Hydrosphere (J. Westall, W. Stumm)
Chemical Oceanography (P.J. Wangersky)
Chemical Aspects of Soil (E.A. Paul, P.M. Huang)
The Oxygen Cycle (J.C.G. Walker)
The Sulfur Cycle (A.J.B. Zehnder, S.H. Zinder)
The Phosphorus Cycle (J. Emsley)
Metal Cycles and Biological Methylation (P.J. Craig)
Natural Organohalogen Compounds (D.J. Faulkner)

Volume 1B: Basic Concepts of Ecology (S.W.F. van der Ploeg)
Natural Radionuclides in the Environment (R. Fukai, Y. Yokoyama)
The Nitrogen Cycles (R. Söderlund, T. Rosswall)
The Carbon Cycle (A.J.B. Zehnder)
Molecular Organic Geochemistry (P.A. Schenck, J.W. de Leeuw)
Radiation and Energy Transport in the Earth Atmosphere System (H.-J. Bolle)

Volume 1C: Humic Substances, Structural Aspects, and Photophysical, Photochemical and Free Radical Characteristics (G.G. Choudhry)
Organic Material in Sea Water (P.J. Wangersky)
Marine Gelbstoff (M. Ehrhardt)
The Surface of the Ocean (L.W. Lion)
Atmospheric Nitrogen. Chemistry, Nitrification, Denitrification, and their Interrelationships (R.D. Hauck)
Carbon Dioxide: A Biogeochemical Portrait (E.T. Degens, S. Kempe, A. Spitzy)

Volume 1D: The Cycles of Copper, Silver and gold (H.J.M. Bowen)
Modelling the Global Carbon Cycle (G. Kratz)
Chemical Limnology (T. Frevert)
Environmental Microbiology (W.D. Grant, P. E. Long)

Volume 1E: The Thermodynamics of Ecosystems (L. Johnson)
Environmental Systems (G.H. Dury)
Global Transport Processes in the Atmosphere (J.R. Holton)
The Atmosphere: Physical Properties and Climate Change (R. Eiden)

Volume 2 series: Reactions and Processes

Volume 2A: Transport and Transformation of Chemicals: A perspective (G.L. Baughman, L.A. Burns)
Transport Processes in Air (J.W. Winchester)
Solubility, Partition Coefficients, Volatility, and Evaporation Rates (D. Mackay)
Adsorption Processes in Soil (P.M. Huang)
Sedimentation Processes in the Sea (K. Kranck)
Chemical and Photo Oxidation (T. Mill)
Atmospheric Photochemistry (T.E. Graedel)
Photochemistry at Surfaces and Interphases (H. Parlar)
Microbial Metabolism (D.T. Gibson)
Plant Uptake, Transport and Metabolism (I.N. Morisson, A.S. Cohen)
Metabolism and Distribution by Aquatic Animals (V. Zitko)
Laboratory Microecosystems (A.R. Isensee)
Reaction Types in the Environment (C.M. Menzie)

Volume 2B: Basic Principles of Environmental Photochemistry (A.A.M. Roof)
Experimental Approaches to Environmental Photochemistry (R.G. Zepp)
Aquatic Photochemistry (A.A.M. Roof)
Microbial Transformation Kinetics of Organic Compounds (D.F. Paris, W.C. Steen, L.A. Burns)
Hydrophobic Interactions in the Aquatic Environment (W.A. Bruggeman)
Interactions of Humic Substances with Environmental Chemicals (G.G. Choudhry)
Complexing Effects on Behavior of Some Metals (K.A. Daum, L.W. Newland)
The Disposition and Metabolism of Environmental Chemicals by Mammalia (D.V. Parke)
Pharmacokinetic Models (R.H. Reitz, P.J. Gehring)

Volume 2C: OECD Fate and Mobility Test Methods (A.W. Klein)
Biodegradation and Transformation of Recalcitrant Compounds (A.H. Neilson, A.-S. Allard, M. Remberger)
Biodegradation of Water-Soluble Compounds (H.A. Painter, E.F. King)
The Fugacity Concept in Environmental Modelling (S. Paterson, D. Mackay)

Volume 2D: Hydrology (R. Herrmann)
Outdoor Ponds: Their Construction, Management, and Use in

Experimental Ecotoxicology (N.O. Crossland, C.J.M. Wolff)
Hydrolysis of Organic Chemicals (Th. Mill, W. Mabey)
Exchange of Pollutants and Other Substances Between the Atmosphere and the Oceans (M. Waldichuk)
Root-Soil Interactions (P.B. Tinker, P.B. Barraclough)
Reaction Types in the Environment (C.M. Menzie)

Volume 3 series: Anthropogenic Compounds

Volume 3A: Mercury (G. Kaiser, G. Tölg)
Cadmium (U. Förstner)
Polycyclic Aromatic and Heteroaromatic Hydrocarbons (M. Zander)
Fluorocarbons (J. Russow)
Chlorinated Paraffins (V. Zitko)
Chloroaromatic compounds Containing Oxygen (C. Rappe)
Organic Dyes and Pigments (E.A. Clarke, R. Anliker)
Inorganic Pigments (W. Funke)
Radioactive Substances (G.C. Butler, C. Hyslop)

Volume 3B: Lead (L.W. Newland, K.A. Daum)
Arsenic, Beryllium, Selenium and Vanadium (L.W. Newland)
C_1 and C_2 Halocarbons (C.R. Pearson)
Halogenated Aromatics (C.R. Pearson)
Volatile Aromatics (E. Merian, M. Zander)
Surfactants (K.J. Bock, H. Stache)

Volume 3C: Aromatic Amines (L. Fishbein)
Phosphate Esters (D.C.G. Muir)
Phthalic Acid Esters (C.S. Giam, E. Atlas, M.A. Powers, Jr., J.E. Leonard)
Thallium (J. Schoer)

Volume 3D: Cellulose Production Processes (C.C. Walden, D.J. McLeay, A.B. McKague)
Asbestos (P.E. Ney)
Carbon Black (D. Rivin)
Creosote (G. Sundström, A. Larsson, M. Tarkpea)
Elemental Phosphorus (R.F. Addison)
Molybdenum (G.A. Parker)

Volume 3G: Isocyanates (F. Brochhagen)
Nitro Derivatives of Polycyclic Aromatic Hydrocarbons (NO_2-PAH) (H. Fiedler, W. Mücke)

Chlorinated Ethanes: General Sources, Biological Effects, and Environmental Fate (J. Konietzko, K. Mross)
Organic Explosives and Related Compounds (D.H. Rosenblatt, E.P. Burrows, W.R. Mitchell, D.L. Parmer)

Volume 4 series: Air Pollution

Volume 4A: Air Pollution in Perspective (A. Wint)
Halogenated Hydrocarbons in the Atmosphere (P. Fabian)
Formation, Transport and Control of Photochemical Smog (H. Güsten)
Atmospheric Distribution of Pollutants and Modelling of Air Pollution Dispersion (H. van Dop)
The Mathematical Characterization of Precipitation Scavenging and Precipitation Chemistry (J.M. Hales)

Volume 4B: Peroxyacyl Nitrates (Pans): Their Physical and Chemical Properties (J.S. Gaffney, N.A. Marley, E.W. Prestbo)
Semivolatile Organic Compounds in the Atmosphere (R. Harkov)
Arctic Haze (G.E. Shaw, M.A.K. Khalil)
Air Pollution and Materials Damage (F.W. Lipfert)
Air Pollution Control Equipment (H. Brauer)

Volume 4C: Lichens as Indicators of Air Pollution (T. H. Nash, C. Gries)
Morbidity Associated with Air Pollution (M. Lippmann)
Mortality and Air Pollution (F. W. Lipfert)

Volume 5 series: Water Pollution

Volume 5A: Epidemiologic Studies of Organic Micropollutants in Drinking Water (G.F. Craun)
Water Quality Genesis and Disturbances of Natural Freshwaters (M. Falkenmark, B. Allard)
Eutrophication of Lakes, Rivers and Coastal Seas (H.L. Golterman, N.T. de Oude)
Mathematical Models for Describing Transport in the Unsaturated Zone of Soils (W.T. Piver, T. Lindström)

Subject Index

Activated sludge, carboxymethyl cellulose, 335
Adenosine triphosphate (ATP), 183
AES (alkyl ether sulfate)
 –, biodegradability, 61, 62
 –, effects on environment, 65–68
 –, fate, 61–64
 –, fish toxicity, 66–68
 –, half-life, 48
 –, metabolic pathways, 64–65
 –, sewage concentration, 61
 –, – effluents, 61
 –, simulation tests, 63
 –, synthesis, 59
 –, uses, 59
Alcohol ethoxy sulfate *see* Alkyl ether sulfate (AES)
Alcohol ethoxylate, biodegradability, 120–127
 – –, chromatographic methods, 107
 – –, dinitrobenzoate derivatives, 108
 – –, HBr cleavage, 107
 – –, metabolic pathways, 122–127
 – –, phenyl isocyanate derivatives, 108
 – –, sewage effluent concentrations, 111
 – –, structure, 93
Alcohol sulfate *see* Alkyl sulfate (AS)
Alcohols, branched, 97
 –, OXO process, 95
Aliphatic sulfonate *see* Alkane sulfonate (SAS)
Alkane sulfonate (SAS), 8, 69–75
Alkene sulfonate *see* α-Olefine sulfonate (AOS)
Alkyl aryl sulfonate *see* LAS
Alkyl benzene sulfonate *see* ABS
Alkyl ether sulfate (AES), 7, 58–69
Alkyl glycosides, structure, 93
Alkyl isothionates, 8
Alkyl sulfate (AS), 7, 41–57
Alkyl taurides, 8
Alkylbenzene, 96
Alkyloxybenzene sulfonate, 291
Alkylphenol ethoxylate, analysis, 108–109
 –, biodegradability, 128–130
 –, chromatographic methods, 108
 –, degradation products, 117
 –, metabolic pathways, 128–129
 –, sewage effluent concentrations, 112–114
 –, structure, 93

Alumina, activated, 197, 199
Analcime, 209
Anionic surfactants, types, 7–8
AOS, biodegradability, 78, 79
 –, effects of environment, 80–82
 –, fate, 78–80
 –, fish toxicity, 81
 –, half-life, 48
 –, metabolic pathways, 80
 –, sewage effluents, 78
 –, synthesis, 76–77
 –, uses, 77
Apatite, 181, 182, 185, 186
AS, biodegradability, 47–49
 –, concentration, biological filter effluent, 46
 –, effects on environment, 51–57
 –, fate, 46–51
 –, fish toxicity, 54–56
 –, half-life, 48
 –, homologues in effluents, 46
 –, metabolic pathways, 51
 –, removal through sewage treatment, 49–50
 –, sewage concentration, 46
 –, sewage effluents, 46
 –, simulation tests, 49–50
 –, synthesis, 42–43
 –, uses, 43–44
Azomethine-H method, 293

BiAS, 106
Bleach, environmental effects of, 294–315
Borax *see* Boron
Boron, analytical methods for 293
 –, concentrations and water quality, 292–313
 –, effects on aquatic life, 294–305
 –, – – crops, 305–310
 –, – – fish, 294–300
 –, – – human health, 312–315
 –, – – plants, 305
 –, levels in diet, 314–315
 –, toxicity to aquatic life 294–305
 –, – – higher plants 305–310
 –, water quality standards 304, 305
Butoxylates, 58, 64

Carboxymethyl cellulose (CMC), 329–336
  – –, biodegradation, aerobic/anaerobic, 333
  – –, degree of polymerization, 330
  – –, – – substitution, 330–333
  – –, ecotoxicity, 335
  – –, enzymatic hydrolysis, 331–333
  – –, laundry, 330, 331
  – –, water soluble polymers, 329–330
CATS, 107
Cellulase, 331, 332, 334
  –, endoglucanase, 332
  –, exoglucanase, 332
Citrate, aquatic chemistry 235
  –, biodegradation, 236–238
  –, detergent applications, 232, 233
  –, environmental acceptability, 233–235
  –, eutrophication, 240
  –, production, 231, 232
  –, properties, 230–233
  –, safety, 238, 239
CMC see Carboxymethyl cellulose
Cobalt thiocyanate, 107
Coenzyme A, 183
Creatine phosphate, 183
Curcumin method, 293

DAED, 321, 322
Detergents, environmental effects of, 294–305
  –, laundry, 7
  –, phosphate-free, 338
Dialkyltetralins, 38
Diamino stilbene, 352–355, 357
Diatoms, interactions with anthropogenic silicates, 375
Die-away tests, 30, 31, 47, 48, 158–159
Distyryl biphenyl, 352
Disulfine blue-active substances (DBAS), 152–153
DNA, 183

EDTA, absorption spectra, 250
  –, acceptable daily intake, 257
  –, adsorption, 247
  –, biodegradation, 248, 249
  –, chromatographic analysis, 246, 247
  –, concentrations in drinking water, 248
  –, consumption, 244
  –, domestic waste water treatment, 248, 249, 252, 253
  –, levels in the environment, 251, 252
  –, manufacture, 244
  –, metal mobilization, 254, 255
  –, production, 244
  –, removal from waste waters, 247
  –, toxicity in mammals, 255–257
  –, uses, 244–246

  –, water solubility, 247
EDTA-iron II complex, photodegradation, 250
  – – –, quantum efficiency, 250
EDTA-metal chelates, stability constants, 255
Ethoxylation reaction, 96
Eutrophication, anthropogenic silicates, 375
  –, citrate, 240

Fatty acids, ethanolamides, 93
  – –, ethoxylate, 93
  – –, polyol esters, 93
  – –, sorbitol esters, 93
Faujasite, 209
Fluorescent whitening agents, animal studies, 363–364
  – – –, bioaccumulation, 361–362
  – – –, biological availability, 361–362
  – – –, carcinogenic effects, 363–364
  – – –, chromatographic analysis, 356–357
  – – –, contamination of surface waters, 358–359
  – – –, degradation, 360
  – – –, diamino stilbene, 352–355, 357
  – – –, ecotoxicology, 362–363
  – – –, E,Z-isomerization, 352, 357
  – – –, in drinking water, 363
  – – –, in fish, 361
  – – –, lethal concentration (LC), 362–363
  – – –, liquid detergents, 356
  – – –, photoallergy, 363
  – – –, photocarcinogenic effects, 363
  – – –, production plant effluents, 358
  – – –, o-sulfobenzoic acid, 360
  – – –, toxicology, 362–363
  – – –, triazolyl stilbene, 352, 355
  – – –, uptake in bean plants, 362
  – – –, UV light, 351, 352
  – – –, washing powder, 356
Foaming, 6–7, 13, 15
Freundlich isotherm, 14
FWAs see Fluorescent whitening agents

Hampshire test, 340
Hydroxy alkane sulfonate see α-Olefine sulfonate

Interfactants, 16
Irrigation water, 304–310

LAS (linear alkyl benzene sulfonate)
  –, biodegradability, 12, 20, 29–33
  –, detergency, 13
  –, effects on the environment, 35–40
    –, – – sediment organisms, 40
    –, – – sewage treatment, 35–36

Subject Index

–, fate, 27–35
–, fish toxicity, 39–40
–, half-life, 30, 34–35
–, homologues, 29–30, 38, 39
–, isomers, 29
–, metabolic pathways, 27–28
–, per capita amounts of, 20–21
–, removal through sewage treatment, 32–34
–, sewage concentration, 20–21
–, simulation tests, 31–32
–, synthesis, 10
–, uses, 11
LAS/MBAS, concentration in tapwater, 6, 26
– –, content, digested sludge, 23
– –, estuarine sediments, 26
– –, sediment content, 25–26
– –, sewage effluents, 22
– –, tapwater, 26
Linear alkyl benzene sulfonate (LAS), 7–68
Lysimeter tests, 346

MBAS, 6, 7, 16
Methylene blue, 12, 15, 16, 18
Micelle formation, 13
Molybdatophosphate, 184

Nonionic surfactants, adsorption isotherms, 100
– –, biodegradability, 119–130
– –, cloud point, 99
– –, CMC values, 98
– –, ecotoxicology, 131–139
– –, groundwater concentrations, 118
– –, hydrophilic structures, 95
– –, hydrophobic routes, 94
– –, Krafft point, 98
– –, liquid-liquid extraction, 105
– –, manufacture, 95
– –, marine concentrations, 118
– –, phase diagram, 100
– –, sewage treatment, 110–115
– –, solid phase extraction, 104
– –, surface tension, 98
– –, – water concentrations, 116–118

Octamethylcyclotetrasiloxane (OMCTS), 386, 388, 391
OECD, 12
 , confirmatory test, 343
α-Olefine sulfonate (AOS), 8, 75–82

Paraffin sulfonate see Alkane sulfonate (SAS)
PEG, biodegradability, 127
Pentaacetylglucose, 291
Pentanatriumtriphosphate (STPP), 188

Perborate, chemical/physical characteristics, 288–290
–, discharge into the environment, 292
–, manufacture, 290
–, sodium monohydrate/tetrahydrate, 288–292
–, use of, 291, 292
Phillipsite, 209
Phosphate, algae, 184, 192–194
–, apatite, 181, 182, 185, 186
–, bacteria, 193, 195, 197, 198
–, biological elimination, 197, 198
–, bones, 181–183
–, condensed, 184, 195
–, creatine, 183
–, detergent, 186, 187, 189, 191, 193, 200, 201
–, drinking water, 184, 186
–, erosion, 189–192
–, esters, 8
–, excrement, 182, 190
–, fertilizer, 181, 182, 186, 187, 190–192, 199
–, filtration, 196, 197
–, food, 181, 184–186, 188–190
–, hyperkinetic syndrome, 185
–, magnetic separation, 197
–, molybdato-, 184
–, pentanatriumtri-, 188
–, photosynthesis, 183
–, precipitation, 184, 195–199
–, rock/ore, 181, 182, 185–187
–, sewage, 182, 184, 189, 190, 193–202
–, sludge, 195–200
–, super-, 187, 199
–, teeth/bones, 181–183
–, Thomas, 188
–, threshold effect, 188
–, trophic state, 194
Phosphonates, aquatic toxicity, 279–282
–, bioconcentration, 278
–, degradation, microbial, 268–271
–, – soil, 271
–, drinking water purification, 278
–, environmental fate, 264–268
–, – partitioning, 265–267
–, – safety assessment, 282–283
–, metal complexation 274–277
–, – mobilization, 273–274
–, microbial degradation, 268–271
–, persistence, 267–273
–, photochemical degradation, 271–273
–, physicochemical properties, 265
–, production, 262
–, sewage treatment, 268–269
–, structures, 263–264
–, uses, 261–263
Phosphoric acid, 187, 188
Polyacrylates, 338
Polycarboxylates, 338–346
–, adsorption, 340

Polycarboxylates (*cont.*)
 –, anaerobic biodegradation, 344
 –, analytical methods, 341
 –, aquatic toxicity, 344
 –, behavior during phosphate treatment, 343
 –, – – sludge digestion, 244
 –, – in soil, 345
 –, – – surface water, 344
 –, bioaccummulation, 348
 –, biodegradation, 342
 –, consumption, 339
 –, concentration in the environment, 346
 –, effect of drinking water preparation, 346
 –, elimination during sewage treatment, 342
 –, heavy metal remobilization, 344
 –, incrustation-inhibiting effect, 340
 –, market share, 339
 –, precipitation with calcium, 340
 –, production, 338
 –, sedimentation 342
 –, toxicity on plants, 346
 –, toxicology, 348
 –, use in cleaning agents, 339
Polydimethylsiloxane (PDMS), 384–393
Polyelectrolyte, 339
Polyethermethylsiloxanes (PEMS), 385–386, 389–391
Polyol, esterification, 96
Potassium picrate, 107
Propoxylates, 58, 64
Propoxylation reaction, 96

QAS (Quaternary ammonium surfactants)
 –, analytical methods, 152–153, 174–175
 –, antistatic agents, 148
 –, aquatic toxicity, 164–169, 174
 –, biodegradation, 157–160, 174
 –, categorization, 146
 –, clinical toxicology, 169
 –, consumption, 149, 174
 –, discharge to environment, 149, 174
 –, ecotoxicity, 164–169, 174
 –, fabric softeners, 148
 –, fate in receiving waters, 157–161, 174–175
 –, germicides, 148
 –, microbial toxicity, 161–164
 –, monitoring, 169–173
 –, octanol/water partition coefficient, 150
 –, photodegradation, 160–161
 –, physicochemical properties, 150–152, 174
 –, production methods, 146–149
 –, sewage, 169, 171–173
 –, sorption, 150–151, 174
 –, structures, 146–147
 –, synthesis, 146–149
 –, toxicology, 169
 –, uses, 148–149

Respirometric tests, 342

SAS, biodegradability, 72
 –, effects on environment, 73–75
 –, fate, 72–73
 –, fish toxicity, 74
 –, metabolic pathways, 73
 –, sewage concentration, 71
 –, – effluents, 71
 –, synthesis, 69
 –, uses, 70
SCAS-test, 343
Secondary alkane sulfonate *see* Alkane sulfonate (SAS)
Silica, anion distribution and reactivity, 371
 –, cycle, natural, 374
Silicates, anthropogenic, aluminum inhibition, 378
 –, –, animal toxicity, 376
 –, –, biological interactions, 374
 –, –, diatoms, 375
 –, –, eutrophication, 375
 –, –, hazardous waste treatment, 378
 –, –, land treatment, 380
 –, –, plant uptake, 374
 –, –, soluble silica, 370
 –, –, structural classes, 368
 –, –, uses, 378
 –, –, water treatment, 379
 –, –, zeolites, 372
Silicone, atmosphere, 388–391
 –, bioaccumulation, 392
 –, fluids, 384–393
 –, Henry's constant, 391–392
 –, hydrophobic, 388–389
 –, interfaces, 389–391
 –, sediment, 388–391
 –, soil, 390
 –, waste waters, 386, 389, 390
Siloxanes, 384–393
Sodalite, 209
Sodium dodecyl benzene sulfonate standard, 16
Stilbenes, 352–355, 360
α-Sulfo fatty acids and methyl esters, 8
Sulfo succinamates, 8
Sulfo succinate esters, 8
o-Sulfobenzoic acid, 360
3(4'-Sulfophenyl)butyrate, 19
Sulfophenyl aklanoic acids, 27–29, 38–40
 –, undecanoic acid, 29
Superphosphate, 187, 199
Surface tension, 15

TAED, biodegradability, 322–324
 –, chemical structure, 319
 –, hydrolysis, 322

Subject Index

–, level in environment, 324, 325
–, partition coefficient, octanol/water, 321
–, perhydrolysis, 320
–, production, 320
–, sewage treatment, 328
–, toxicity to aquatic organisms, 326, 327
–, – – mammals, 327, 328
–, water solubility, 321
Tetraacetyl ethylenediamine, 291
Tetrapropylene benzene sulfonate (TPBS), 6, 9
Thomas phosphate, 187, 199
Triazolyl stilbene, 352, 355
Trophic state (lake), 194
Tween, biodegradability, 130

Zahn-Wellens test, 343
Zeolite A, aquatic organisms, 223, 224
   –, detection, 220, 221
   –, detergency properties, 212, 213
   –, field trials, 221, 222
   –, ion exchange selectivity, 212
   –, plant growth, effects on, 224
   –, sewage treatment, effects in, 222, 223

Zeolite NaA, 211
Zeolite NaX, 209
Zeolite NaY, 209
Zeolites, acid-clay process, 216
–, adsorption materials, 213
–, – properties, 211
–, anthropogenic silicates, 372
–, building units, 207, 208
–, cage, 210
–, β-cage, 208
–, catalysts, 213, 214
–, chemical properties, 210, 211
–, composition, 206, 208
–, crystal morphology, 211
–, history, 206
–, identification, 220
–, ion exchange, 212
–, pore diameter, 211
–, precipitation process, 216
–, preparation, 216
–, production volumes, 214, 215
–, regulations, 226
–, secondary building unit (SBU), 208
–, structure, 206–210
–, toxicity, 224–226

Environmental Toxin Series

Edited by S. Safe, O. Hutzinger

Volume 3

S. Safe, Texas A&M University, College Station, TX;
O. Hutzinger, University of Bayreuth;
T. A. Hill, Washington, DC (Eds.)

Polychlorinated Dibenzodioxins and -furans (PCDDs/PCDFs)

Sources and Environmental Impact, Epidemiology, Mechanisms of Action, Health Risks

1990. IX, 145 pp. 8 figs. 32 tabs. Hardcover DM 174,–
ISBN 3-540-15552-X

Polychlorinated Dibenzodioxins and -furans (PCDDs and PCDFs) are potent environmental toxins. Environmental exposures to these compounds in part-per-billion (ppb) and part-per-trillion (ppt) concentrations are of particular interest. These exposures may arise from bleached paper products, paper production process sludge, effluent waste water from paper plants, consumption of food and water from contaminated sources and contact to paper products.

The Handbook of Environmental Chemistry

Edited by O. Hutzinger

Volume 5

Water Pollution

Part A

1991. XI, 264 pp. 25 figs. 29 tabs.
Hardcover DM 198,– ISBN 3-540-51599-2

Contents: *G.F. Craun, Cincinnati, OH:* Epidemiologic Studies of Organic Micropollutants in Drinking Water. – *B. Allard, Linköping, Sweden; M. Falkenmark, Stockholm, Sweden:* Water Quality Genesis and Disturbances of Natural Freshwaters. – *H.L. Golterman, Arles, France; N.T. de Oude, Strombeek-Bever, Belgium:* Eutrophication of Lakes, Rivers, and Coastal Seas. – *W.T. Piver, Research Triangle Park, NC; T. Lindstrom, L. Boersma, Corvallis, OR:* Mathematical Models for Describing Transport in the Unsaturated Zone of Soils.

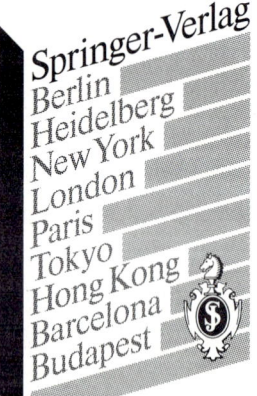

Springer-Verlag
Berlin
Heidelberg
New York
London
Paris
Tokyo
Hong Kong
Barcelona
Budapest